AIM HIGH
FOR THE TOP GRADES

AQA GCSE

Additional Science

Muriel Claybrook • Peter Gale • Richard Grime • Keith Hirst
Sue Kearsey • Nigel Saunders • Martin Stirrup

Series editor
Nigel English

Longman is an imprint of Pearson Education Limited, Edinburgh Gate, Harlow, Essex, CM20 2JE.

www.pearsonschoolsandfecolleges.co.uk

Edited by Stephen Nicholls
Typeset by Tech-Set Ltd, Gateshead

Illustrated by Tech-Set Ltd, Geoff Ward, Tek-Art
Cover design by Wooden Ark
Cover photo: A snowflake photographed through a light microscope.
© PhotoDisc: Photolink.

First published 2011

15 14 13 12 11
10 9 8 7 6 5 4 3 2 1

British Library Cataloguing in Publication Data
A catalogue record for this book is available from the British Library

ISBN 978 1 408253 71 7

Printed in Spain by Grafos S.A.

Acknowledgements
The authors and publisher would like to thank the following individuals and organisations for their kind permission to reproduce photographs:

(Key: b - bottom; c - centre; l - left; r - right; t - top)

x Pearson Education Ltd: Trevor Clifford. **2–3** iStockphoto: siun. **6** Photolibrary.com. **7** Science Photo Library Ltd: Power and Syred. **8** Shutterstock.com: sgame. **9** Shutterstock.com: Neil Webster (t); KariDesign (b). **10** Pearson Education Ltd: Trevor Clifford (l, c, r). **12** Science Photo Library Ltd: Dr Yorgas Nikas. **14** Science Photo Library Ltd: Cordelia Molloy (br). Shutterstock.com: Marie C. Fields (t); knin (bl). **16** Shutterstock.com: martellostudio (t); Dmitriy Kuzmichev (b). **18** Science Photo Library Ltd: Dr Jeremy Burgess. **19** Shutterstock.com: SamTan (b); nito (t). **20** Shutterstock.com: cybervelvet (r); corepics (l). **21** Shutterstock.com: Brett Nattrass. **22** Alamy Images: Nigel Cattlin. **23** Alamy Images: Bernhard Classen. **28** Shutterstock.com: Thomas Payne. **29** PhotoDisc: David Buffington. **30** Corbis: Willard Culver / National Geographic Society. **38–39** Science Photo Library Ltd: Herve Conge, ISM. **40** Pearson Education Ltd: Jules Selmes. **44** Pearson Education Ltd: Jules Selmes (t). Shutterstock.com: Jubal Harshaw (b). **46** Alamy Images: Dynamic Graphics. **48** Imagestate Media: John Foxx Collection (r). Shutterstock.com: Carmen Steiner (l). **50** Digital Vision (bl). Guillaume Dargaud (br). Pearson Education Ltd: Tudor Photography (tl); Malcolm Harris (tr). **51** Shutterstock.com: MichaelTaylor. **53** Science Photo Library Ltd: BSIP, Laurent / B. Hop AME. **54** iStockphoto: Florian Batschi. **55** Shutterstock.com: Ivan Histand. **58** Science Photo Library Ltd: CNRI. **62** Alamy Images: Momentum Creative Group. **68** Corbis: Jorge Z. Pascual / epa (b). Science Photo Library Ltd: Simon Fraser / RVI, Newcastle-upon-Tyne (t). **70** Photos.com: Jupiterimages (bc). Science Photo Library Ltd: Silkeborg Museum, Denmark / Munoz–Yague (tc). Shutterstock.com: Snowshill (t); mikeledray (b). **71** image courtesy of NASA Earth Observatory: Ames Research Centre. **72** Digital Vision. PhotoDisc: InterNetwork Media Inc. (b). **74** Shutterstock.com: jordache (t); Andreas Gradin (b). **75** Ardea: Pat Morris (b). Shutterstock.com: Vishnevskiy Vasily (t). **79** Photos.com: Jupiterimages. **86–87** PhotoDisc: Photolink. **88** Science Photo Library Ltd: Sheila Terry. **89** Pearson Education Ltd: Trevor Clifford (br); Trevor Clifford (bl). Peter Gould (tl). **92** Pearson Education Ltd: Trevor Clifford (t); Trevor Clifford (b). **96** Shutterstock.com: mirounga. **97** CERN Geneva (t). Shutterstock.com: jordache (b). **98** Science Photo Library Ltd: George Steinmetz. **99** Pearson Education Ltd: Trevor Clifford. **100** Shutterstock.com: Blaz Kure. **102** Alamy Images: Steve Hamblin (t). Shutterstock.com: Craig Jewell (b). **103** Shutterstock.com: Kevin Britland. **104** Shutterstock.com: Elnur. **105** Science Photo Library Ltd: Pascal Goetgheluck. **106** Pearson Education Ltd: Mark Bassett. **108** Science Photo Library Ltd: Susumu Nishinaga (t); Eye of Science (b). **114** Pearson Education Ltd: Stefan Grippon. **118** Pearson Education Ltd: MindStudio (t). Peter Gould. **121** Shutterstock.com: Antonio S. **122** Shutterstock.com: Yenyu Shih. **130–131** Alamy Images: David Wall. **132** Martin Stirrup. **134** iStockphoto: redmonkey8 (t). Martin Stirrup (b). **136** Peter Gould. **138** Peter Gould. Shutterstock.com: Losevsky Pavel (t). **141** Corbis: Document General Motors / Reuter R (t). Shutterstock.com: Sergey Peterman (b). **142** Peter Gould. Science Photo Library Ltd: Charles D. Winters (c); Andrew Lambert Photography (b). **143** Creatas (l). Pearson Education Ltd: Debbie Rowe (r). **144** Alamy Images: Art Directors & Trip (tl). Martin Stirrup. **145** Photolibrary.com: Stockbyte / George Doyle. **146** Martin Stirrup. Peter Gould. **147** Heat in a Click (t). iStockphoto: Daniel Laflor (b). **151** Pearson Education Ltd: Trevor Clifford. **153** Pearson Education Ltd: Trevor Clifford (l, r). **154** Alamy Images: ScotStock. **156** Peter Gould. **158** FLPA Images of Nature: Nigel Cattlin (b). Pearson Education Ltd: Trevor Clifford (t). **161** Shutterstock.com: terekhov igor. **162** Science Photo Library Ltd: Sheila Terry (b). Shutterstock.com: Ingvar Bjork (t). **164** Science Photo Library Ltd: Martin Bond. **167** PhotoDisc: Photolink. **168** Heat in a Click. **178–179** PhotoDisc: Karl Weatherly. **180** PhotoDisc (t); StockTrek (b). **182** Science Photo Library Ltd: Prof. Harold Edgerton (t). Shutterstock.com: Philip Date (b). **185** PhotoDisc: Photolink. **186** PhotoDisc: Photolink. **190** Pearson Education Ltd: Jules Selmes (l). Science Photo Library Ltd: Takeshi Takahara (r). **191** Shutterstock.com: Kondratenkov Vadim. **193** Alamy Images: Paul Bernhardt. **195** Shutterstock.com: Craig Jewell. **197** Shutterstock.com: Selena. **202** Shutterstock.com: daseaford. **203** Science Photo Library Ltd: TRL Ltd. **210–211** PhotoDisc: StockTrek. **217** Shutterstock.com: Feng Yu. **226** Pearson Education Ltd: Gareth Boden. **227** Pearson Education Ltd: Trevor Clifford (r). Shutterstock.com: StudioSmart (l). **231** Pearson Education Ltd: Trevor Clifford. **234** Pearson Education Ltd: Trevor Clifford. **238** PhotoDisc: StockTrek. **244** Science Photo Library Ltd: Ria Novosti. **248** image courtesy of NASA Earth Observatory. **250** PhotoDisc: StockTrek. **252** PhotoDisc: StockTrek. **254** Alamy Images: Paul Bernhardt. **261** Shutterstock.com: StudioSmart (r); stopwarnow (l)

All other images © Pearson Education

Every effort has been made to contact copyright holders of material reproduced in this book. Any omissions will be rectified in subsequent printings if notice is given to the publishers.

Introduction

This student book has been written by experienced examiners and teachers who have focused on making learning science interesting and challenging. It has been written to incorporate higher-order thinking skills to motivate high achievers and to give you the level of knowledge and exam practice you will need to ensure you get the highest grade possible.

The book follows the AQA 2011 GCSE Science specification, the first examinations for which are in November 2011. It is divided into three units, B1, C1 and P1, covering biology, chemistry and physics. Within each unit there are two sections, each with its own section opener page. Each section is divided into chapters, which follow the organisation of the AQA specification.

There are lots of opportunities to test your knowledge and skills throughout the book: questions on each spread, ISA-style questions, questions to assess your progress and exam-style questions. There is also plenty of practice in the new-style question requiring longer-text answers.

There are several different types of page to help you learn and understand the skills and knowledge you will need for your exam:

- Section openers with learning objectives and a check of prior learning.
- 'Content' pages with lots of challenging questions, examiner tips, skills boxes, and 'Route to A*' boxes.
- 'GradeStudio' pages with examiner commentary to help you understand how to move up the grade scale to achieve an A*.
- 'ISA-style' pages to give you practice with the types of questions you will be asked in your controlled assessment.
- Assess yourself question spreads to help you check what you have learnt.
- End of unit Exam-style questions to provide thorough exam preparation.

This book is supported by other resources produced by Longman:

- an ActiveTeach (electronic copy of the book) with BBC video clips, games, animations, and interactive activities
- an online exam booster for independent study, which takes you through exam practice tutorials focusing on the new exam questions requiring longer answers, difficult science concepts and questions requiring some maths to answer them.
- revision books

In addition there are Teacher and Technician Packs and Activity Packs, containing activity sheets, skills sheets and checklists, as well as Teacher Books.

The next two pages explain the special features we have included in this book to help you learn and understand the science and do the very best in your exams. At the back of the book you will also find an index and a glossary.

Contents

How to use this book

These two pages illustrate the main types of pages in the student book and the special features in each of them. (Not shown are the end-of-topic Assess yourself question pages and the Examination-style question pages.)

Section opener pages – an introduction to each section

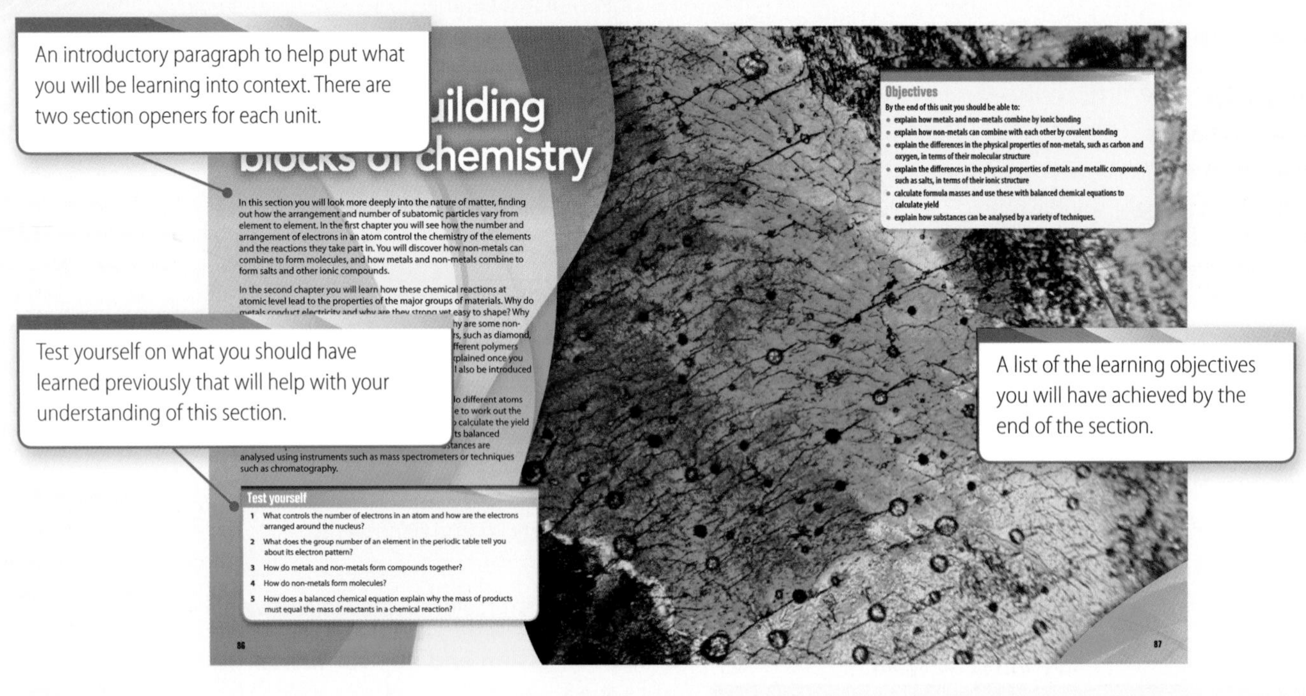

Content pages – covering the AQA specification

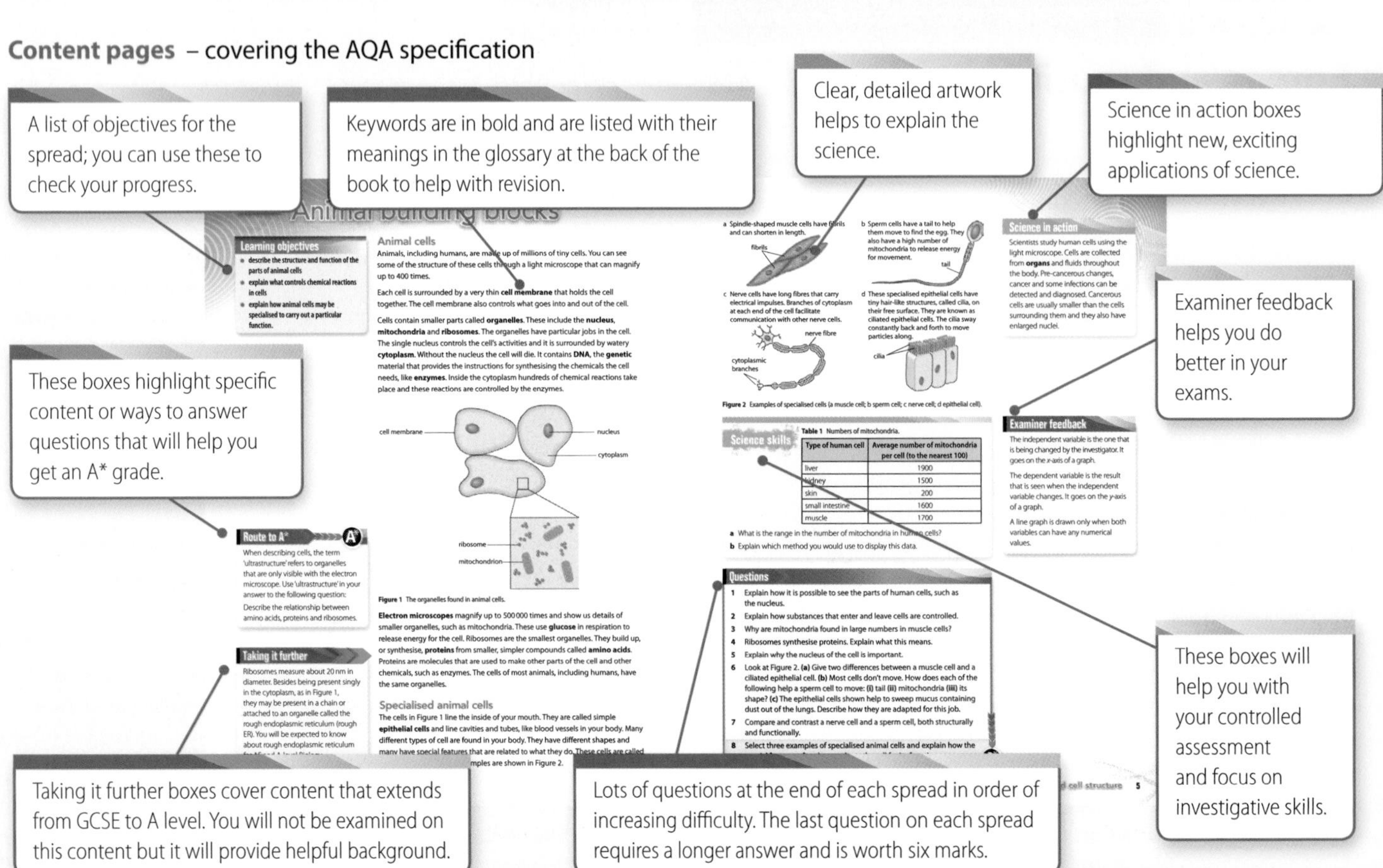

ISA practice pages – to help you with your controlled assessment

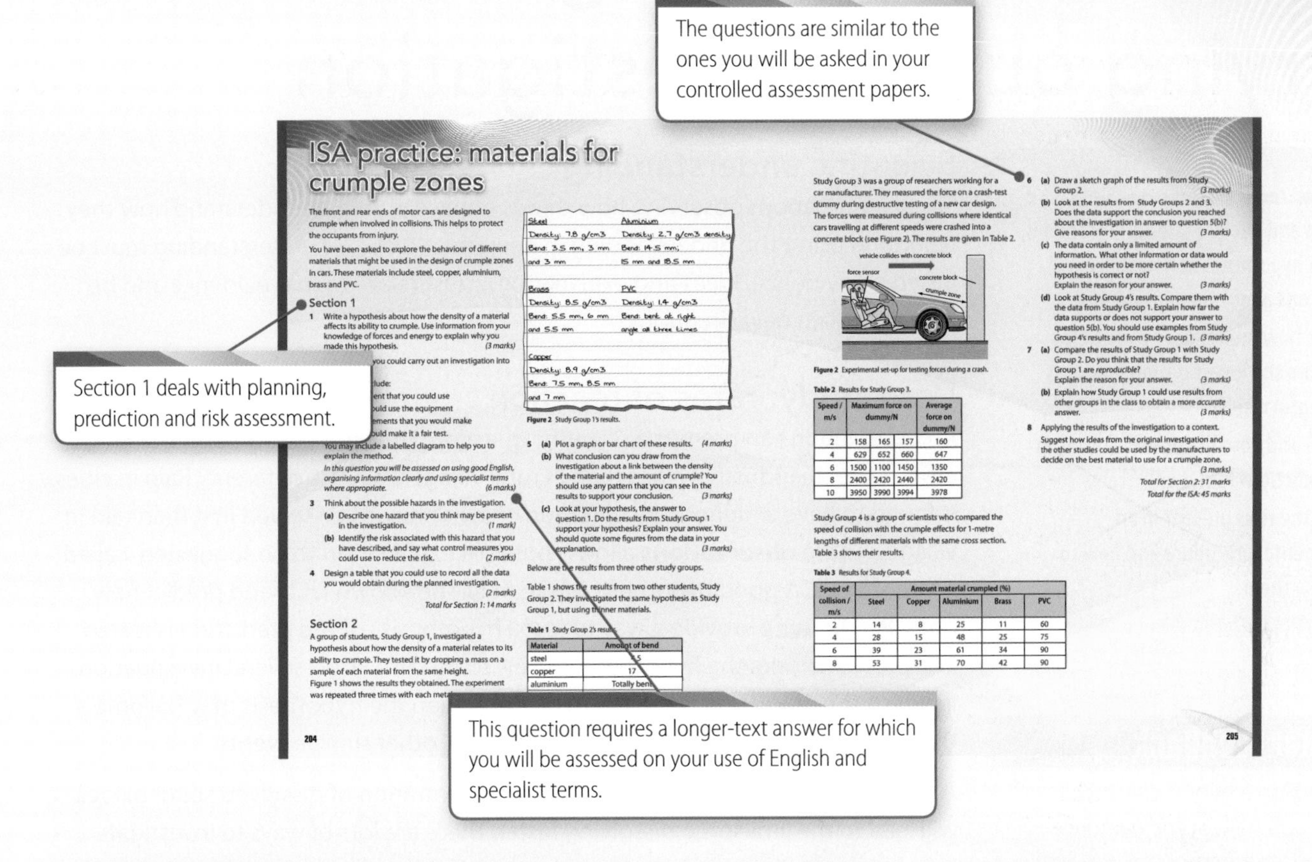

GradeStudio pages – helping you achieve an A*

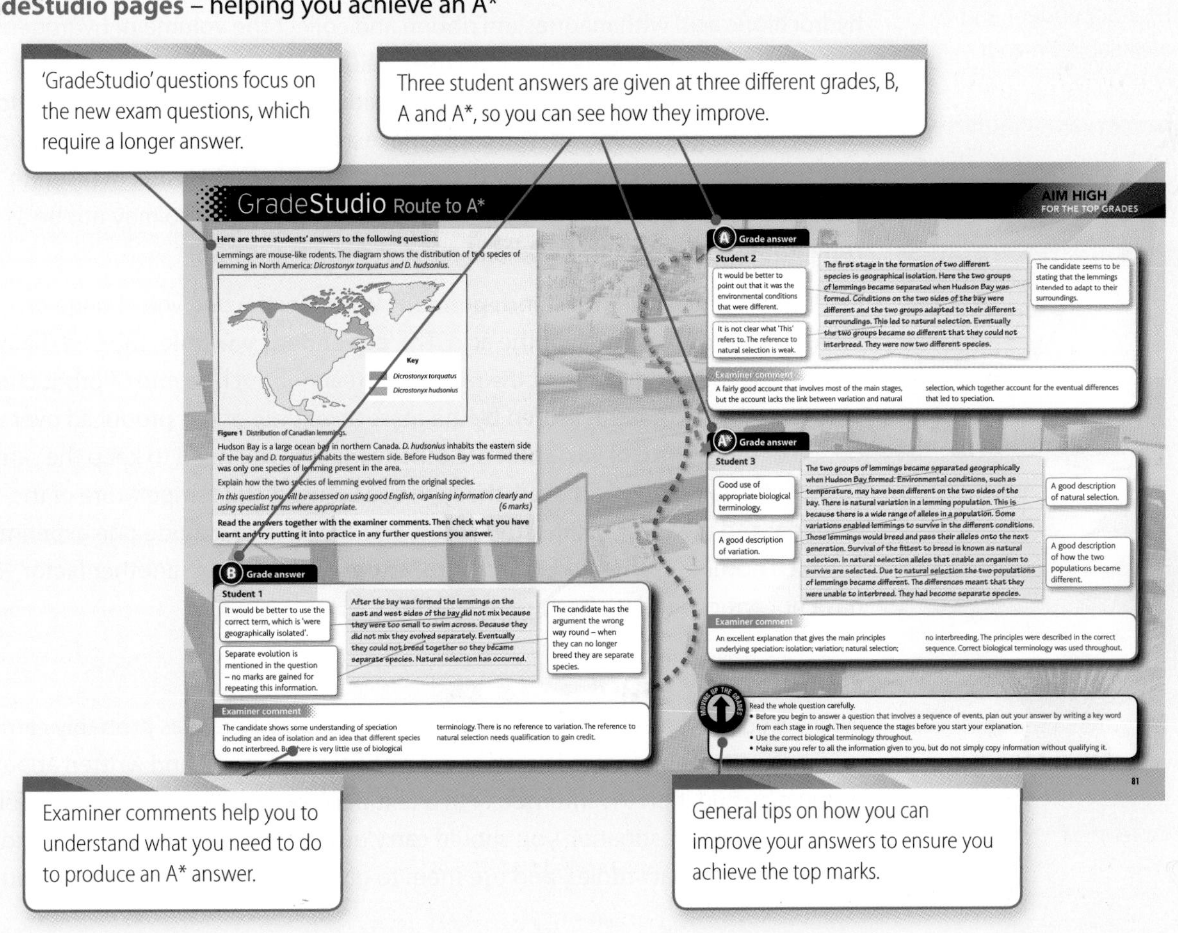

Researching, planning and carrying out an investigation

Learning objectives

- research and analyse scientific problems, suggest hypotheses and make predictions about these problems
- describe how preliminary work helps make sure the investigation produces meaningful results
- research and describe practical ways to test predictions fairly
- explain the risks present in an experimental procedure and how to minimise them.

Examiner feedback

When making a hypothesis make sure that you have some evidence to back up your hypothesis. Identify clearly both the independent and the dependent variables in your hypothesis.

One way of measuring the rate of a reaction is to time the mass loss as gas is given off.

Scientific understanding

Science is all about observing how things behave, trying to understand how they work, and using the understanding in new situations. This understanding must be based on evidence. Ideas and explanations must relate to that evidence and be able to explain new situations.

Case study: rates of reaction

An observation can prompt a scientific question. Limestone statues on buildings erode faster in industrial cities than in rural villages. What is different? Rain in cities is found to have a higher concentration of sulfur oxides dissolved in it than rain in villages. These **observations** allow a possible explanation to be suggested, called a **hypothesis**. A good hypothesis has to fit all the known facts and predict new ones. **Predictions** provide a way to test a hypothesis – if the prediction is tested and proved wrong, the hypothesis is likely to be false. After several investigations in which the prediction is proved to be right, then the hypothesis may become a theory, and be used to make predictions about other similar events.

A prediction is made that increasing the concentration of dissolved sulfur oxides will lead to the limestone dissolving faster. There are lots of ways to investigate how concentration affects the rate of a chemical reaction. You could react hydrochloric acid with magnesium ribbon and collect the volume of hydrogen gas given off. Alternatively you could use calcium carbonate and hydrochloric acid and measure the amount of carbon dioxide produced by measuring the loss of mass as the gas is evolved. You could make a choice by between the methods by considering the likely accuracy of the two methods. In this case, measuring gas volumes can be difficult, owing to leaks in the apparatus, so may not be as **repeatable** as measuring the loss of mass as the gas escapes.

In testing this prediction, the **independent variable**, the one you change or select, is the concentration of the acid. The **dependent variable**, the one that you measure, is some indication of the rate of the reaction such as rate of production of carbon dioxide gas, measured by the mass or volume of gas produced over a period of time. **Control variables** are other variables you need to keep the same, such as the volume of acid used, the type of acid used, the temperature of the reaction, and the mass of limestone used. You should also include one experiment in which nothing is changed, the **control experiment**, in case another factor you have not anticipated is affecting the results.

Preliminary work

No investigation is ever entirely new. Someone, somewhere, has probably carried out an investigation that is related to the one you plan to do, and written about it. You may find it on the Internet, or in a text book or a science magazine. Before you start your investigation you should carry out some **preliminary work**: find two or three similar studies, and use them to give you ideas about the way you could investigate your problem. One method will seem better than another, perhaps because it's quicker or you have the apparatus needed.

All scientific investigations have **hazards**, things that can go wrong with the experiment and cause injury to people or objects. The biggest hazard in this investigation is the risk of acid burns. To minimise the **risk**, or the chance of it happening, **control measures** are used. A control measure is something that reduces the hazard to a level of risk that is acceptable. The control measures could be to wear gloves and eye protection, and to be aware that any liquid may be acidic and should be mopped up immediately.

Validity, repeatability, precision and accuracy

It may be that your preliminary work helps you improve the **validity** of your investigation. Will your investigation really provide the information you need to prove or disprove the prediction? The results you obtain may suggest that there are more variables that need to be controlled than you originally thought if you are to adequately test the prediction.

To be able to draw a conclusion from a set of results, your data must be **repeatable**. This means that if you repeated the investigation the new data obtained would give the same or similar results.

If you change the method or use different equipment, or if someone else does the investigation, and the results are still similar, then we say that the results are **reproducible**.

The closer your measurements are to each other and to the **mean** of the results, the greater the **precision** of the results. This does not mean that your results are **accurate**. For your results to be accurate they must be close to the true value, the value you would obtain if there were no errors (for example **zero errors**) in your measurements – see the explanation of error types on page x in the Science student book.

You can also check whether the findings are valid by looking for similar evidence from classmates or on the Internet, or by trying a different method to see if you get the same answer to the investigation.

Questions

1 Suggest a hypothesis to explain why limestone statues on buildings erode faster in industrial cities than in rural villages.

2 Give brief details of two methods to investigate how concentration affects the rate of a chemical reaction.

3 Describe in detail one of the two methods you have given in question 3. You should state the equipment you would use, the measurements to make and why your method will produce a fair test.

4 Explain the difference between hazards, risks and control measures.

5 You have carried out your investigation, and to confirm your findings you ask another group to repeat your method. They find the same trend as you did. Are the results from your investigation repeatable or reproducible? Explain your answer.

A*

Examiner feedback

When researching, don't always turn to the Internet first. A good textbook may describe a range of methods that work, and you can find them quickly using the index. Much of the information online may be too advanced to be of much use to your investigation.

Decide which method you to plan to use and be able to say why it is better than the other method(s) you found. You should be able to briefly describe two of the methods.

Presenting, analysing and evaluating results

Learning objectives

- describe how to report and process your experimental data
- evaluate the data collected, identifying errors
- analyse the evidence from your data and other data
- use research to confirm whether the findings are valid.

Recording and displaying results

It is best to collect your results in a table that you have already prepared. Table 1 shows a table for recording loss of mass in an investigation into concentration and rate of reaction. The independent variable, that is, the one you change, is usually recorded in the left-hand column. In this case, though, time is more conveniently placed in the left-hand column, and the independent variable – the concentration of acid – is placed above each column containing the value of the dependent variable.

Table 1 Recording the results of an investigation into rates of reaction.

Time/s	Carbon dioxide lost/g at a given acid concentration/moles per dm^3		
	0.5	**1.0**	**2.0**
0	0	0	0
30	0.06	0.12	0.16
60	0.11	0.21	0.30
90	0.16	0.28	0.34
120	0.20	0.32	0.34
150	0.23	0.34	0.34

Patterns and relationships

Sometimes you can spot a trend or pattern in a set of results from the table, but more likely you will need to see them as a bar chart, if the independent variable is **categoric**, or as a line graph, for **continuous variables**.

The data in Table 1 is continuous, so three curves of best fit should be drawn as a graph. If you were investigating the effect of the size of the limestone pieces used, the data would be categoric, and a bar chart should be drawn.

When plotting a line graph you should first plot the points, and then look for a pattern in the points. Sometimes there are **anomalous results**. These show up as points that would not lie on a best-fit line or curve. Plot the best line or curve you can, ignoring any anomalous results, and passing approximately equal numbers of the other points on each side of your line to take account of the **random errors** in any experimental data. Your graph will help you to identify the relationship between the two variables.

Straight lines indicate a linear relationship. In Figure 1, the top graph shows a **positive linear** relationship, whilst the middle graph shows a **negative linear** relationship. If the line goes through the origin, where the axes meet, then the relationship may be **directly proportional**. This is only the case when the origin of the graph is truly zero on both axes.

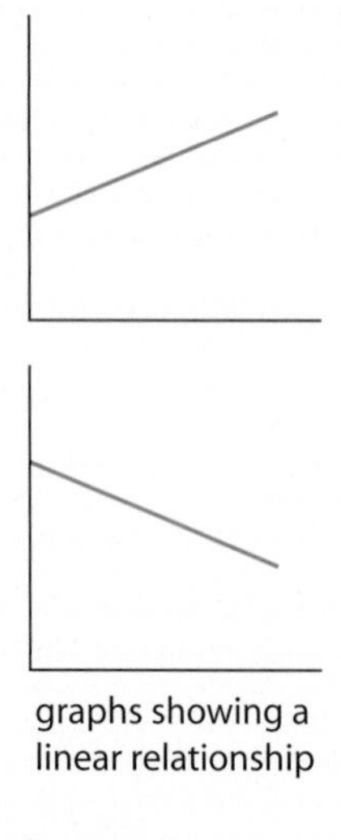

graphs showing a linear relationship

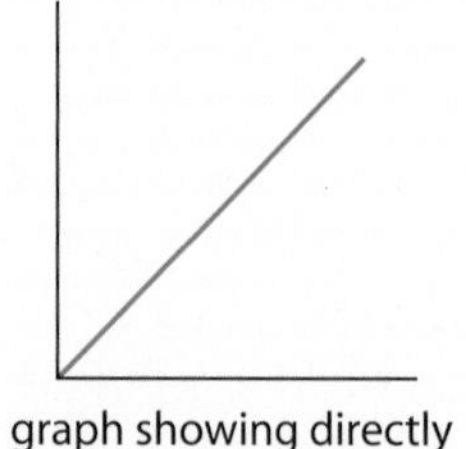

graph showing directly proportional relationship

Figure 1 Examples of line graphs.

Analysing the evidence

Figure 2 shows the results from the table. In this case the curved lines show a more complex relationship between the variables than those in Figure 1.

Your conclusion must relate to the investigation. The lines on the graph show that as the concentration increases, the mass of carbon dioxide lost increases, so the rate of reaction is greater in more concentrated solutions. The curves are a result of the drop in concentration of the acid throughout each experiment as the acid reacts with the limestone.

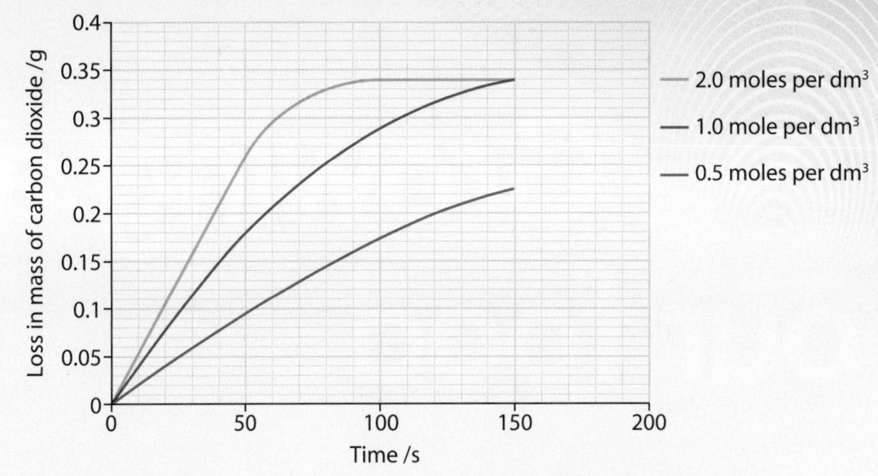

Figure 2 Three curves plotted from the results in Table 1 showing the complex relationship between concentration and rate of reaction over time.

Using other evidence

Other evidence can help you to establish whether your conclusion is likely to be correct. A comparison is best undertaken between the trends, rather than the actual data, because it is unlikely that your data will numerically match the other data. Do the data give the same conclusion as yours? If so, then your conclusion is reinforced. If not, then you should look at other data to establish which trend is better supported. Sometimes you find data that is not relevant to your investigation. This would be the case if you found an investigation into temperature and the dissolving of limestone. The shape of the curves on the graph would appear to support your trend, but because your investigation is into concentration and not temperature the data are not valid for your purpose, and cannot be used to support your hypothesis.

Examiner feedback

In science you should only plot your line to the data you have. Unless you have a value for the origin, do not plot it or connect your line to it. Not all graphs meet the axes at the origin. For example, remember that in winter the temperature drops to less than 0 °C. This means that 0 °C is not zero for a temperature axis. In fact, the lowest possible temperature, the true value for zero, is −273.15 °C.

Improving and extending the investigation

Your results and the other data you may find may give you ideas for further work, or an improvement to the design of your investigation. The graph in Figure 3 suggests several further questions that could be answered by extending the investigation, for example how much carbon dioxide gas will be produced by 1.0 mole per dm^3 of hydrochloric acid? Will it be the same as 2.0 moles per dm^3?

Examiner feedback

When looking at data provided, make sure that anomalous results were discarded before the mean was calculated. The data may not show a trend if anomalous results are included. If you spot that this has happened, then say that the conclusions are not valid, or that the data do not support the conclusion, but this fact is masked by the failure to remove the anomalous result.

Questions

1. Explain the difference between an anomalous result and a random error.
2. Make a sketch of Figure 3. Add onto your sketch the line you would expect for 0.25 moles per dm^3 of hydrochloric acid.
3. In the investigation, a student noticed that all the limestone dissolved for the acid at 1.0 and 2.0 moles per dm^3. Suggest the question that the student should ask about 0.5 moles per dm^3 of acid.
4. Suggest, with reasons, what alterations should be made to the method to enable the student to investigate the question you wrote for question 3.
5. The student formed the hypothesis '*doubling the concentration of the acid would halve the time taken for the reaction to complete*'. Use Table 1 and the graphs to explain whether the hypothesis is supported by the data.
6. Turn in the book to the pages on rates of reaction. Find additional data or information to support the hypothesis.

A*

B2 Growing and using our food

Cells of animals, plants and bacteria have a characteristic structure, typical of their type. Multicellular organisms, like humans, are made up of cells that have differentiated to perform different functions. The digestive system, for example, includes many specialised cells grouped into different types of tissues. Organs like the stomach and small intestine are each made up of several different tissues.

Green plants use light energy to make their own food during photosynthesis. This is the basis of our agriculture. Some of our food crops are grown in greenhouses or polytunnels where the environment can be controlled. Light, temperature and carbon dioxide concentration can be manipulated to enhance the crop's growth.

In natural environments, physical factors, such as amount of light and availability of water, affect where we find certain plants. Fieldwork using quadrats and transects can provide quantitative data about the types and number of plants growing in a habitat. A tree trunk's surface, the ground in the shade of a tree, or a field can be investigated.

Test yourself

1. Where are chromosomes found in plant and animal cells?
2. Describe the reproduction rate of bacteria and explain whether bacteria reproduce faster in the human body or at a room temperature of 20 °C.
3. Explain what is meant by a healthy diet.
4. Why are organisms able to survive best in conditions in which they normally live?

Objectives

By the end of this unit you should be able to:

- describe the structure and function of animal, plant, bacterial and yeast cells
- explain what is meant by diffusion and how it is affected by the difference in concentration between two areas
- list the hierarchy of specialised cell organisation into organ systems, such as the digestive system
- explain that the salivary glands, stomach, pancreas and small intestine contain glandular tissue that produces digestive juices
- describe how the limiting factors of photosynthesis can be controlled to enhance crop growth in artificial environments
- tabulate the uses of glucose in plants and algae
- interpret the distribution of organisms in a community by referring to the physical factors that may affect them.

Animal building blocks

Learning objectives

- describe the structure and function of the parts of animal cells
- explain what controls chemical reactions in cells
- explain how animal cells may be specialised to carry out a particular function.

Animal cells

Animals, including humans, are made up of millions of tiny cells. You can see some of the structure of these cells through a light microscope that can magnify up to 400 times.

Each cell is surrounded by a very thin **cell membrane** that holds the cell together. The cell membrane also controls what goes into and out of the cell.

Cells contain smaller parts called **organelles**. These include the **nucleus**, **mitochondria** and **ribosomes**. The organelles have particular jobs in the cell. The single nucleus controls the cell's activities and it is surrounded by watery **cytoplasm**. Without the nucleus the cell will die. It contains **DNA**, the **genetic** material that provides the instructions for synthesising the chemicals the cell needs, like **enzymes**. Inside the cytoplasm hundreds of chemical reactions take place and these reactions are controlled by the enzymes.

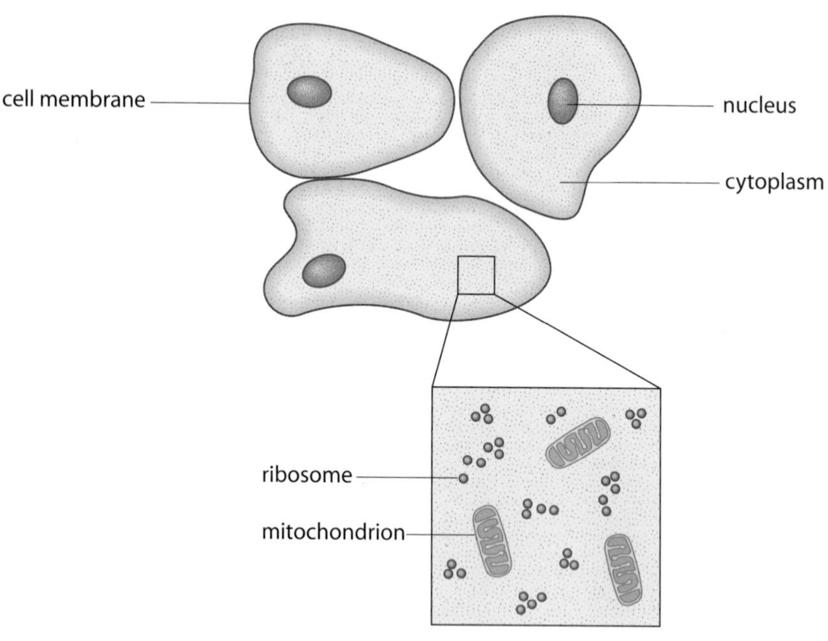

Figure 1 The organelles found in animal cells.

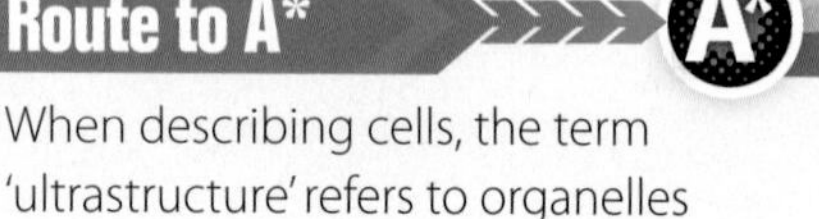

Route to A*

When describing cells, the term 'ultrastructure' refers to organelles that are only visible with the electron microscope. Use 'ultrastructure' in your answer to the following question:

Describe the relationship between amino acids, proteins and ribosomes.

Electron microscopes magnify up to 500 000 times and show us details of smaller organelles, such as mitochondria. These use **glucose** in respiration to release energy for the cell. Ribosomes are the smallest organelles. They build up, or synthesise, **proteins** from smaller, simpler compounds called **amino acids**. Proteins are molecules that are used to make other parts of the cell and other chemicals, such as enzymes. The cells of most animals, including humans, have the same organelles.

Taking it further

Ribosomes measure about 20 nm in diameter. Besides being present singly in the cytoplasm, as in Figure 1, they may be present in a chain or attached to an organelle called the rough endoplasmic reticulum (rough ER). You will be expected to know about rough endoplasmic reticulum for AS and A-level Biology.

Specialised animal cells

The cells in Figure 1 line the inside of your mouth. They are called simple **epithelial cells** and line cavities and tubes, like blood vessels in your body. Many different types of cell are found in your body. They have different shapes and many have special features that are related to what they do. These cells are called **specialised cells**. Some examples are shown in Figure 2.

a Spindle-shaped muscle cells have fibrils and can shorten in length.

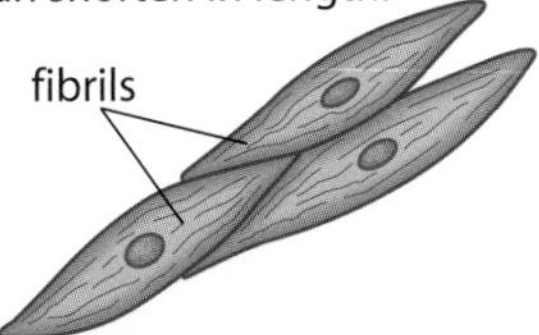

b Sperm cells have a tail to help them move to find the egg. They also have a high number of mitochondria to release energy for movement.

tail

c Nerve cells have long fibres that carry electrical impulses. Branches of cytoplasm at each end of the cell facilitate communication with other nerve cells.

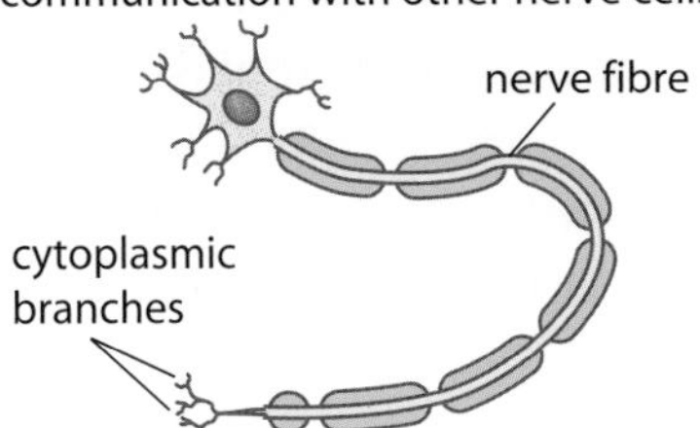

d These specialised epithelial cells have tiny hair-like structures, called cilia, on their free surface. They are known as ciliated epithelial cells. The cilia sway constantly back and forth to move particles along.

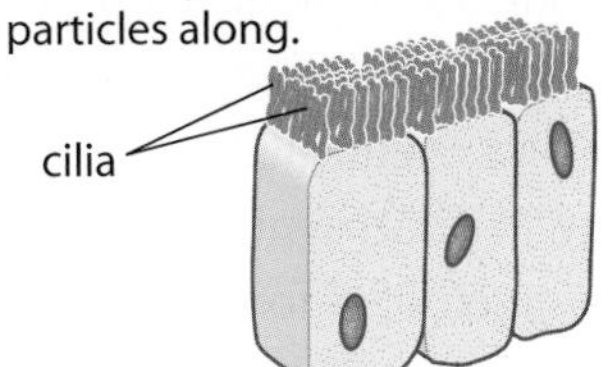

Figure 2 Examples of specialised cells (a muscle cell; b sperm cell; c nerve cell; d epithelial cell).

Science in action

Scientists study human cells using the light microscope. Cells are collected from **organs** and fluids throughout the body. Pre-cancerous changes, cancer and some infections can be detected and diagnosed. Cancerous cells are usually smaller than the cells surrounding them and they also have enlarged nuclei.

Science skills

Table 1 Numbers of mitochondria.

Type of human cell	Average number of mitochondria per cell (to the nearest 100)
liver	1900
kidney	1500
skin	200
small intestine	1600
muscle	1700

a What is the range in the number of mitochondria in human cells?

b Explain which method you would use to display this data.

Examiner feedback

The independent variable is the one that is being changed by the investigator. It goes on the *x*-axis of a graph.

The dependent variable is the result that is seen when the independent variable changes. It goes on the *y*-axis of a graph.

A line graph is drawn only when both variables can have any numerical values.

Questions

1. Explain how it is possible to see the parts of human cells, such as the nucleus.
2. Explain how substances that enter and leave cells are controlled.
3. Why are mitochondria found in large numbers in muscle cells?
4. Ribosomes synthesise proteins. Explain what this means.
5. Explain why the nucleus of the cell is important.
6. Look at Figure 2. **(a)** Give two differences between a muscle cell and a ciliated epithelial cell. **(b)** Most cells don't move. How does each of the following help a sperm cell to move: **(i)** tail **(ii)** mitochondria **(iii)** its shape? **(c)** The epithelial cells shown help to sweep mucus containing dust out of the lungs. Describe how they are adapted for this job.
7. Compare and contrast a nerve cell and a sperm cell, both structurally and functionally.
8. Select three examples of specialised animal cells and explain how the special features of each one adapts the cell for its function.

B2 1.2

Plant and alga building blocks

Learning objectives

- describe the structure and function of the parts of plant cells
- explain how plant cells may be specialised to carry out a particular function
- compare and contrast animal and plant cells.

Route to A*

Leaves contain two different types of mesophyll cells that you can distinguish between. Figure 1 shows palisade mesophyll cells, which are adapted, by the presence of many chloroplasts, to trap light energy for photosynthesis.

Find out the name of the other type of mesophyll cells in a leaf and learn what their function is.

Plant cells

Like animal cells, plant cells usually have a cell membrane, nucleus, mitochondria, ribosomes and cytoplasm. Unlike animal cells, plant cells also have a **cell wall**, made of a carbohydrate called **cellulose**. The cellulose in the cell wall is in the form of tiny fibres. Together, these fibres are very strong so the cell wall supports the cell and strengthens it. Algae are also made of cells that have a cell wall made of cellulose. Examples of algae include seaweed and microscopic, single-celled algae that grow on tree trunks or in fish tanks, giving the water a green colour. Some plant cells also have organelles called **chloroplasts** in their cytoplasm. Inside the chloroplasts is a green pigment called **chlorophyll**. Chlorophyll is a chemical that plants use in **photosynthesis** to absorb the Sun's light energy. In photosynthesis, light energy is converted to chemical energy in the form of glucose, as food for the plant.

In the centre of many plant cells there is a large, permanent, liquid-filled space called a **vacuole**. The liquid in the vacuole is called **cell sap** and it contains sugars, salts and water. When it is full the vacuole supports the cell, making it firm. If the vacuole is less full, the cell is not so firm.

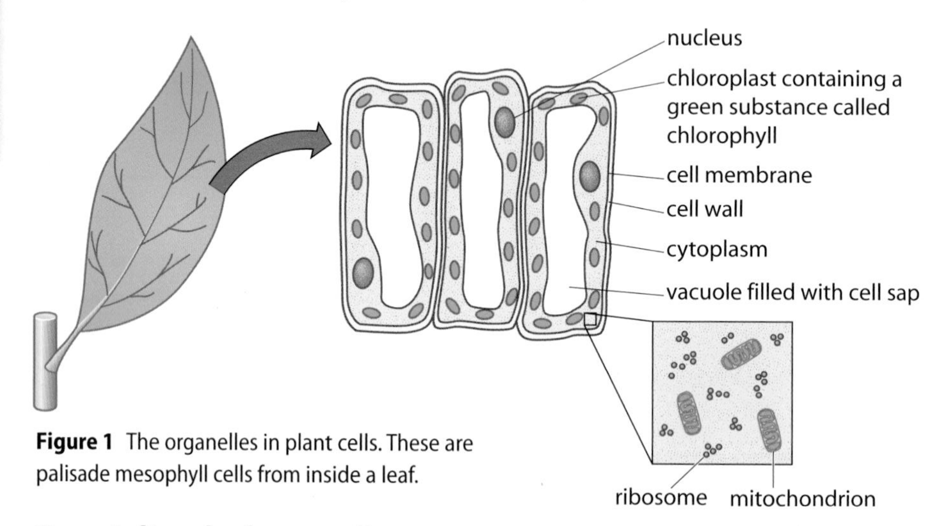

Figure 1 The organelles in plant cells. These are palisade mesophyll cells from inside a leaf.

Specialised plant cells

Many plant cells are specialised to carry out particular jobs.

- **Palisade mesophyll cells** are found in the leaf and are packed with chloroplasts. They are the main photosynthetic cells.
- **Root hair cells** have extensions into the soil to absorb water and dissolved mineral ions. These extensions are the actual root hairs. They are long and narrow so they can fit between soil particles. A thin film of water surrounds each soil particle and it contains the dissolved mineral ions.

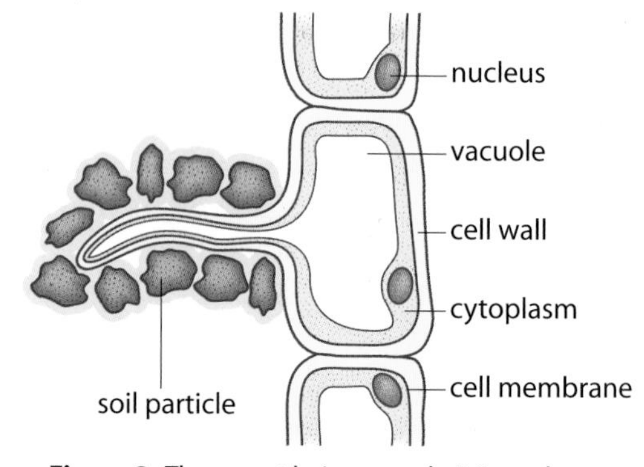

Figure 2 These root hairs are only 0.1 mm long. Each root end has thousands of these cells.

Root hairs on a germinating seed.

- **Xylem vessels** are made up of empty dead cells, arranged as long tubes of cell wall only, with no end walls between them. The cell wall has various chemicals added to it in xylem, so it does not rot away. Xylem vessels transport water from the roots, up through the stem to the leaves.

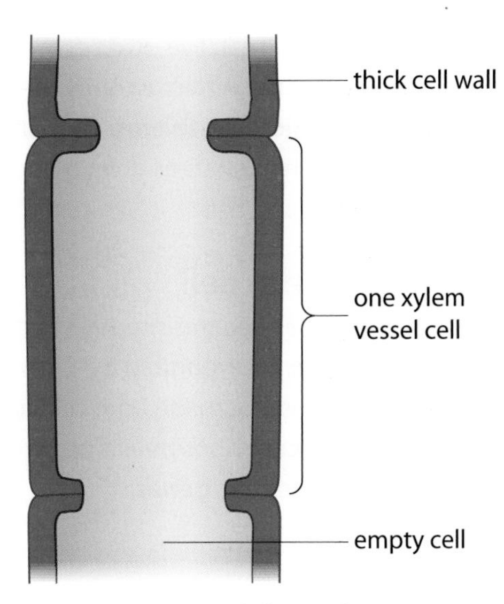

Figure 3 Xylem vessels from a plant stem.

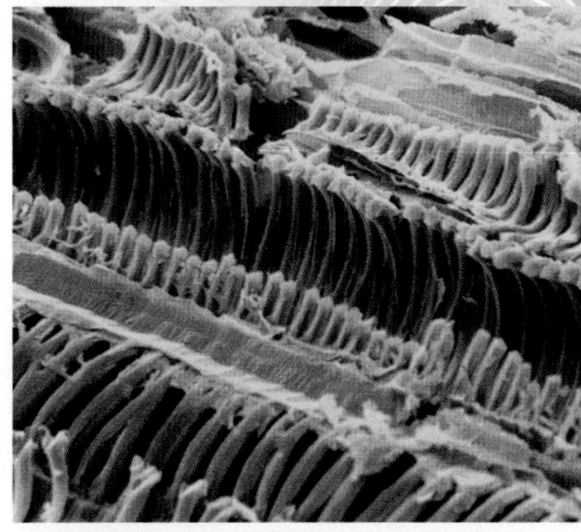

Scanning electron microscope photo of a section through a rhubarb stem, showing a xylem vessel cut open. Magnification ×290.

Science skills

Two students were measuring cells with a microscope. They used a clear plastic ruler, calibrated in mm, clipped to the microscope. The magnification of the microscope was ×10.

a Suggest how many cells they should measure to give a reliable result. Explain your answer.

b The students reported that one type of cell was 0.3 mm in diameter. Comment on the accuracy of the result bearing in mind the resolution of the ruler.

Practical

Figure 4 You can make slides of plant cells and look at them through a light microscope.

Questions

1. Why is the cell wall important in an algal cell and what is the cell wall made from?
2. Explain why plants need: **(a)** a nucleus **(b)** mitochondria **(c)** a cell membrane **(d)** ribosomes.
3. List three facts about chloroplasts.
4. What fills the large, permanent space in a plant cell? What chemicals does it contain and what is its function in the cell?
5. Explain fully the advantage of palisade cells having lots of chloroplasts.
6. Suggest which parts of a plant might not contain chloroplasts. Explain your answer.
7. Compare and contrast the structures of animal and plant cells.
8. Look at Figures 2 and 3. Explain how root hair cells and xylem vessels are adapted for their function in the plant.

A*

Examiner feedback

Distinguish carefully between the cell wall and cell membrane. A cellulose cell wall is present in plant, but not animal, cells and it allows all substances in solution to pass through it. A cell membrane is present in every cell and allows some substances to pass through, but not others. Cellulose is only found in plant and algal cells.

B2 1.3 Bacteria and yeast cells

- describe the structure of a bacterial cell
- describe the structure of a yeast cell
- compare and contrast the structure of a bacterial cell and yeast cell.

The most abundant cells on Earth

Tens of billions of bacteria may be present in a handful of soil. You have more bacteria in your intestines and on your skin than cells in your body. Bacteria are **unicellular organisms**. Each single cell can live on its own and carry out all the seven characteristics of living organisms. A microscope is needed to see bacteria, so they are known as **microbes** or **microorganisms**. They are found in and on plants and animals, and worldwide in habitats as diverse as deserts, deep oceans, snow and boiling mud.

Examiner feedback

The mnemonic 'Mrs Gren' will help you to remember the seven characteristics of living things. They are: movement, reproduction, sensitivity, growth, respiration, excretion and nutrition.

The structure of a bacterial cell

Under the electron microscope we can see the internal structure of a bacterial cell. A bacterium consists of cytoplasm surrounded by a membrane and an outer cell wall. The cell wall is semi-rigid and not made from cellulose. There is no nucleus, and no other organelles, except ribosomes. The cytoplasm contains a loop of DNA that contains most of the cell's **genes**.

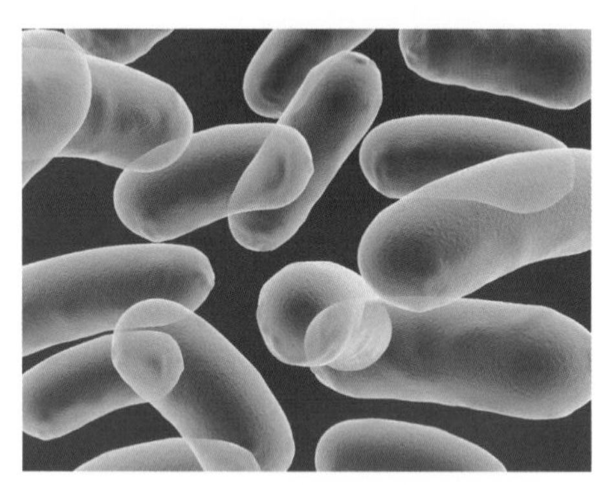

Bacteria under the electron microscope.

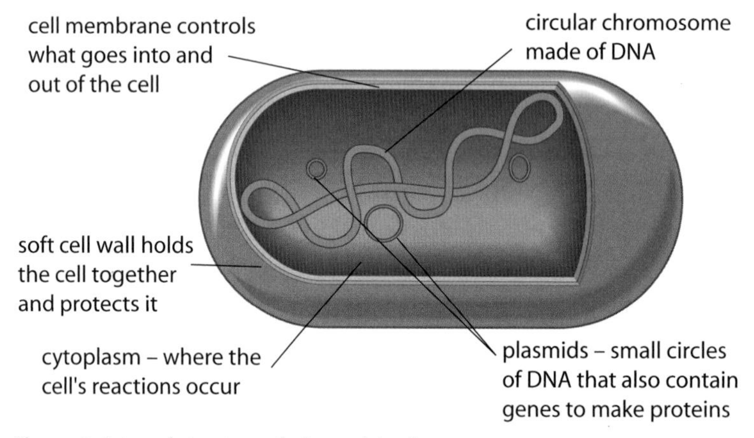

Figure 1 Internal structure of a bacterial cell.

Taking it further

Bacterial cells have no nucleus and no other organelles surrounded by a membrane. They are known as prokaryotic cells.

Yeast cells have a membrane-bound nucleus and other organelles. They are known as eukaryotic cells. Animal and plant cells are also eukaryotic.

Science skills

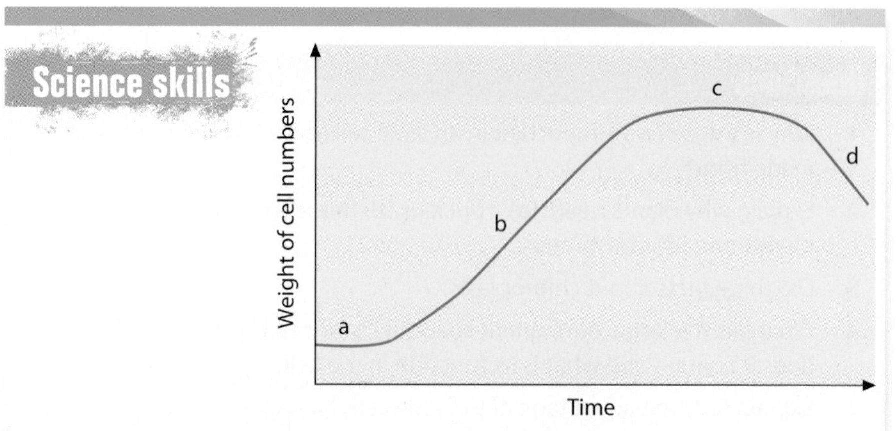

Figure 2 Bacterial cells grow quickly in certain conditions.

a Suggest which region of the graph corresponds to ideal growth conditions.

b Give a reason for your answer to part **a**.

c Suggest what is happening to the bacteria in region 'd' of the graph.

Yeast is a single-celled microscopic fungus

Some fruits – like grapes, plums and apples – often have a pale grey 'bloom' on their surface. This is partly due to naturally occurring yeast. If you have ever polished a plum or apple by rubbing it on your clothes you have removed the natural yeast. Yeast occurs on plant leaves, flowers and in the soil. Yeasts are also found in dust, water, milk and even on some of the inside surfaces of our body, such as the linings of the body cavities and various tubes.

Yeast growing on fruit skins.

Structure of a yeast cell

You can buy a block of fresh yeast for baking bread. Each square centimetre of it contains millions of individual yeast cells.

Each yeast cell is oval or spherical and has a nucleus, cytoplasm, mitochondria, a vacuole and a cell membrane surrounded by a cell wall. Yeast cells are about 10 times bigger than bacterial cells.

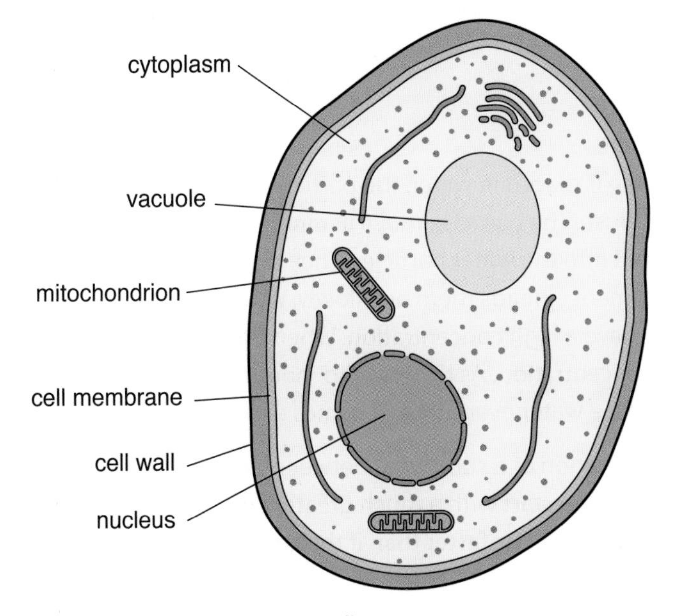

Figure 3 Section through a yeast cell.

Compressed yeast. Each 1-cm^3 block contains millions of yeast cells.

Science in action

Breweries use vast amounts of yeast. It is sent to them in granulated form from suppliers called maltsters. Yeast cells survive even when they are dehydrated. Yeast produces the alcohol in beer.

Questions

1. Why do we refer to bacteria as microorganisms?
2. Name and describe the only organelle found in bacterial cells.
3. Where is the genetic material found in bacterial cells and what is it made of?
4. What type of organism is a yeast cell?
5. Name and describe the organelle that contains genes in a yeast cell.
6. **(a)** Give three similarities in structure between bacterial and yeast cells.
   (b) Give two differences in structure between bacterial and yeast cells.
   (c) How much bigger are yeast cells than bacterial cells?
7. Give two ways that yeast can be used in the home and describe two forms that yeast could be in when it is bought.
8. Explain, as fully as you can, where bacteria and yeast are found naturally and why they are usually found in such large numbers.

Route to A*

A question that asks you to compare two organisms, such as bacteria and yeast, requires you to give similarities *and* differences between the two. Use the mark allowance to help you to supply a suitable number of marking points in your answer.

B2 1.4

Getting in and out of cells

Learning objectives

- describe the process of diffusion
- explain diffusion through cell membranes
- describe how the difference in concentration can affect the rate of diffusion
- explain what a partially permeable membrane is.

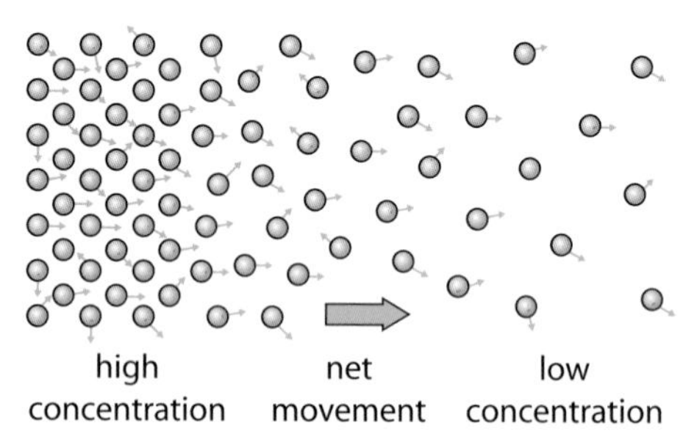

Figure 1 Particles move from areas of high concentration to areas of low concentration.

Route to A*

When defining diffusion, include the term 'net movement' and finish by saying that diffusion carries on until the particles in an area are equally distributed.

Diffusion of gases

You can smell the perfume released by a flower because smelly particles spread through the air. The smell can spread many metres. The movement of the perfume particles through the particles of air is called **diffusion**.

As you get closer to the flower the smell gets stronger. This is because there are more smelly particles of gas near the flower. We say that the **concentration** of perfume particles is higher nearer the flower. The sense of smell detects chemicals in the form of a gas. Gas molecules diffuse rapidly through the air.

Science in action

In England in July 2007, smoking was banned in all enclosed public places and workplaces to prevent inhalation of smoke by non-smokers. This is called passive smoking and occurs due to diffusion of the particles in cigarette smoke through the gases in the air.

Diffusion of liquids

When a **soluble** substance is placed in water, the particles that make up the substance will start to diffuse. The particles move in random directions, and bump into each other and into the water particles. They start all clumped in one place, but this movement spreads them out slowly. When the particles are clumped together, they have a high concentration. When they are more spread out, they have a lower concentration. As they spread more, the concentration of the particles throughout the water eventually becomes equal.

A difference in the concentration of a substance between two areas is called a **concentration gradient**. If you start with a much greater concentration in one place than the other, diffusion will be faster than if the concentrations in the two places are nearly the same.

Practical

(a)

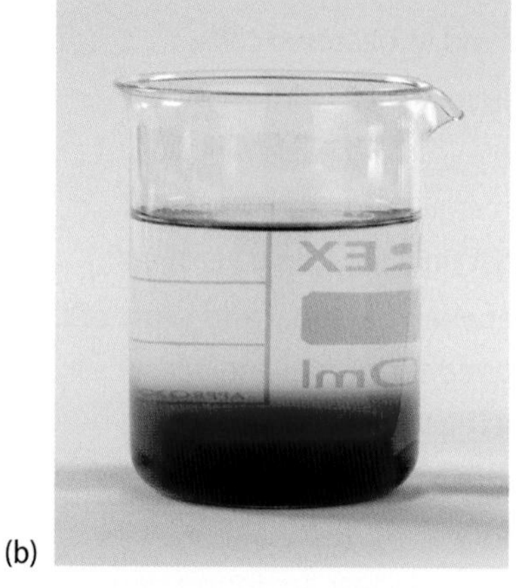

(b)

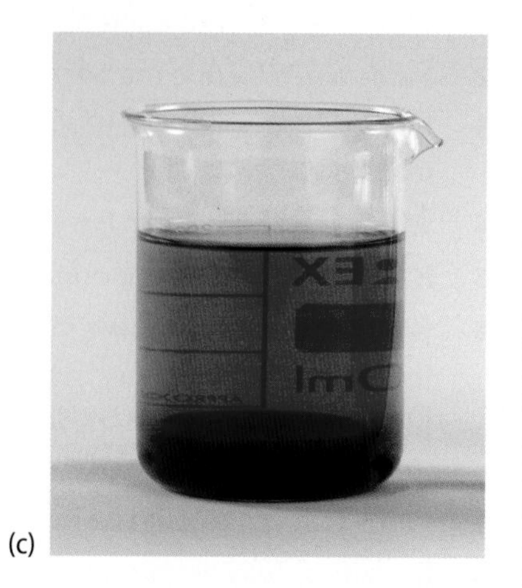

(c)

A slice of boiled beetroot is put in water at 12 noon (a). Diffusion is shown at 6 am (b) and 6 pm (c).

Diffusion through cell membranes

All plant and animal cells have a cell membrane. The cell membrane has tiny holes through which small particles can pass by diffusion. We say that cell membranes are **partially permeable membranes** because large particles cannot get through.

If the concentration of small particles on each side of a membrane is different, then more particles will diffuse through the membrane from the concentrated solution to the dilute solution. The overall or **net movement** of particles is from a higher concentration to a lower concentration. Diffusion results in an even distribution of particles on both sides of a membrane.

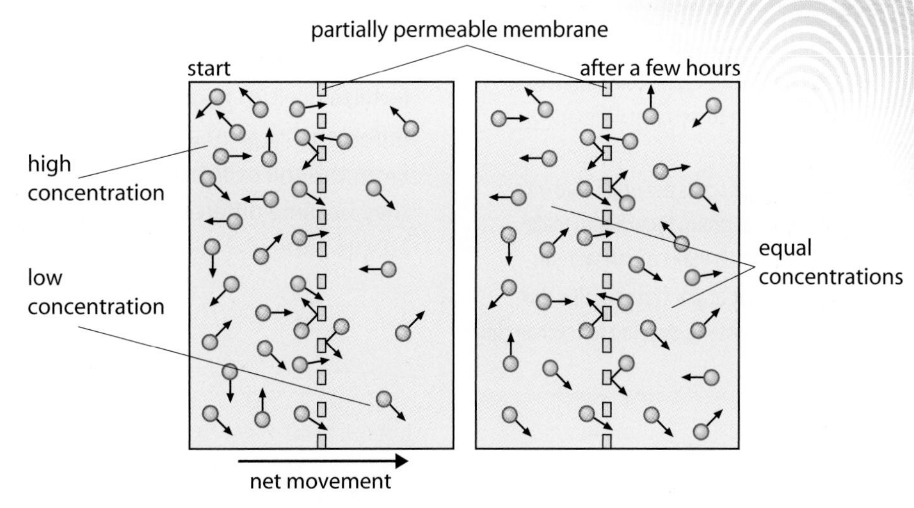

Figure 2 When the concentrations are different, the net movement of particles is from the higher concentration to the lower concentration.

Examiner feedback

It is important to remember that even when there is a big difference in concentration, diffusion occurs in both directions. There is a *net* movement from high to low concentration.

Route to A*

Remember that plant and algal cells require oxygen for respiration too, but because their rate of respiration is low, compared with animals, the rate of diffusion of oxygen into their cells is lower too.

All living organisms need oxygen for respiration. Oxygen molecules are small molecules that can pass through cell membranes by diffusion. There are five cell membranes for oxygen to pass through from an air sac, or **alveolus**, in the lungs, to a red blood cell that carries the oxygen round the body. The rapid rate of diffusion keeps the cells alive. It enables oxygen to get into the blood fast enough to be transported to body cells.

Questions

1. Name one sense that a butterfly uses to find a flower that is far away.
2. Explain how the perfume of the flower reaches the butterfly.
3. Where is the concentration of perfume particles highest: near the flower or far from it? Explain your answer.
4. Explain what we mean by 'partially permeable membrane'.
5. Explain how cell membranes control which particles pass through them by diffusion.
6. Explain what we mean by 'net movement'.
7. If milk is poured into a mug of hot, black coffee and not stirred, explain, using the word 'diffusion', why the coffee changes colour from black to medium brown.
8. Cells need oxygen for respiration. Explain how oxygen can get into cells.

Taking it further

A short diffusion distance increases the rate of diffusion.

From the inside of an alveolus to the inside of a red blood cell the distance is 5 × the width of a cell membrane. Find out what this distance is.

B2 1.5

Specialised organ systems

Learning objectives

- explain that cells differentiate, which allows them to perform different functions
- describe how cells are organised into tissues that group together to make organs, which make organ systems
- explain that organ systems, like the digestive system, develop for exchanging materials.

Differentiation from a single cell

Your multicellular body has developed from one fertilised cell, or **zygote**. This one cell divided repeatedly to form a tiny ball of identical cells. From this the cells began to **differentiate**: they became different by specialisation. This adapts particular cells for a specific function.

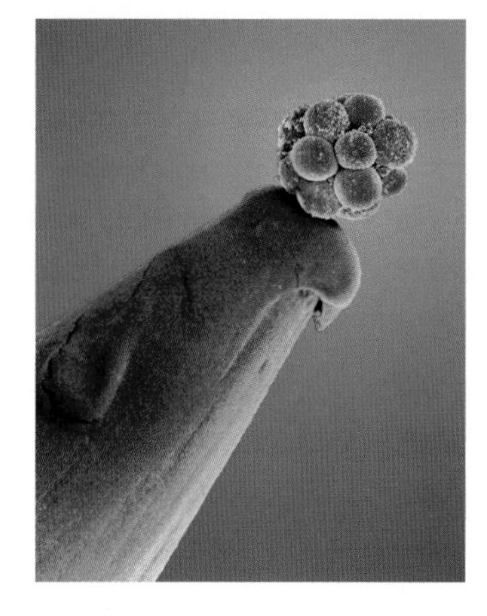

An early human embryo on a pin head, before the cells start to differentiate.

Figure 1 Humans have many different types of tissue.

Similar cells make a tissue

Some living things are made up of only one cell. Other living things are made from millions of cells. In more complex organisms, specialised cells of the same type group together to form **tissues**. A tissue is a group of cells with similar structure and function. In animals a tissue might make a thin sheet of cells, like the epithelial cells that make up linings inside the body. Cells in other tissues group together, like muscle cells that make muscle tissue. Muscle tissue contracts to bring about movement. Cells in glandular tissue produce and release particular chemicals such as enzymes and hormones.

Plants also have tissues. Epidermal tissues cover the plant. Mesophyll tissue carries out photosynthesis in the leaf and xylem and phloem tissues transport substances around the plant.

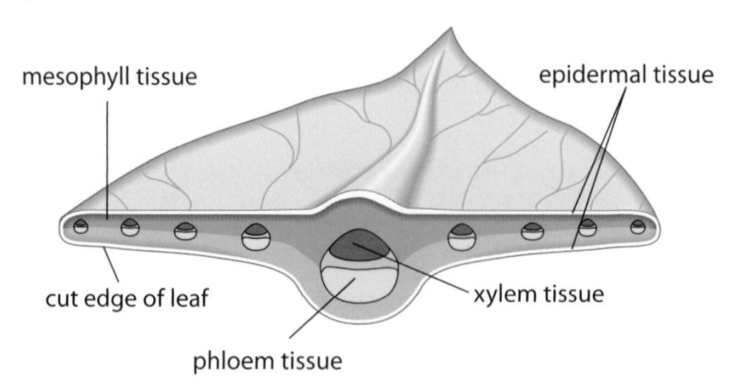

Figure 2 Section through a leaf showing some tissues.

Figure 3 A section through the stomach's tissues.

Several different tissues make an organ

Groups of tissues join together to make more complicated structures that are called organs. For example your stomach is an organ. It has three layers of muscular tissue to churn or mix up the stomach contents as the muscles contract and relax. The inner wall contains glandular tissue that produces digestive juices. Epithelial tissue covers the outside and inside of the stomach. Other organs in your body include your heart, brain, liver and lungs. Organs have a specific function. The function of the stomach is the storage and digestion of food.

Several different organs make a specialised organ system

Systems are groups of organs that perform a particular function. The digestive system in humans and other mammals is one example of a system where substances are exchanged with the environment. Food enters the body, is broken down and absorbed into the blood. Undigested food with some added waste chemicals is returned to the environment.

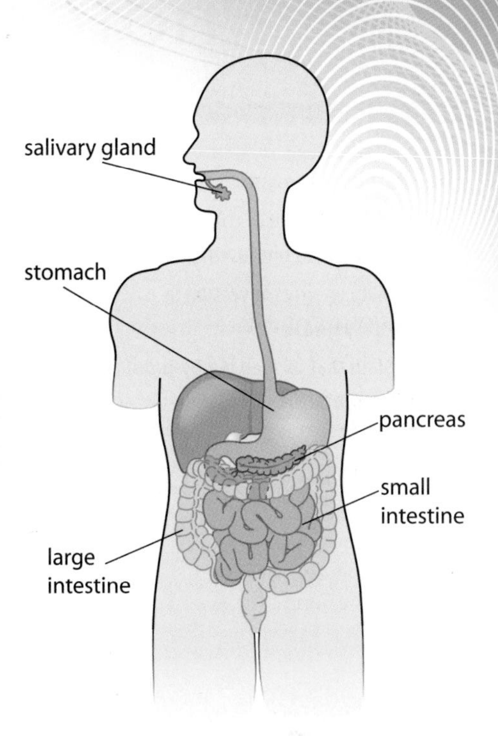

Figure 4 Some organs in the digestive system.

Science skills

Read the information below about the digestive system and then construct a table to show it.

The digestive system includes glands such as the pancreas and salivary glands that produce digestive juices. Digestion occurs mostly in the stomach and small intestine. Absorption of soluble food occurs in the small intestine. In the large intestine, two processes occur: water is absorbed into the blood from the undigested food and faeces are produced.

Plant organs

You have just been studying the hierarchy of cell organisation in animals. You may have realised, when looking at Figure 2, that a leaf is a plant organ, as it contains several different types of tissue and has the specific function of photosynthesis in the light. Two other plant organs are roots and stems. Roots absorb water and minerals. The stem holds the leaves in a good position to catch as much light as possible. Leaves are where photosynthesis occurs in plants when they are in the light.

Questions

1. What is the function of: **(a)** epithelial tissue **(b)** glandular tissue?
2. Define an organ and write down six organs in your body.
3. Arteries are blood vessels carrying blood away from the heart. Suggest what the functions of the following tissues are in an artery: **(a)** epithelial tissue **(b)** muscular tissue.
4. Draw a table listing four plant organs and their functions. Suggest why plants need organs.
5. Outline how you could see plant cells for yourself, in a leaf you were given in the laboratory.
6. Using the stomach as an example of an organ, describe three types of tissue it contains and the function of each tissue's cells.
7. Name a system in humans where substances are exchanged with the environment. Explain how three organs are involved in the example you have given.
8. Explain the terms 'differentiate' and 'specialise', by referring to human cells and tissues.

A*

Taking it further

Xylem, the main supporting tissue in plants, is distributed differently in stems and roots. In stems it is found as distinct oval patches arranged in a circle, while in roots it is found as a central cylinder.

Suggest what forces a plant stem and root are subject to as they grow and survive in the air and soil respectively.

How will the arrangement of xylem in the root and shoot equip them for their survival?

Route to A*

Suggest one other organ system in humans and mammals, in addition to the digestive system, where substances are exchanged with the environment.

Photosynthesis

Learning objectives

- describe photosynthesis in an equation
- describe the role of chloroplasts and chlorophyll in photosynthesis
- explain energy conversion in photosynthesis
- explain that oxygen is a by-product of photosynthesis.

A life-giving chemical reaction

The word 'synthesis' means to combine or to join together to create something new. 'Photo' means light. Photosynthesis uses light energy and two simple molecules: carbon dioxide from the air and water from the soil, to make a more complex molecule called glucose. Oxygen is also released as a **by-product**. Photosynthesis happens in a series of reactions that can be summarised by the equation:

$$\text{carbon dioxide} + \text{water} \xrightarrow{\text{light energy}} \text{glucose} + \text{oxygen}$$

Glucose is a larger molecule than carbon dioxide and water, and contains more energy in its bonds. This energy can be used for growth. Plants make glucose by photosynthesis.

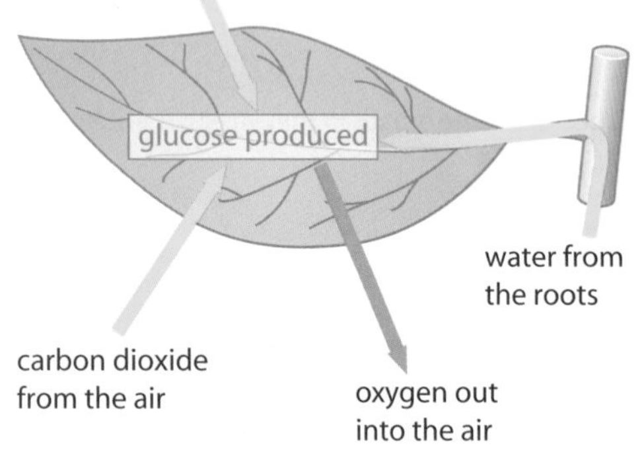

Figure 1 Photosynthesis produces sugar.

Photosynthesis takes place in chloroplasts

Chlorophyll is a green pigment that is found in chloroplasts. Chloroplasts are found mainly in palisade cells in the upper layer of leaves. The chlorophyll absorbs light energy, which is used to convert carbon dioxide and water into glucose.

A variegated plant.

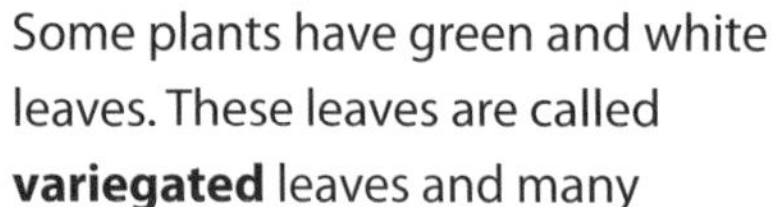

Some plants have green and white leaves. These leaves are called **variegated** leaves and many cultivated plants have them. Only the green parts contain chlorophyll. Photosynthesis can only take place in the green parts of the leaves.

Some algae can photosynthesise

Green algae are organisms that have some different characteristics to plants. They do, however, have chloroplasts and can photosynthesise. You may have seen green algae in ponds and streams in summer. Also, seaweeds are algae. **Plankton** contains the most numerous types of algae. They are single celled and microscopic and are found in the surface layers of lakes, rivers and oceans.

These algae can photosynthesise.

Investigating photosynthesis

One of the easiest ways to see if a plant is photosynthesising is to test it for **starch**. Any excess sugar produced during photosynthesis is stored as the insoluble product starch and this will stain blue-black with the iodine test. In the photograph of the two leaves, the brown leaf is showing the brown iodine stain because no starch is present in it. It has not photosynthesised. The other leaf is stained blue-black because it contains starch. This

The iodine test for starch. If starch is present the leaf turns blue-black.

tells us that the leaf has photosynthesised. Before iodine is added to the leaves they are plunged into boiling water and decolourised by heating in ethanol.

You can also check that oxygen is produced in photosynthesis by collecting and testing it with a glowing splint. The easiest way to do this is by using pondweed, a plant that lives under water.

Practical

Figure 2 Collecting the gas given out by an aquatic plant.

Science skills

The easiest way to measure the rate of photosynthesis is to measure the rate at which oxygen is produced. The two sets of apparatus shown in Figure 3 do this in different ways.

a What two measurements do you need to make in order to calculate the rate at which oxygen is produced?

b Which set of apparatus in the diagram would give the more reliable data: A or B? Explain the reason for your answer.

c Table 1 shows results obtained from this experiment. Suggest the most suitable method of displaying the results.

Table 1 Results from experiment.

Light intensity/ arbitrary units	Rate of movement of meniscus/mm in 5 min
0	0
2	3
4	6
6	9
8	12
10	12

d How many times should the experiment be repeated to make the results reliable?

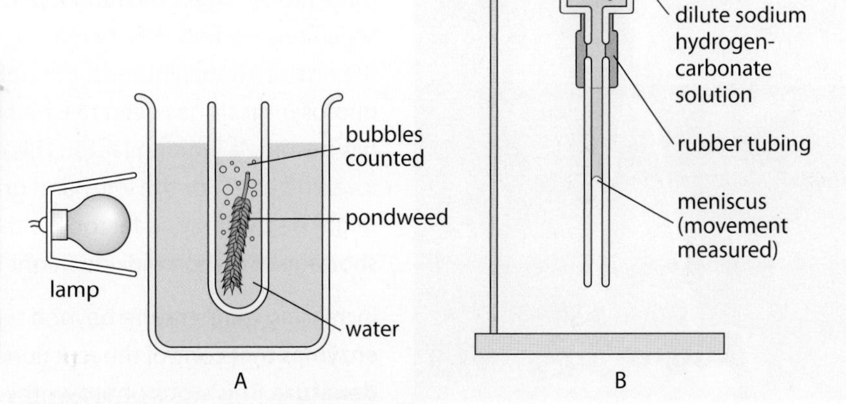

Figure 3 Measuring the rate of photosynthesis.

Questions

1 **(a)** What are the reactants in photosynthesis? **(b)** Where do they come from? **(c)** Why is light needed for photosynthesis?

2 What are the products of photosynthesis and how are they useful to the plant?

3 Explain the role of chlorophyll in photosynthesis.

4 A farmer forgets to water the crops when the weather is dry. What effect will this have on photosynthesis?

5 Describe the appearance of variegated leaves. If you had two leaves of the same type and size, that had been kept in the same conditions but one was variegated and the other was not, suggest with reasons which would have the higher rate of photosynthesis.

6 How could you test a leaf from a plant that has been kept in a dark cupboard for 3 days to see if it had been photosynthesising? What result would you expect from the test?

7 Some tiny green algae grow on the surface of ponds. If a pond gets completely covered by these, the larger plants underneath them die. Suggest why.

8 Pondweed is a plant that lives under water. Explain how it obtains glucose.

B2 2.2 Limiting factors

Learning objectives

- explain what is meant by a limiting factor
- describe the factors that may limit photosynthesis
- describe how factors that limit the rate of photosynthesis can interact with one another
- interpret a graph showing the interaction of limiting factors.

Crop plants photosynthesising.

What limits the rate at which crops grow?

Growing tomatoes is big business. Tomato plants produce our food through photosynthesis. Growers need to know the conditions in which photosynthesis works fastest if they are to harvest the largest possible crop.

Rate of photosynthesis

The rate of photosynthesis is the speed at which photosynthesis takes place. It is affected by the environment. The factors that affect it most are temperature, **light intensity**, availability of carbon dioxide and availability of water. The rate of photosynthesis is limited by low temperature, shortage of carbon dioxide and shortage of light.

If the level of one or more of these is low, the rate of photosynthesis will be slowed down, or limited. The factor that is reducing the rate of photosynthesis is called the **limiting factor**. A limiting factor is something that slows down or stops a reaction even when other factors are in plentiful supply.

How do variations in the amount of light, carbon dioxide and temperature affect the rate of photosynthesis?

If we are to advise growers we need to investigate how variations in these three factors affect the rate of photosynthesis. When we investigate each factor separately we find that increasing the amount of the limiting factor will increase the rate of photosynthesis, but only up to a certain value. After this rate of photosynthesis has been reached there is no further increase. Some other factor has become a limiting factor. This can be seen in Figure 1 where each graph levels off at 'X'. For the left-hand graph low temperature or low light intensity might be the limiting factor. For the right-hand graph low temperature or shortage of carbon dioxide might be the limiting factor.

Increasing temperature beyond the optimum value for a plant causes the enzymes that control the reactions of photosynthesis to break down or **denature**. This stops photosynthesis.

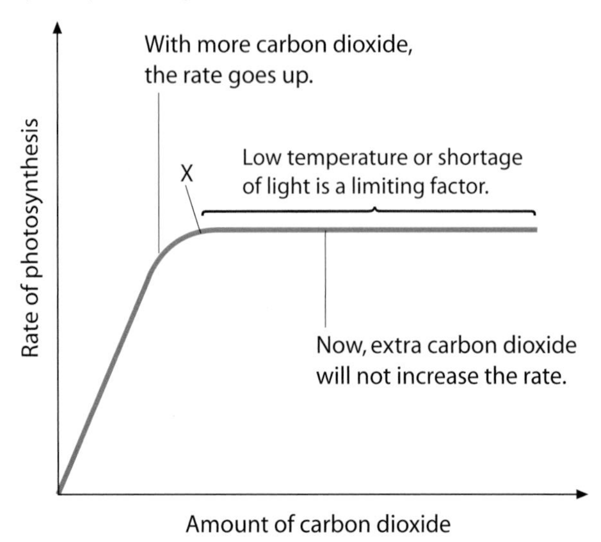

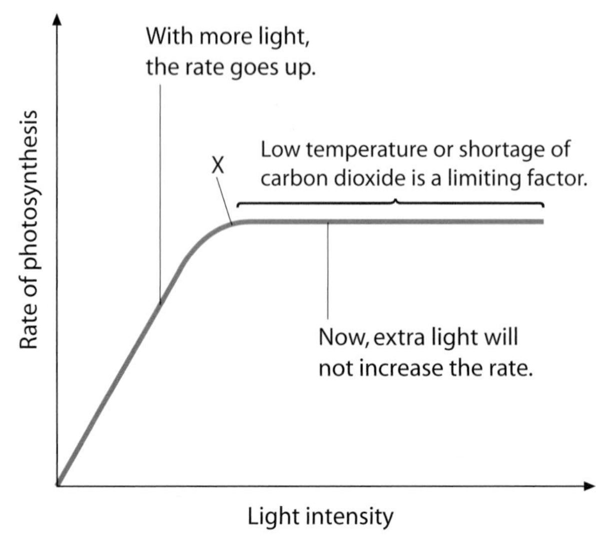

Figure 1 Carbon dioxide and light intensity affect the rate of photosynthesis.

Interaction of limiting factors

In practice, light intensity, temperature and the level of carbon dioxide interact to affect the rate of photosynthesis. Any one of them might be the limiting factor at a specific time of day. For instance at dawn and in the evening the temperature may be low, or in rainy weather thick cloud may reduce the light intensity. The Earth's atmosphere has a very low concentration of carbon dioxide of about 0.04% and carbon dioxide can be a limiting factor when plants grow densely together as crops or in tropical rainforests.

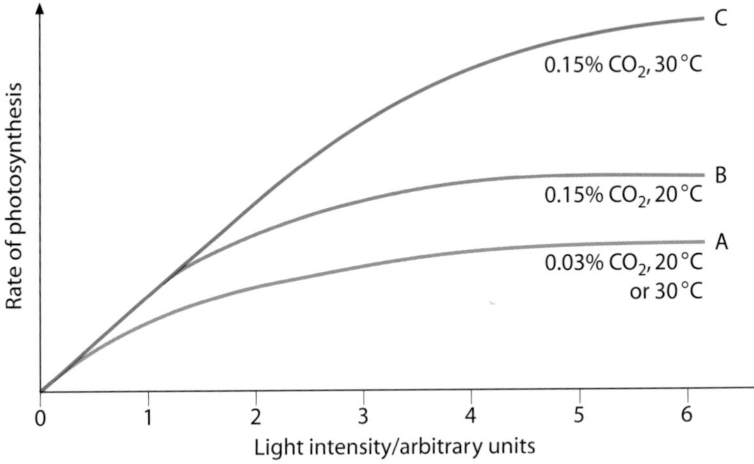

Figure 2 Interaction of limiting factors affects the rate of photosynthesis.

Questions

1 What is meant by the rate of photosynthesis?

2 List three factors that affect the rate of photosynthesis.

3 Explain what a limiting factor is.

4 **(a)** Using the three limiting factors discussed on this page, construct a table and suggest, using ticks, which of them you think may be limiting in the following locations: the Arctic, a hot desert, a tropical rainforest. **(b)** Which other substance in short supply will limit or prevent photosynthesis?

5 Explain the financial implications for a grower of ignoring the concept of limiting factors.

6 Look at the graphs in Figure 1. **(a)** Suggest why increasing the amount of carbon dioxide increases the rate of photosynthesis. **(b)** Suggest why increasing the amount of light can increase the rate of photosynthesis. **(c)** Explain why both graphs level off as the factor continues to increase. **(d)** The percentage of carbon dioxide in the air is about 0.04%. On a warm, sunny day, suggest which factor is limiting the rate of photosynthesis in the middle of a crop. Give a reason for your answer.

7 Look at Figure 2. **(a)** Which curve shows the highest rate of photosynthesis at 6 units of light intensity? **(b)** Which curve shows the lowest concentration of carbon dioxide? **(c)** Explain why curve C is much higher than curve B. **(d)** Suggest why you get curve A even if the temperature is increased from 20 °C to 30 °C.

8 You have been asked to experiment and find out the rate of photosynthesis of a specimen of pondweed provided by your teacher. Outline what you would do, giving reasons for your proposed method.

Route to A*

Curves with a similar shape to those in Figure 1 are very common in biology. You need to be able to explain why the curves go up to start with and why they then level out.

Taking it further

Plants are **autotrophic** in their nutrition – this means they feed themselves. Plants build up organic molecules from simple inorganic molecules, using light as a source of energy.

The biochemical pathway involved has two main stages: the light-dependent reaction and the light-independent reaction. Suggest which of these stages is mostly limited by temperature.

B2 2.3

Uses of glucose produced in photosynthesis

Learning objectives

- describe how glucose may be converted to insoluble starch, fat or oil and stored in cells
- explain that cells use some glucose for respiration
- explain that glucose is also used to synthesise cellulose and proteins.

Energy foods

Many of our energy foods come from parts of plants that are stores of carbohydrate. Potatoes and cereals are part of our **staple** diet. Both store plenty of carbohydrate. These stores are not for the benefit of humans. Stored carbohydrate helps the plants to survive over the winter and to support the growth of new plants the following spring.

Glucose releases energy during respiration

During photosynthesis plants make glucose:

$$\text{carbon dioxide} + \text{water} \xrightarrow{\text{light energy}} \text{glucose} + \text{oxygen}$$

The glucose a plant makes does not stay as glucose for long. Some of it is used for respiration:

$$\text{glucose} + \text{oxygen} \longrightarrow \text{carbon dioxide} + \text{water (+ energy)}$$

The energy released in respiration in plant and animal cells is used in many processes. It is used to build new cells for growth, or for repair of damaged cells. Some energy is used in chemical reactions to change some materials into others. More energy is used to move materials around inside the organism.

Examiner feedback

Remember to use the word 'release' when you are writing about the energy from glucose that is available after respiration. This is because energy cannot be created or destroyed.

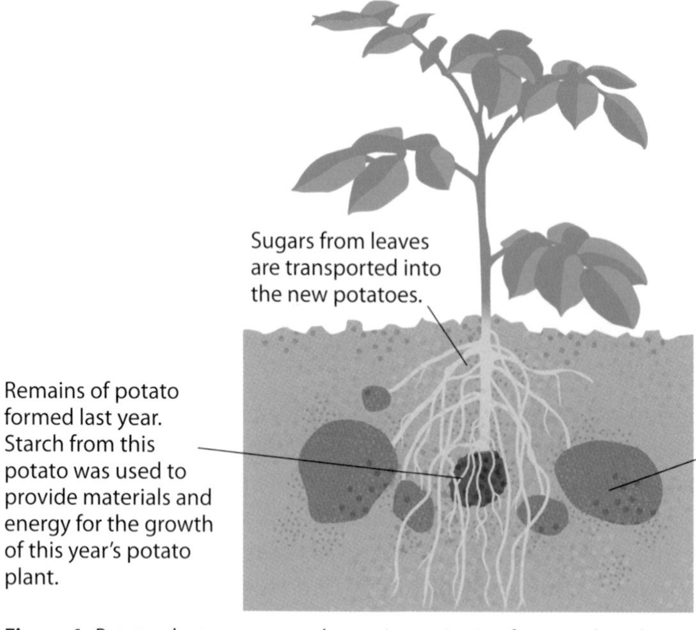

Figure 1 Potato plants use some glucose in respiration for growth and some to make potatoes.

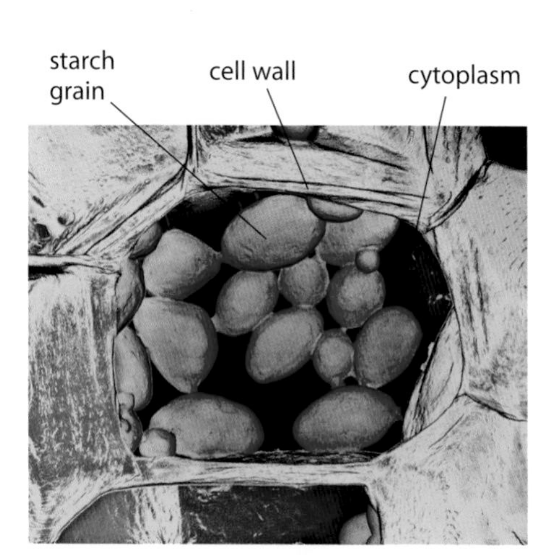

Starch is stored in plant cells as starch grains in the cytoplasm. Magnification ×309.

Conversion of glucose to starch

A plant won't use all the glucose it makes straight away. Some glucose needs to be converted to starch and stored for times when the plant can't make enough glucose, for instance when it is too cold.

Starch is useful for storage because it is **insoluble** and doesn't react easily with other chemicals in the cytoplasm. The starch is an **energy store**, because the plant can change it back to glucose when it needs more energy from respiration.

Most plant cells contain some starch, but some cells are specially adapted to store a lot of it. For example, potatoes are stem tubers full of starch that grow underground. In the spring, the potatoes provide the energy stores needed to make new potato plants. These grow from buds on the old potatoes. Seeds, such as those of wheat and rice plants, also store a lot of starch. These stores are used to provide the energy needed to make new leaves when the seed germinates.

Conversion of glucose to fat or oil

Some plant seeds, plants, and algae store energy as fat or oil droplets in the cytoplasm. Fats and oils belong to a group called **lipids**. One gram of lipid has a higher energy content than 1 gram of carbohydrate. Lipids provide more glucose for respiration than carbohydrates, such as starch, when cells need it for respiration.

High-energy lipids are formed in some plant and algal cells after photosynthesis.

Synthesis of cellulose and proteins from glucose

When large molecules of cellulose are synthesised, many molecules of glucose are linked together to form strong fibres. This makes cellulose a useful structural material in plant and algal cell walls. It strengthens the cell wall and prevents the cell from bursting when it absorbs water.

All living cells need proteins to form enzymes and cell membranes. Plant and algal cells can synthesise proteins from glucose and other raw materials, such as **nitrates**. Soluble nitrates can be absorbed through cell membranes from soil, or water if the plant is aquatic.

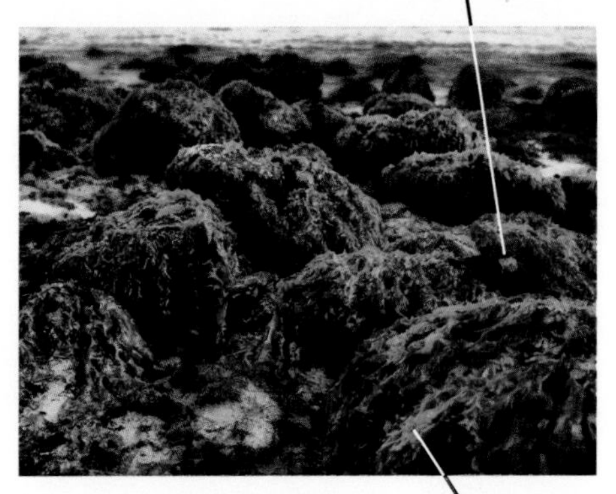

Seaweed uses glucose and nitrates to build its cells.

Questions

1. Look at the equations for photosynthesis and respiration. **(a)** What are the similarities in the two equations? **(b)** What are the differences in the two equations? **(c)** Animals also respire. Where do they get their glucose from for respiration?
2. Give three processes for which a plant needs the energy from the respiration of glucose.
3. Explain what we mean by energy store.
4. Name: **(a)** one plant where starch is stored in stem tubers **(b)** two plants where starch is stored in the seeds **(c)** two fruits that store lipid **(d)** two seeds that store lipid.
5. Explain why potato plants and plants growing from seed need food stores.
6. A wheat grain contains a lot of starch. **(a)** Explain where the starch comes from. **(b)** Explain why the starch is stored in the grain. **(c)** Write down what the starch will be used for by the plant.
7. Why is the formation of cellulose dependent on photosynthesis?
8. Describe the uses of glucose produced in photosynthesis that is not used for respiration or stored as starch.

Taking it further

Cereal crops, which are grown for their seeds, require large amounts of nitrate fertilisers.

Explain the links between seed formation, photosynthesis and nitrates in wheat crops.

B2 2.4

Enhancing photosynthesis in greenhouses and polytunnels

Learning objectives

- explain how greenhouses and polytunnels enable crops to be grown throughout the year
- explain how enhancing limiting factors for photosynthesis can increase crop yield
- evaluate the financial benefits of changing environments to improve growth rates.

All-year salad

In the nineteenth century, most people in this country could only buy lettuces in summer. Now you can eat lettuce grown in the UK at any time of year. Artificially controlling the environment around a crop makes it possible to grow it at times of year when it wouldn't grow as well outside, or in places where it wouldn't grow well, such as very dry places. The conditions outside in fields cannot be easily controlled. Growing crops in **greenhouses** or **polytunnels** makes it possible to control the environment around the plants, including the soil.

Maximising the conditions for photosynthesis.

Healthy eating at any time of the year.

Crops grow when the plants are photosynthesising. The faster the rate of photosynthesis, the better the growth of the plants; the better the growth of plants, the higher the **yield**. Yield is the amount of crop a plant produces. Growers can now get plants to photosynthesise all year round. They can do this by enhancing light, carbon dioxide concentration and temperature.

Taking it further

There are several aspects of light quality that are relevant to photosynthesis and plant growth. In addition to light intensity and duration of the light period, the wavelength of the light is important for photosynthesis.

Find out which wavelengths of light are most useful to photosynthesising green plants.

Enhancing light

Light is rarely a limiting factor in summer, but some growers supplement natural light with artificial light in winter. In industrial greenhouses the problem with enhancing light is making sure that all the plants receive the extra light. Tall plants shade each other if they are not adequately spaced. Supplementary light is usually only economic when the plants are small and then they can be grown more densely.

Enhancing carbon dioxide

The very low concentration of carbon dioxide in the atmosphere (about 0.04%) can limit the rate of photosynthesis when plants are grown closely together. Carbon dioxide levels in a greenhouse can be increased by burning a fuel such as propane or adding the gas through PVC tubing. Growers find that there is a limit to increasing the rate of photosynthesis by increasing the concentration of carbon dioxide. Another factor, such as temperature or light, may then become limiting.

Controlling temperature

Keeping a greenhouse at a temperature of between 25 °C and 30 °C in winter requires heating. Most greenhouses are heated in winter by a boiler and radiators. In summer the temperature in the greenhouse will sometimes exceed 30 °C. This is because heat waves from the Sun are trapped in the greenhouse by being continually reflected. This causes the rate of photosynthesis to decrease. Vents can be opened in the greenhouse roof and blinds can be used to lower the temperature.

Science in action

When propane is burned, carbon dioxide is produced and heat is released. Liquid carbon dioxide is pure carbon dioxide and is stored in special cylinders. After vaporisation the carbon dioxide is delivered to the plants by PVC tubing with a hole punched in the tubing near each plant.

Scottish strawberries growing in a polytunnel.

Science skills

Rate of photosynthesis

0 10 20 30 40 50

Temperature/°C

Figure 1 How temperature affects the rate of photosynthesis in a greenhouse crop.

a What is the best temperature for growing this crop?

b At what temperature would you open the roof vents? Explain your answer.

c Describe the change in rate of photosynthesis between 30 °C and 40 °C.

Science in action

Pollination is essential to fruit formation. When strawberries are grown in greenhouses or polytunnels, growers import live bees to pollinate the flowers. These imported bees must not be released outside as they may affect the health of British bees.

Questions

1 Give three factors that can be controlled in polytunnels and greenhouses to increase the rate of photosynthesis.

2 **(a)** For each of the factors in your answer to question **1**, explain how a grower can increase that factor in a greenhouse. **(b)** Why would a grower want to increase these factors?

3 Give two advantages of growing crops in controlled environments.

4 Give two disadvantages of growing crops in controlled environments.

5 Suggest why a grower might choose to grow crops in a controlled environment.

6 UK-grown lettuces are available at Christmas time. Explain how this is possible and any disadvantages for growers and consumers.

A*

Examiner feedback

When extracting data from a graph, it is sensible to draw a construction line from the curve of the graph to the relevant axis. This will enable you to read off a value and give an accurate answer.

Manipulating the environment of crop plants

Learning objectives

- explain the advantages and disadvantages of manipulating environmental conditions in which plants grow
- evaluate the practice of manipulation in greenhouses
- interpret data showing how environmental factors affect the rate of photosynthesis.

Propane burner for carbon dioxide enrichment, installed in a greenhouse.

Science in action

All of the carbon dioxide released in commercial greenhouses, as described in this chapter, has been produced by burning fossil fuels. The carbon dioxide-enriched atmosphere boosts crop growth. However, carbon dioxide released into the atmosphere by the same combustion process is regarded as one of the main contributors to global warming. Globally, this may reduce crop growth in future decades.

Economics of enhancing photosynthesis

High energy costs for supplementary heating, lighting and carbon dioxide concentration prohibit some potential growers from using greenhouses and polytunnels. In addition there are one-off costs for building and equipment to set up this type of agriculture. However, it results in a bigger yield, and crops grown out of season usually sell for more money.

Getting carbon dioxide to greenhouse crops

Three ways of supplying carbon dioxide to greenhouse crops are described below. The information in each section is for 4 hectares of greenhouse maintaining a carbon dioxide concentration of 1300 parts per million (ppm).

Propane burners

When propane is burned, carbon dioxide is produced and heat is released. Propane is derived from fossil fuel and these fuels tend to contain sulfur as an impurity. If there is sulfur in the fuel, sulfur dioxide will also be released. About 1.4 kg of water is released for each cubic metre of propane burned. The one-off cost of installing the burners is £32 600 and the daily cost of propane is £217.

Carbon dioxide from flue gases

Natural gas is burned in a microturbine, which is used to generate electricity. The heat released during combustion is used to heat water. This can be circulated immediately throughout the greenhouse by pipes, or stored in large tanks for use at night. The carbon dioxide in the flue gas is distributed to the crops through a pipework system. The one-off cost of the equipment is approximately £118 000 and the natural gas fuel for the microturbine costs £84 per day.

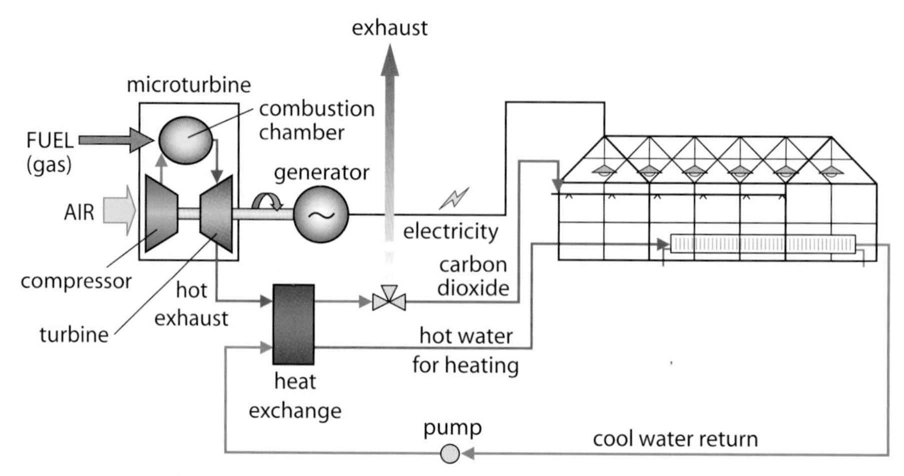

CO_2-enrichment from a combined heat and power (CHP) unit.

Liquid carbon dioxide

Liquid carbon dioxide is pure carbon dioxide. It is delivered in bulk by tankers and stored in special cylinders. The liquid carbon dioxide is vaporised then delivered to the plants by PVC tubing with a hole punched near each plant. The equipment for storing and vaporising the carbon dioxide is rented for £6900 per year and the daily cost of the carbon dioxide is £234. This method of providing carbon dioxide is used more on mainland Europe than in the UK. Once installed, equipment can be expected to last for at least 10 years.

Bulk storage of liquid carbon dioxide.

Science skills

During the day, plants both photosynthesise and respire. The relationship between gross photosynthesis, net photosynthesis and respiration is given in the equation:

gross photosynthesis = net photosynthesis + respiration

Table 1 The rates of gross photosynthesis and net photosynthesis for a cereal crop at different temperatures.

Temperature/°C	Rate of gross photosynthesis/ arbitrary units	Rate of net photosynthesis/ arbitrary units
12	12	10
19	26	24
26	40	37
34	34	27
41	26	11

a Plot a graph of the data in Table 1. Choose suitable scales for the axes. Label each of the curves.

b Describe the effect of temperature on the rate of gross photosynthesis.

c Which factor is limiting the rate of gross photosynthesis between 19 °C and 26 °C? Explain the reasons for your answer.

d The rate of gross photosynthesis is the same at 19 °C as it is at 41 °C. The cereal crop grows more slowly at 41 °C than at 19 °C. Suggest an explanation for this.

Cucumbers are now grown mainly in greenhouses.

Table 2 The yield of cucumbers grown in a well-lit greenhouse under different conditions.

Temperature/°C	Yield of cucumbers/kg per 10 plants	
	0.13% carbon dioxide	0.04% carbon dioxide
12	12	10
19	26	24
26	40	37
34	34	27
41	26	11

e In which conditions did the cucumbers give the greatest yield?

f Would the grower make most profit by using these conditions? Explain the reasons for your answer.

Questions

1. In a table, summarise the advantages and disadvantages of each of the methods of supplying carbon dioxide described above.
2. Imagine you are a grower. Which method would you use? Explain the reasons for your answer.

Assess yourself questions

1 The drawing shows part of a plant as seen through an electron microscope.

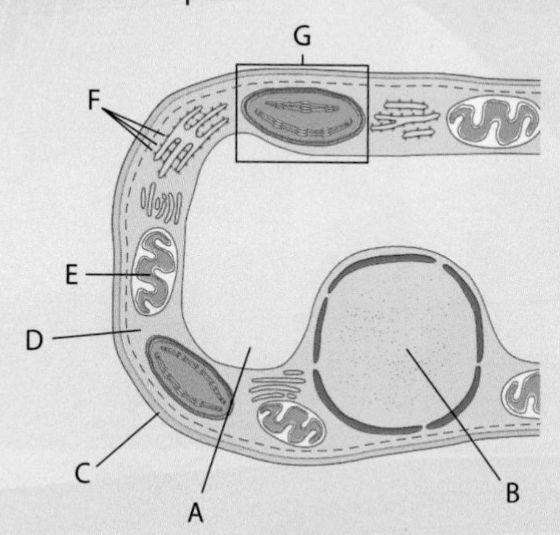

Figure 1 Part of modified plant cell.

(a) Name the structures labelled A–G. *(7 marks)*

(b) Give the function of the part labelled:

(i) E **(ii)** F **(iii)** G *(3 marks)*

(c) 1 μm is 1/1000 mm. The length of five of the structures labelled E were measured. Their lengths were as follows:

5.1 μm 5.5 μm 5.8 μm 5.4 μm 5.7 μm

Calculate the mean length of structure E. *(1 mark)*

2 Figure 2 shows the structure of the type of muscle that moves our limbs.

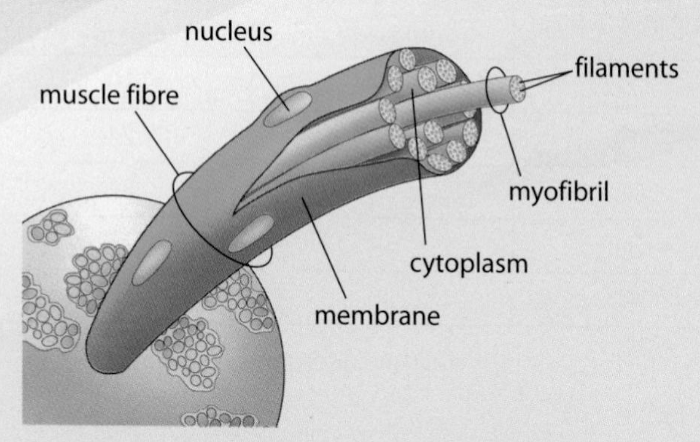

Figure 2 Muscle fibres and filaments.

(a) Give a difference between a muscle fibre and a typical animal cell related to the nucleus. *(2 marks)*

(b) There are large numbers of mitochondria in a muscle fibre. Explain why the muscle fibre needs so many mitochondria. *(2 marks)*

(c) **(i)** Suggest the function of the filaments. *(1 mark)*

(ii) Suggest the advantages of having many filaments in a fibre. *(1 mark)*

3 **(a)** Explain what is meant by diffusion. *(2 marks)*

(b) Figure 3 shows four ways in which molecules may move into and out of a cell. The dots show the concentration of molecules.

Which arrow, A, B, C or D, represents the movement of:

(i) carbon dioxide during photosynthesis?

(ii) carbon dioxide during respiration?

(2 marks)

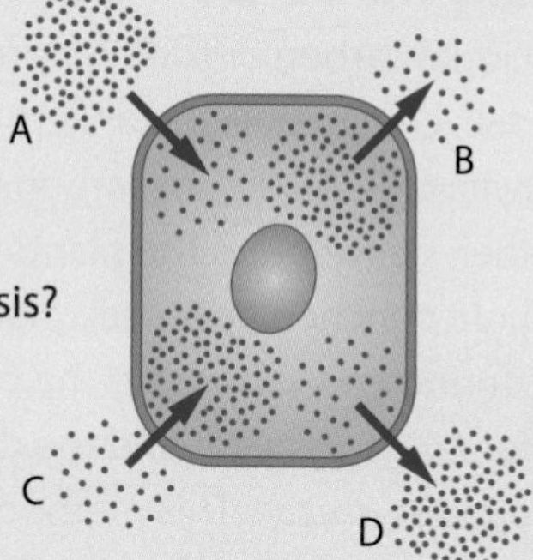

Figure 3 Cell and molecules.

4 **(a)** What is meant by a limiting factor? *(1 mark)*

(b) Figure 4 shows the effect of light intensity, carbon dioxide concentration and temperature on the rate of photosynthesis.

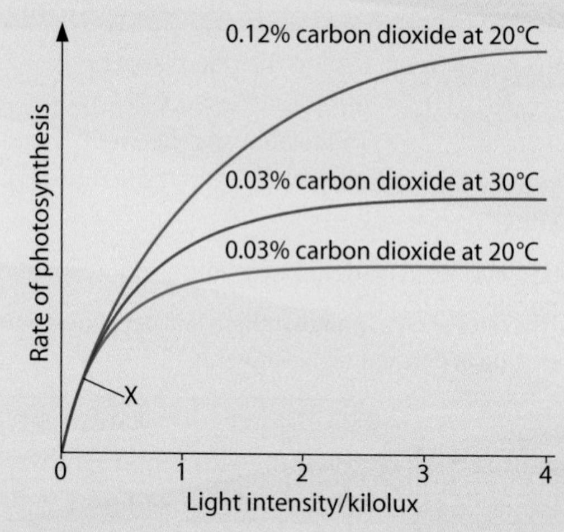

Figure 4 Light intensity versus rate of photosynthesis.

(i) Which factor is limiting the rate of photosynthesis at X? *(1 mark)*

(ii) In a greenhouse in winter the carbon dioxide concentration is 0.03%, the temperature is 20 °C and the light intensity is 3 kilolux.

Using the data on the graph, predict whether increasing the carbon dioxide concentration to 0.12% or the temperature to 30 °C would result in the greater increase in the rate of photosynthesis. Explain your answer as fully as you can. *(2 marks)*

5 **(a)** Explain what is meant by:

(i) digestion **(ii)** differentiation

(iii) specialisation. *(3 marks)*

(b) Use *all* the information in Table 1 to write an educational flyer about the role of a specialised organ system like the digestive system. You must write in full sentences.

In this question you will be assessed on using good English, organising information clearly and using specialist terms where appropriate. *(6 marks)*

Table 1 Information about the digestive system.

	Examples (random order)
Tissues	muscular, glandular, epithelial
Organs	pancreas, stomach, large intestine
Digestive functions	mix stomach contents, digestive juices, absorb water

6 A student studied ivy plants of the same species growing against a fence in her garden. She noticed that the leaves on the plants were not all the same size. She thought there might be a link between the height above the ground and the size of the leaves.

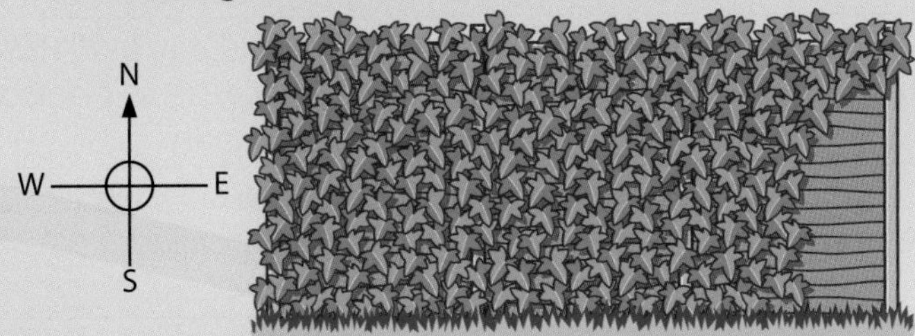

Figure 5 A fence with ivy on it.

She measured the surface area of five leaves at five different heights above the ground.

Table 2 The student's results.

Height above ground/ cm	Area/cm^2					Mean surface area of leaves/cm^2
	Leaf 1	Leaf 2	Leaf 3	Leaf 4	Leaf 5	
30	11	15	14	11	12	13
60	19	20	24	18	26	23
90	35	32	34	32	37	
120	44	41	40	43	40	42
150	57	43	49	52	55	51

(a) Calculate the mean surface area of the leaves collected at 90 cm above the ground. *(1 mark)*

(b) What is the range of size of leaves for 60 cm above the ground? *(1 mark)*

(c) Display the student's results as a graph. *(8 marks)*

(d) Describe how the mean surface area of the leaves is related to their height above the ground. *(2 marks)*

(e) Copy and complete the sentence by choosing the correct word from the box:

precise, reliable, valid, variable

The mean could have been improved by sampling 10 leaves instead of five. This would have made the mean more ___________. *(1 mark)*

(f) The student thought that the further away from the ground the leaves were, the more light they received. How could she measure this in her garden? *(1 mark)*

(g) Suggest *two* other factors that could influence leaf size in the ivy plant. *(2 marks)*

7 A group of students was studying the distribution of daisy and dandelion plants in a field by counting along a transect.

(a) Explain what a transect is. *(1 mark)*

(b) How would sampling be done with quadrats along the transect? *(2 marks)*

(c) **Table 3** The students' results.

Distance along the transect/m	Percentage cover of plants	
	Daisy	Dandelion
5	30	10
10	20	20
15	5	5
20	5	–
25	10	5
30	20	10
35	45	15
40	40	20

Plot the results on a graph. *(6 marks)*

(d) Describe the distribution of the two plants along the transect. *(3 marks)*

(e) The students also measured soil depth at each distance along the transect and the results are shown below.

Table 4 The students' results.

Distance along the transect/m	Soil depth /cm
5	20
10	25
15	8
20	6
25	10
30	15
35	>30
40	>30

Add these data to your graph. *(4 marks)*

(f) Suggest an explanation for the distribution of dandelions along the transect. *(2 marks)*

(g) Suggest two other abiotic factors that could be measured. *(1 mark)*

B2 3.1

Communities of organisms and their environment

Learning objectives

- identify the physical factors that affect the distribution of organisms
- explain how each particular factor affects the distribution of organisms
- calculate the mean, median and mode for environmental data.

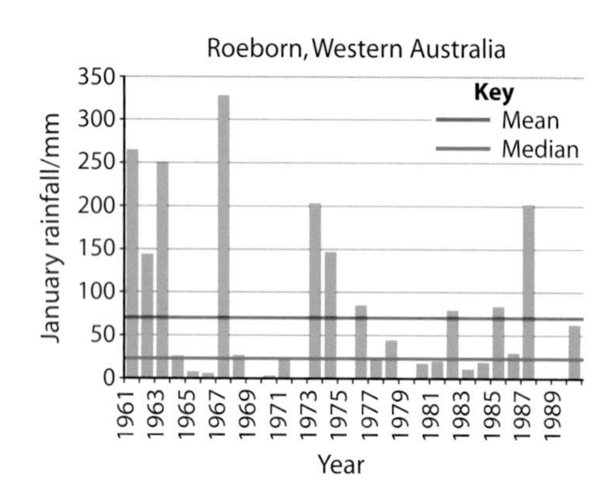

Figure 1 January rainfall for a city in Western Australia.

Examiner feedback

The high rainfall figures in some years are *not* anomalous results – so they cannot be disregarded when calculating the mean. Data can only be disregarded if the measuring technique or equipment is faulty.

Examiner feedback

It is easy to confuse mean, median and mode. Remember:

- mean by writing it as meAn – A = average
- median by writing it as meDian – D = the miDDle letter
- mode by writing it as MOde – MO = MOst common.

Physical factors

Organisms are affected by their environments.

Table 1 Some environmental factors alongside their effects on living organisms, and the methods we use to measure them.

Physical factor	Measurement	Effect on living organisms
temperature	thermometer or probe	In low temperatures metabolism slows right down. This reduces the activities of animals that cannot regulate their body temperatures.
nutrients	chemical analysis	Plants and microorganisms need ions to produce chemicals essential for living processes, so plants do not grow well in nutrient-poor soils.
amount of light	light meter or light sensor	Plants need light for photosynthesis, so only specialised plants can grow well in shade conditions.
availability of water		Most living organisms contain a very high proportion of water, which is essential for all living processes.
availability of oxygen in water or soil	oxygen electrode	Oxygen is essential for aerobic respiration, so only specialised animals can live in water with a low concentration of dissolved oxygen.
availability of carbon dioxide	gas analysis	Carbon dioxide is essential for photosynthesis and may be a limiting factor in some circumstances.

Analysing data

Environmental factors may vary throughout the year or even minute by minute, so before an experiment we must plan data collection and analysis so that the results reflect actual conditions as faithfully as possible.

A student investigating the activity of insects measured the temperature on the school playing field at noon each day for 1 week in summer.

The temperatures, in °C, on the 7 days were

19 21 18 22 18 17 25

To find the **mean** noon temperature for the week, add all seven numbers and divide by seven.

$19 + 21 + 18 + 22 + 18 + 17 + 25 = 140$

$140 \div 7 = 20$

So the mean temperature for the week was 20 °C.

To find the **median** temperature for the week, rearrange the daily temperatures in ascending order and select the one in the middle. If two numbers are left in the middle, add them together and divide by two.

17 18 18 19 21 22 25

So the median temperature for the 7 days is 19 °C.

To find the **mode**, group the data and find out how many there are in each group.

17 (1), 18 (2), 19 (1), 21 (1), 22 (1), 25 (1)

There were 2 days when the temperature was 18 °C and only 1 day for each of the other temperatures, so the mode temperature was 18 °C.

If the values you are working with are fairly close together, then calculating the mean gives useful information. However, if the values for a factor are very variable, the median may be more useful.

Figure 1 shows wide variation in January rainfall. The mean is influenced by the very high rainfall in some years. The median gives a far more typical measure of the rainfall.

Science skills

Plant species	Number of plants growing in a 10 m length of bank	
	South-facing bank	North-facing bank
Cow parsley	7	10
Dandelion	8	4
Groundsel	15	10
Lesser celandine	18	8
Thistle	5	1
White deadnettle	23	0

Figure 2 Which factors affect the growth of plants on the banks by a road?

a Describe the pattern shown by the data for most of the plants growing on the banks.

b Which plant does not fit into the general pattern?

c **i** Suggest the two factors that are most likely to affect the general pattern.

ii Name two processes in plants that could be affected by the factors you have named in **i**.

d Suggest one other factor that might affect the general pattern.

Questions

1 Give one way in which an increase in temperature affects organisms.

2 Which factors are essential for the healthy growth of plants in a habitat?

3 When is it best to calculate: **(a)** the mean value for an environmental factor? **(b)** the median value for an environmental factor?

4 The minimum temperature (°C) on each of 7 successive days in a habitat was:

9 9 11 5
14 2 6

Calculate the mean, the mode and the median for this set of data.

5 The graph in Figure 3 shows changes in the populations of two species of plants in an area of deforestation.

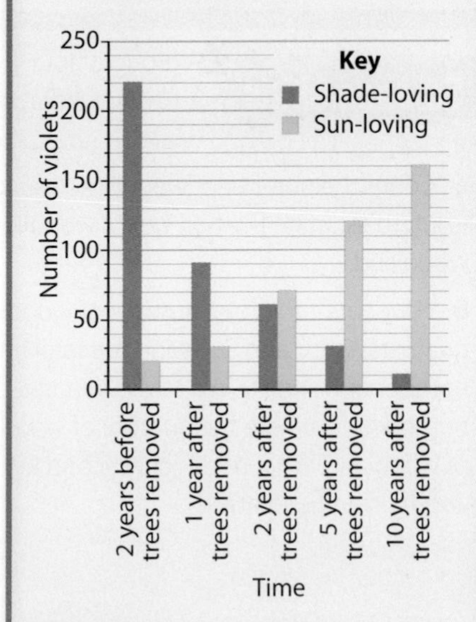

Figure 3 Numbers of violets in an area of deforestation.

Describe and suggest an explanation for the changes in the populations of the two species.

A*

Collecting ecological data

Learning objectives

- explain how quadrats are used to collect information about the distribution of organisms
- explain random and systematic sampling
- analyse information obtained using quadrats.

Monitoring biodiversity

The Government is planning to build a high-speed rail link between London and Birmingham. The outline route passes through countryside that has a large biodiversity. Scientists will survey the wildlife in these areas in order to report on the effects of the proposals on threatened species. Some of the techniques the scientists will use are described below.

How many woodlice live in a wood?

Imagine you want to make a count of the number of woodlice in an area of woodland. Woodlice are tiny, there are very many of them and they are not distributed evenly across their habitat. Because of this it is not easy to find out how many woodlice there are in the wood. Instead of trying to count all the woodlice, we can use **sampling**. We count the numbers in a small area and use this number to estimate the total.

Large numbers of woodlice live in damp woodland soil.

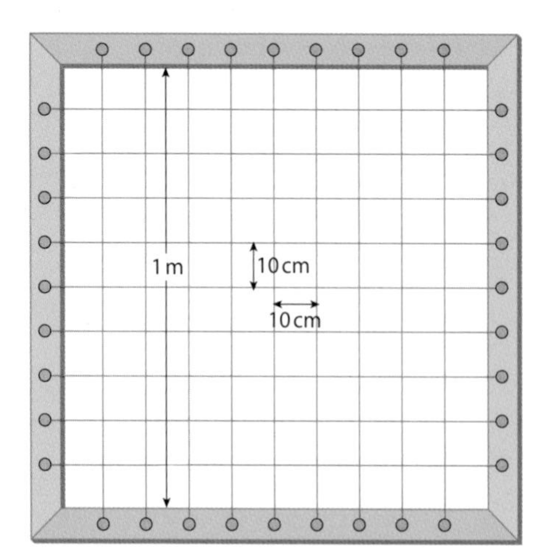

Figure 1 A 1 m quadrat divided into 10 cm squares.

Quadrats

The most common method of sampling organisms is to use a **quadrat**. This is a square frame, usually measuring 10 cm, 50 cm or 1 m along each side. Quadrats are usually subdivided into smaller squares.

Science skills

A group of four students each placed a 10 cm quadrat on the floor of a wood. Figure 2 shows their quadrats.

a Count the number of woodlice in quadrat A. Use this result to estimate the number of woodlice in 1 m^2 of woodland.

b Now count the total number of woodlice in quadrats A, B, C and D. Divide the total to find the mean number of woodlice in a 10 cm quadrat. Use the mean number to estimate the number of woodlice in 1 m^2 of woodland. What does your second estimate tell you about using quadrats?

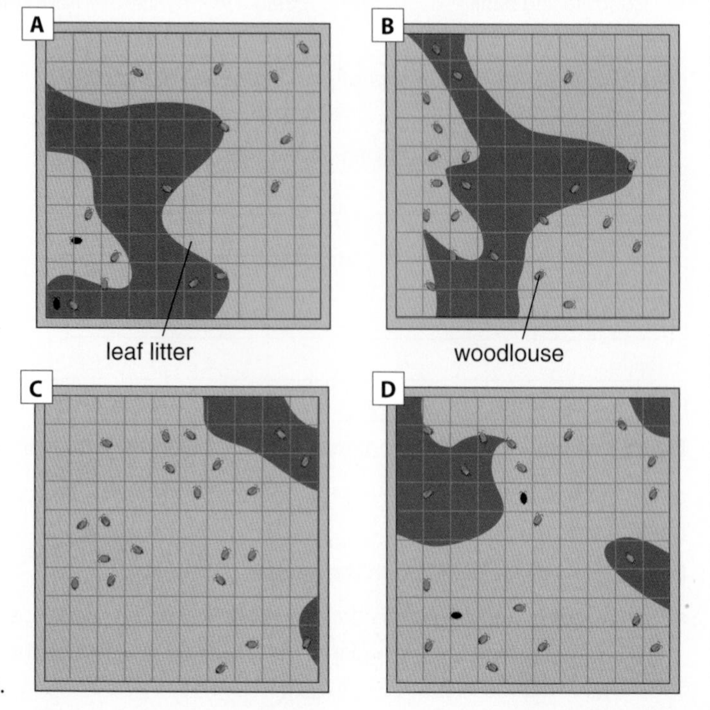

Figure 2 The students' quadrats.

Reliability and validity

The greater the number of quadrat counts that are made, the more reliable the estimate of the size of the population will be. Increasing reliability in this case also increases the validity of the population estimate.

Sampling methods

Not all parts of an area being sampled will be the same – for example, the lawn on the right may have some patches that are full of clover, and some that are almost clover-free. To get more valid results from sampling, quadrats should be placed carefully. There are two approaches to placing quadrats:

- **Random sampling**. A set of random numbers is generated by a computer. The numbers are used as coordinates on a grid as shown in Figure 3a. A similar grid is marked out on the lawn and the quadrat is used at each of the random coordinates.
- **Systematic sampling**. A grid is marked out on the lawn and the quadrat is used at each intersection, as shown in Figure 3b.

Both random and systematic sampling avoid bias in placing the quadrats.

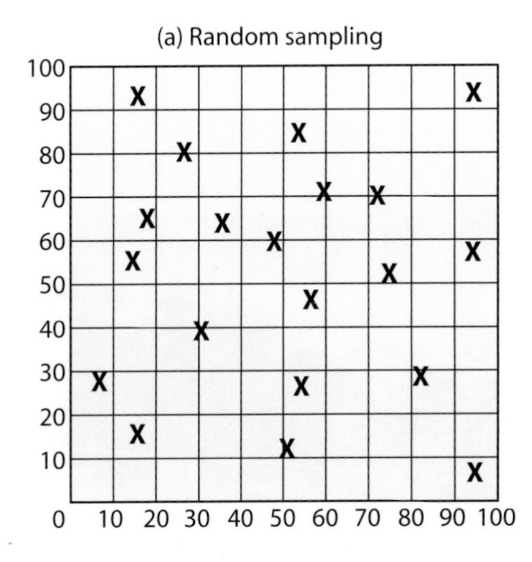

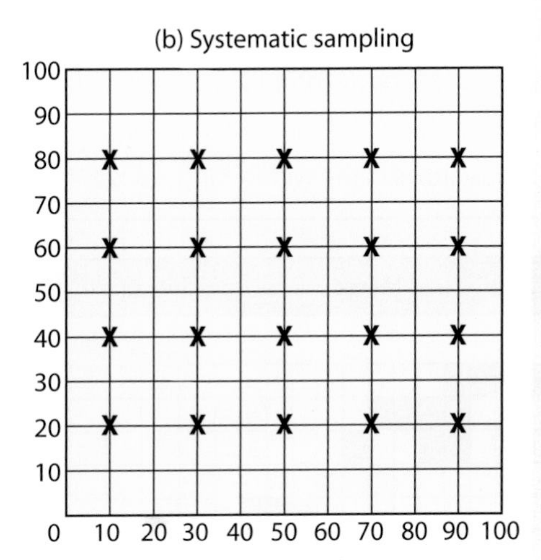

Figure 3 Random and systematic sampling.

Questions

1. How can the reliability of results obtained using quadrats be improved?
2. How can the accuracy of results obtained by using quadrats to estimate cover be improved?
3. Explain how quadrats could be placed randomly to sample organisms in a habitat.
4. Explain how quadrats could be placed systematically to sample organisms in a habitat.
5. Look again at the quadrats labelled A, B, C and D in Figure 2. **(a)** Where are most of the woodlice found? **(b)** Suggest a hypothesis to explain this distribution. **(c)** Design an investigation to test this hypothesis.

A*

Science skills

Quadrats can be used to estimate **ground cover** as well as for counting populations. Clover grows in clumps among grass. A student wanted to find out how much of a lawn was covered by clover. She placed a 50 cm quadrat on the lawn as shown in Figure 4.

Clover growing among grass.

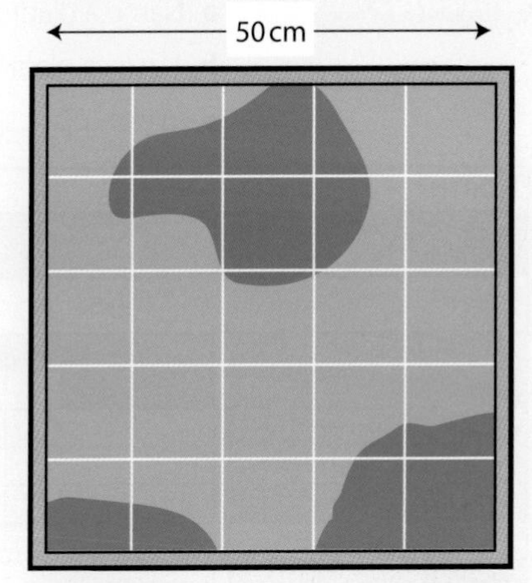

Figure 4 Using a quadrat to estimate cover.

To estimate the area of lawn covered by clover, count any square more than half-covered by clover as a 'clover square'.

c Use the number of 'clover squares' to calculate the percentage of squares covered by clover.

d How could the quadrat be modified to give a more accurate measurement of cover? Give a reason for your answer.

Analysing ecological data

Learning objectives

- explain how transects are used to collect information about the distribution of organisms
- analyse information obtained using transects
- analyse information about succession.

Examiner feedback

Many candidates confuse the use of quadrats in random sampling and in a transect. Remember that quadrats are *not* placed at random on a transect – they are placed at regular intervals.

Using transects

Transects are used to investigate changes in populations from one area to another, for example down a rocky seashore or along sand dunes or marshland. A measuring tape, or a cord marked at regular intervals, is laid out across the area. Organisms are sampled by placing a quadrat at regular intervals along the line as shown in Figure 1.

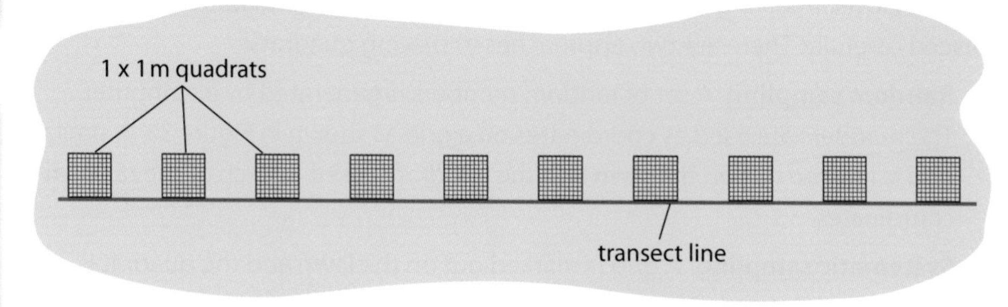

Figure 1 Quadrats along a transect.

Science skills

Look at the distribution of plants along a wetland transect.

a Name a plant species that grows only in open water and swamp.

b Name a plant species that grows mainly on dry land.

c Which plant species grows in the widest range of habitats?

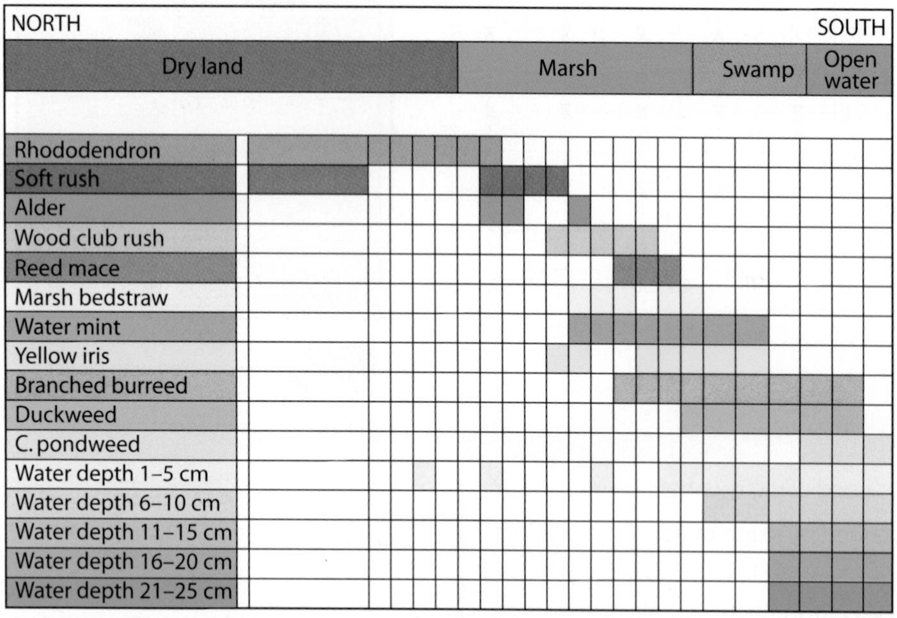

NORTH — SOUTH

Dry land | Marsh | Swamp | Open water

Rhododendron
Soft rush
Alder
Wood club rush
Reed mace
Marsh bedstraw
Water mint
Yellow iris
Branched burreed
Duckweed
C. pondweed
Water depth 1–5 cm
Water depth 6–10 cm
Water depth 11–15 cm
Water depth 16–20 cm
Water depth 21–25 cm

Figure 2 Distribution of plants along the wetland transect.

The position of a transect across a piece of wetland.

Succession

Each plant species is adapted to a particular set of environmental conditions, but a habitat is not static; the environmental factors gradually change. In a wetland, the stems and roots of the plants trap sediment, which gradually builds up and forms soil. The water around these plants therefore gets shallower. Shallow water is more suitable for other species suited to drier habitats, and these dry-land species move in to replace the marshland plants. Eventually all traces of open water will disappear and only the dry-land plants will survive. This process is called **succession**.

Succession in hay infusions

A hay infusion is a habitat for microorganisms that can be set up in a laboratory. Water from a pond or stream is mixed with cut pieces of hay in a jar, and left to stand.

Bacteria from the fresh hay cause the infusion to decay, and a large population of bacteria develops over a period of about 2 weeks. Towards the end of the 2 weeks other unicellular organisms appear. These develop from spores on the fresh hay. The unicellular organisms feed mainly on the bacteria. Some of these organisms are shown in Figure 3.

As the hay infusion ages its pH changes. Different organisms appear as the pH changes. This shows succession in populations of unicellular organisms.

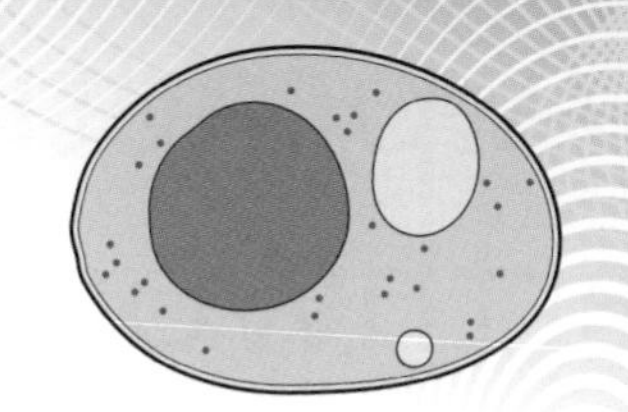

Holophyra

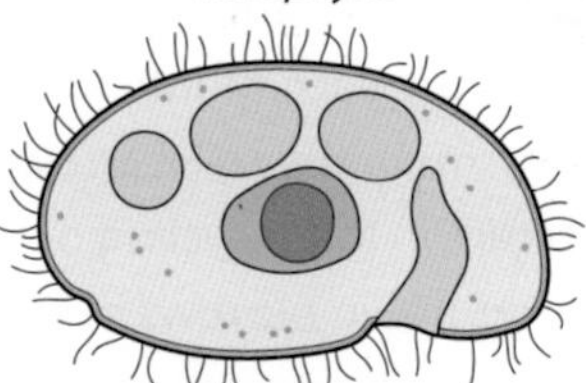

Plagiopyla

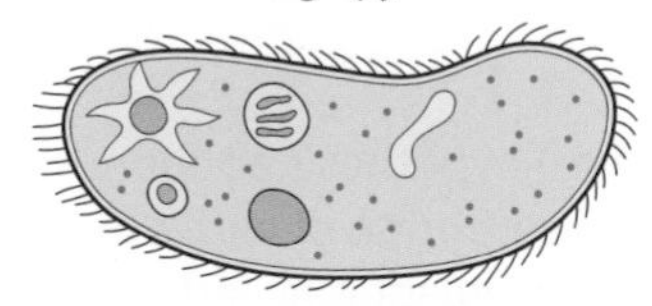

Colpidium

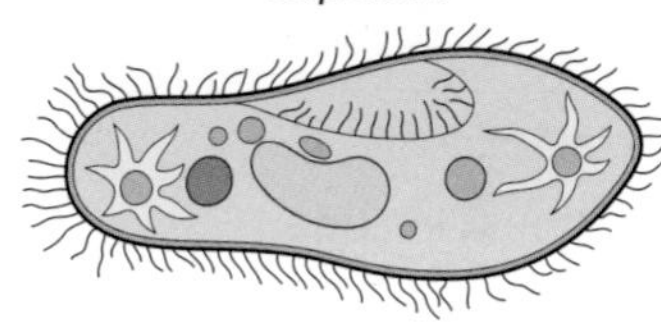

Paramecium

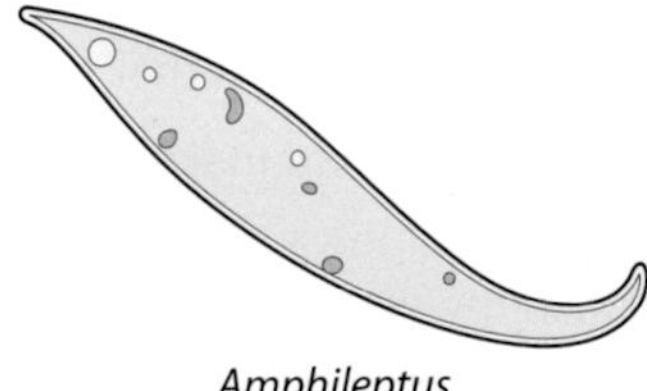

Amphileptus

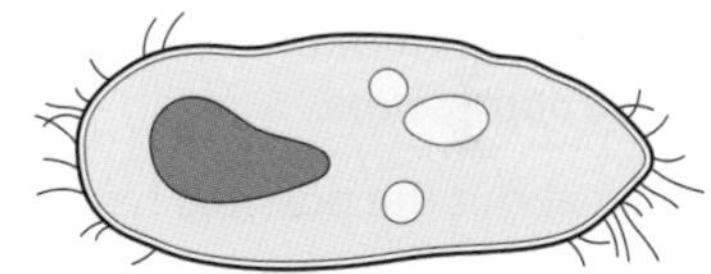

Gastrostyla

Figure 3 Some of the organisms that appear in hay infusions.

Science skills

Figure 4 shows the changes in pH of a hay infusion.

d Describe the changes in pH using the terms acid, alkaline and neutral.

e Suggest an explanation for the change in pH between days 3 and 8.

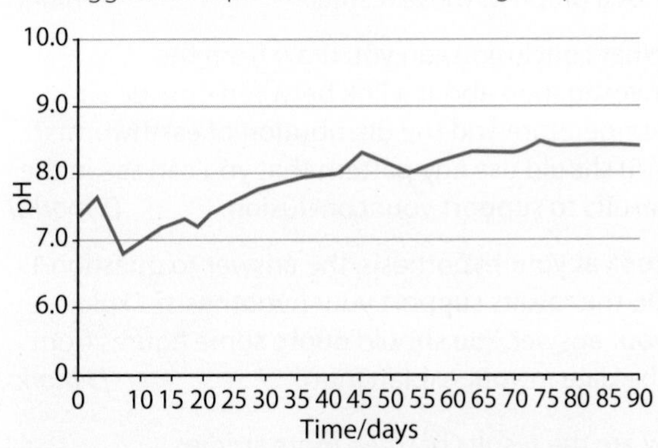

Figure 4 Graph showing changes in pH in a hay infusion.

Table 1 Optimum pH range of different unicellular organisms.

Organism	Optimum pH range
Amphileptus	7.3–7.5
Colpidium	7.0–7.2
Holophyra	6.5–7.4
Paramecium	7.5–7.7
Plagiopyla	6.6–7.5

f Using information from Figure 4 and Table 1, draw a graph to show the days on which the five organisms would live in the hay infusion.

Questions

1. When is it appropriate to use a transect?
2. Describe how a transect and a quadrat are used to sample populations.
3. Explain what is meant by succession.
4. Why would you want to set up a hay infusion in the lab?
5. Describe how you would set up a hay infusion.
6. Which factor influences succession in a hay infusion?
7. Figure 5 shows succession from abandoned farmland to mature forest. Suggest reasons for the changes in populations shown in the drawing.

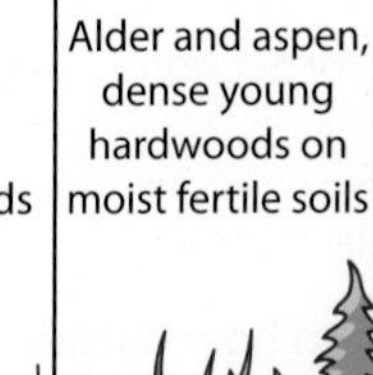
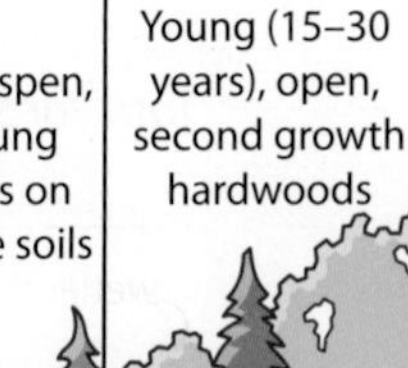

Figure 5 Woodland succession.

ISA practice: earthworm distribution

A farmer has asked students to investigate how the outside air temperature each month affects the distribution of earthworms in the soil. Earthworms are important in agriculture in the recycling of nutrients in the soil. Earthworms live in burrows in soil. The depth at which they live depends mainly on physical factors in the soil.

Section 1

1 Write a hypothesis about how air temperature affects the distribution of earthworms. Use information from your knowledge of earthworm behaviour to explain why you made this hypothesis. *(3 marks)*

2 Describe how you could carry out an investigation into this factor.

You should include:

- the equipment that you could use
- how you would use the equipment
- the measurements that you would make
- how you would make it a fair test.

You may include a labelled diagram to help you to explain the method.

In this question you will be assessed on using good English, organising information clearly and using specialist terms where appropriate. *(6 marks)*

3 Think about the possible hazards in the investigation.

(a) Describe one hazard that you think may be present in the investigation. *(1 mark)*

(b) Identify the risk associated with this hazard that you have described, and say what control measures you could use to reduce the risk. *(2 marks)*

4 Design a table that you could use to record all the data you would obtain during the planned investigation. *(2 marks)*

Total for Section 1: 14 marks

Section 2

Two students, Study Group 1, investigated how outside air temperature affects the distribution of earthworms in the soil. They measured the air temperature and counted the number of worms in 1 m^2 of soil. Their results are shown in Figure 1.

January
3°C 22 worms

February
1°C 8 worms

March
2°C 12 worms

April
5°C 38 worms

May
8°C 92 worms

June
15°C 12 worms

July
20°C 5 worms

August
16°C 18 worms

September
12°C 30 worms

October
9°C 52 worms

November
8°C 78 worms

December
6°C 52 worms

Figure 1 Study Group 1's results.

5 **(a)** Plot a graph of these results. *(4 marks)*

(b) What conclusion can you draw from the investigation about a link between outside air temperature and the distribution of earthworms? You should use any pattern that you can see in the results to support your conclusion. *(3 marks)*

(c) Look at your hypothesis, the answer to question 1. Do the results support your hypothesis? Explain your answer. You should quote some figures from the data in your explanation. *(3 marks)*

Below are the results of three more studies.

Figure 2 shows the results from another two students, Study Group 2.

Study Group 2

January
3°C 22 worms

February
1 °C 2 worms

March
2°C 6 worms

April
5°C 30 worms

May
8°C 68 worms

June
15°C 13 worms

July
20°C 12 worms

August
16°C 17 worms

September
12°C 40 worms

October
9°C 48 worms

November
8°C 65 worms

December
6°C 59 worms

Figure 2 Study Group 2's results.

A third group of students, Study Group 3, decided that another factor might also be affecting the number of earthworms. They decided to find out the rainfall for the area. Their results are shown in Figure 3.

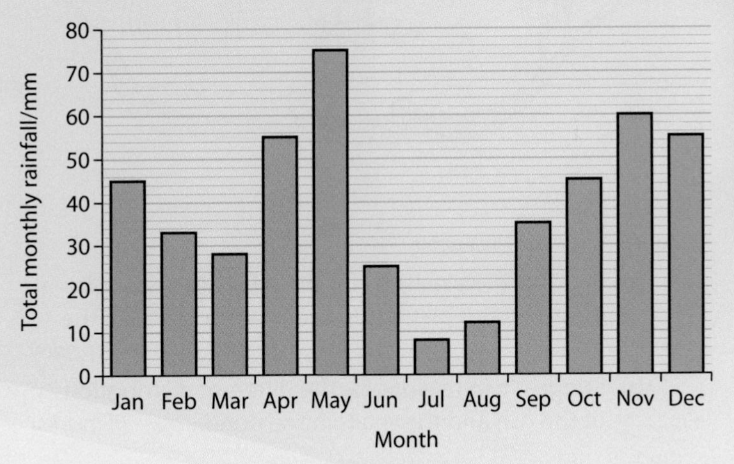

Figure 3 Monthly rainfall data from Study Group 3.

Study Group 4 was a group of scientists in India. They compared earthworm populations in two different national parks, X and Y.

- In each park they sampled eight sites by digging out a piece of soil 25 cm × 25 cm × 20 cm.
- They counted the number of earthworms in each sample.
- They measured the biomass of the worms using an electronic balance.
- They also measured the pH, nitrogen, organic matter, phosphorus, calcium, temperature and moisture content of each sample.

Table 1 shows the mean number and mean mass of earthworms from the two national parks.

Table 1 Mean number and mean mass of earthworms.

Park	Mean number of earthworms/m^3	Mean mass of earthworms / m^3/g
X	82	11.20
Y	17	2.87

Table 2 Average soil characteristics of national parks X and Y.

	Park X	Park Y
pH	6.40	6.02
Total nitrogen as a percentage	0.75	0.58
Organic matter as a percentage	5.13	4.09
Phosphorus as a percentage	0.34	0.23
Potassium as a percentage	0.90	0.63
Nitrate as a percentage	0.23	0.19
Temperature/°C	29.02	29.40
Water as a percentage	26.60	13.66

6 **(a)** Draw a sketch graph of the results from Study Group 2. *(3 marks)*

(b) Look at the results from Study Groups 2 and 3. Does the data support the conclusion you drew about the investigation in answer to question 5(a)? Give reasons for your answer. *(3 marks)*

(c) The data contain only a limited amount of information. What other information or data would you need in order to be more certain whether the hypothesis is correct or not?
Explain the reason for your answer. *(3 marks)*

(d) Look at the results from Study Group 4. Compare the data from Study Group 1 with Study Group 4's data. Explain how far the data shown supports or does not support your answer to question 5(b). You should use examples from Study Group 4 and Study Group 1. *(3 marks)*

7 **(a)** Compare the results of Study Group 1 with Study Group 2. Do you think that the results for Study Group 1 are *reproducible*?
Explain the reason for your answer. *(3 marks)*

(b) Explain how Study Group 1 could use results from other groups in the class to obtain a more *accurate* answer. *(3 marks)*

8 Applying the results of the investigation to a context.

Suggest how ideas from the original investigation and the other studies could be used by the farmer in encouraging growth in the numbers of earthworms in his fields. *(3 marks)*

Total for Section 2: 31 marks

Total for the ISA: 45 marks

Assess yourself questions

1 The map shows the temperatures at noon for 1 day in the UK.

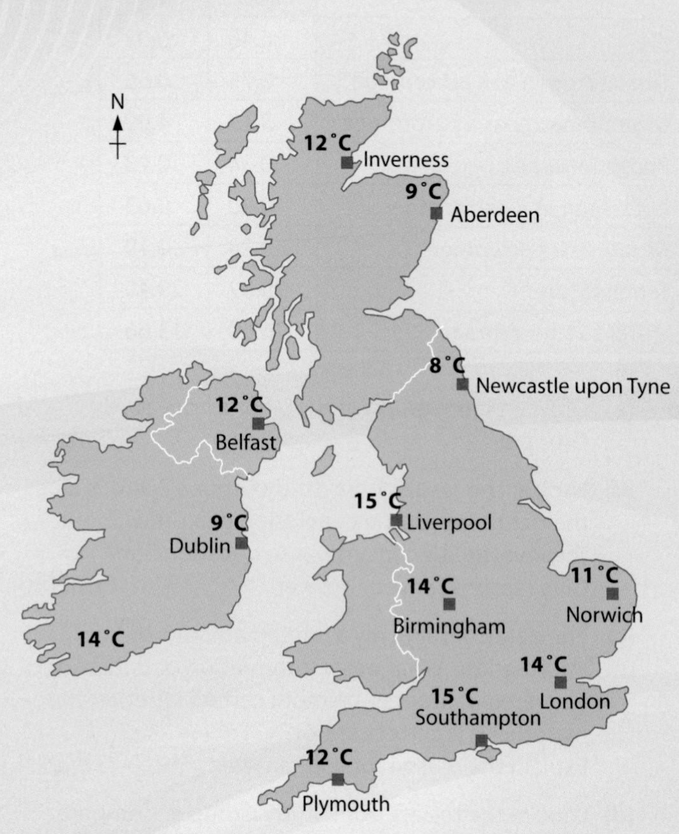

Figure 1 Noon temperatures across the UK.

Calculate:

(a) the median temperature *(2 marks)*

(b) the mode temperature. *(2 marks)*

In each case show your working.

2 The diagram shows three ways in which plant species might be distributed in a field.

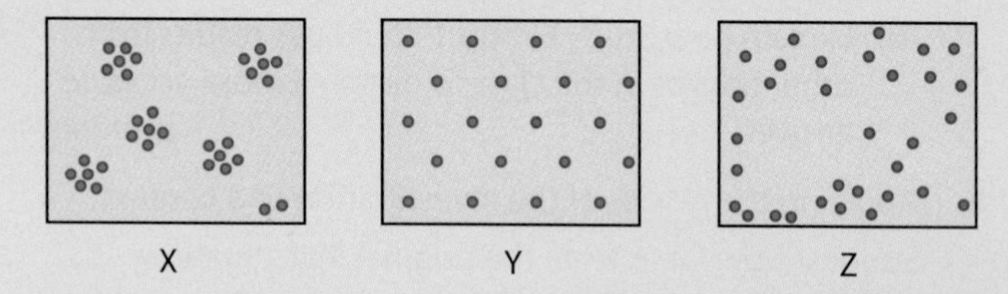

Figure 2 Three distributions of plants in a field.

(a) Describe each of the patterns X, Y and Z. *(3 marks)*

(b) At first glance a species of plant, W, appears to be distributed in a similar way to the plant species in diagram X.

(i) Describe fully how you would investigate the distribution of species W in the field. *(3 marks)*

(ii) Explain how your results would confirm a distribution of species W similar to that in diagram X. *(2 marks)*

3 The diagram shows some of the organisms that live in a pond.

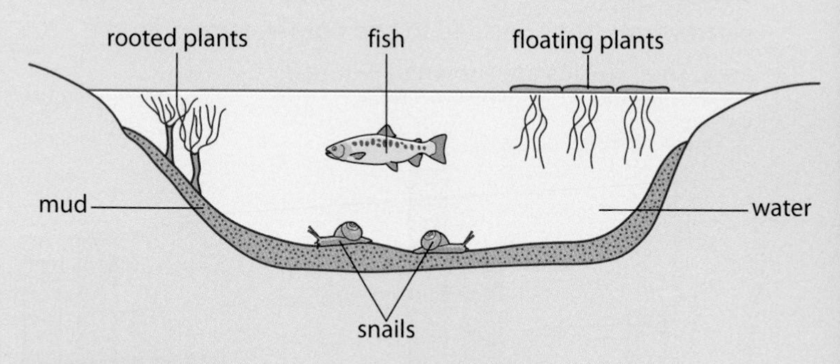

Figure 3 Organisms in a pond.

(a) Suggest *three* reasons for the different distributions of the rooted plants and the floating plants in the pond. *(3 marks)*

(b) Suggest *two* reasons for the different distributions of the fish and the snails in the pond. *(3 marks)*

4 Leafhoppers are small insects that live on leaves in trees. These insects feed on sugars, which they suck from the leaf tissues.

Students sampled the numbers of leafhoppers living on leaves in different conditions in three trees.

Table 1 The students' results.

Tree sampled	Conditions for leaves	Number of leaves examined	Number of leaves with leafhoppers on them
1	sunshine	104	13
1	shade	105	2
2	sunshine	192	44
2	shade	182	18
3	sunshine	93	19
3	shade	0	0

(a) Suggest the hypothesis the students were investigating. *(1 mark)*

(b) The students should have done further calculations on their results.

(i) What should the students have calculated to make the results more valid? *(1 mark)*

(ii) Copy the table then add a further column. Do further calculations and complete your table to give valid results. *(3 marks)*

(c) **(i)** What conclusion can be drawn from these results? *(1 mark)*

(ii) Suggest an explanation for the distribution of the leafhoppers. *(2 marks)*

(d) What could the students have done to obtain more reliable results? *(1 mark)*

5 Students investigated the distribution of weeds in two different lawns.

Table 2 The students' results.

Weed species	Mean population density/ number of plants per m^2	
	Regularly mown lawn	Occasionally mown lawn
Daisy	36.0	18.6
Dandelion	10.8	3.4
Field buttercup	1.2	10.0
Ribwort plantain	4.3	2.8
Greater plantain	0.9	1.5

(a) Describe how the data in the table could be collected. *(3 marks)*

(b) A gardening magazine recommends mowing lawns regularly to keep down weeds. Do the data in the table support this recommendation? Explain the reasons for your answer. *(4 marks)*

6 Mayflies are insects whose nymphs (immature stages) live mainly under stones in streams.

Mayfly nymphs can be sampled by disturbing the stones and collecting the nymphs in a net held downstream of the disturbed area.

Students investigated the distribution of two species of mayfly nymphs, X and Y, in different regions of a stream. They sampled 10 sites in each region.

Table 3 The students' results.

	Species X		Species Y	
	Shallow, fast-running water	Deep, slow-running water	Shallow, fast-running water	Deep, slow-running water
Mean number of nymphs per m^2	2.38	12.88	24.50	6.00
Range	1–3	5–20	18–30	4–8

(a) Suggest how the students used the method described above to collect the data shown in the table. *(3 marks)*

(b) Suggest *two* reasons why the results obtained might not be accurate. *(2 marks)*

(c) Suggest *two* reasons for the different distributions of species X and Y. *(2 marks)*

7 Figure 4 is a kite diagram showing how abundant each organism is at each point on the shore – the broader the 'kite' at that point, the more abundant the organism. High tides reach as far as the shingle. All the organisms are seaweeds except *Arenicola* and *Littorina*.

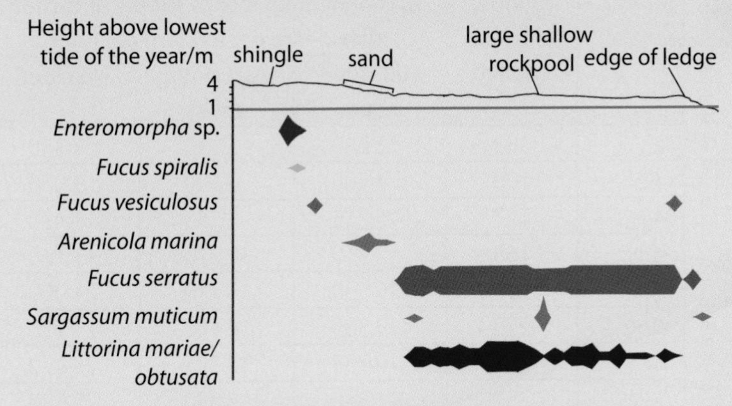

Figure 4 Distribution of some of the organisms living on a seashore.

(a) Suggest one explanation for the difference in the distribution of *Enteromorpha* and *Fucus serratus*. *(2 marks)*

(b) *Arenicola* is a marine worm. Suggest one explanation for the pattern of its distribution. *(2 marks)*

(c) *Littorina* is a snail-like organism. Suggest one explanation for the pattern of its distribution. *(2 marks)*

8 Limpets are snail-like animals that live on rocky shores.

A group of students measured the height and width of ten limpets each from a shore exposed to strong waves and a sheltered shore.

Table 4 Limpets from shore exposed to strong waves.

	1	2	3	4	5	6	7	8	9	10	Mean
Height in cm	1.7	0.9	1.4	1.7	1.2	1.9	1.7	1.2	1.9	0.9	
Width in cm	2.4	1.9	2.8	3.1	2.1	2.2	3.1	2.4	3.1	1.9	

Table 5 Limpets from sheltered shore.

	1	2	3	4	5	6	7	8	9	10	Mean
Height in cm	1.7	1.4	1.4	2.0	2.3	1.9	2.0	3.2	2.7	3.0	
Width in cm	2.8	2.1	2.8	2.3	2.6	1.9	2.5	2.6	2.5	2.9	

(a) Calculate the mean height and width of each group of limpets. *(4 marks)*

(b) Suggest an explanation for any difference in the means. *(2 marks)*

GradeStudio Route to A*

Here are three students' answers to the following question:

The table shows the recommended supply of carbon dioxide to greenhouse crops at different times of the year and in different conditions.

Month	Number of hours applied	Recommended rate of supply of carbon dioxide/kg per hectare per hour		
		Full cloud cover	Part cloud cover	No cloud
Jan	82	3 690	5 330	7 380
Feb	100	4 500	6 500	9 000
Mar	127	5 715	8 255	11 430
Apr	168	7 560	10 920	15 120
May	234	10 530	15 210	21 060
Jun	253	11 385	16 445	22 770
Jul	283	12 735	18 395	25 470
Aug	252	11 340	16 380	22 680
Sep	187	8 415	12 155	16 830
Oct	157	7 065	10 205	14 130
Nov	89	4 005	5 785	8 010
Dec	67	3 019	4 361	6 039

Suggest explanations for different recommended rates of carbon dioxide supply to greenhouse crops. *(6 marks)*

In this question you will be assessed on using good English, organising information clearly and using specialist terms where appropriate.

Read the answers together with the examiner comments. Then check what you have learnt and try putting it into practice in any further questions you answer.

Read the whole question carefully.

- Before beginning to answer a data question, jot down the main trends: in this case the seasonal trends and the sky-cover trends.
- Next, decide which concept the question is addressing: in this case limiting factors for photosynthesis. It is a good idea to introduce this concept in the first part of your answer, then to keep referring back to it as appropriate.
- If the data refer to industry, such as horticulture here, make sure you look for economic and/or environmental issues: in this case the cost of providing carbon dioxide.
- Always use the correct biological terminology.

Grade answer

Student 1

The recommended rate of carbon dioxide supply is low in winter and high in summer. This is because it is colder in winter than in summer so photosynthesis will not be as fast. Also the days are longer in summer. The supply is faster on days when there is no cloud. This is because there is more sun so there will be more photosynthesis.

Always refer to the rate of a process – do not simply state that a process is fast.

Light is the factor that should be referred to – it would be better to state 'there are more hours of daylight in summer'.

'Sun' is far too vague. Plants receive both heat and light from the Sun – it is important to use 'light' and 'heat' rather than 'Sun'.

Examiner comment

The candidate has referred to three patterns – the change in temperature during the year, the change in day length during the year and the change in overhead conditions. These were weakly linked to photosynthesis. There is no reference to the number of hours of carbon dioxide supply. There is no reference to limiting factors or to the economics of supplying carbon dioxide.

Grade answer

Student 2

The number of hours that carbon dioxide is supplied varies with the season. This number is low in winter, rises during spring to a peak in summer, and then falls in winter. This is because day length and air temperature vary seasonally with the same pattern. Increases in light and temperature both increase the rate of photosynthesis, so the supply of carbon dioxide is increased for the maximum rate of photosynthesis. The supply of carbon dioxide is also increased on sunny days to maximise the rate of photosynthesis. But giving too much carbon dioxide would be wasteful.

The candidate has not distinguished between light intensity and duration of light.

The candidate has correctly referred to maximum rate, but has not referred to limiting factors.

Rather than referring to 'sunny' days the candidate should have referred to increased light intensity.

'Wasteful' is ambiguous – it is not entirely clear what the candidate means.

Examiner comment

A good account using correct biological terminology, such as 'rate' and 'light intensity'.

Although the candidate has referred to the maximum rate of photosynthesis this has not been linked to carbon dioxide as a limiting factor. There is an attempt at a reference to economics, but merely stating 'wasteful' is insufficient.

Grade answer

Student 3

The rate of photosynthesis is affected by temperature, carbon dioxide concentration and light intensity. Any of these factors may limit the rate of photosynthesis. If one factor is limiting, then increasing other factors will have no effect on the rate. It is expensive to provide carbon dioxide to glasshouses, so the amount supplied is linked to the other limiting factors so that carbon dioxide is never the limiting factor.

Crop production is also affected by day length, if no artificial light is available, since the longer the day, the longer the plants can photosynthesise.

Crop plants in greenhouses receive light and heat from sunlight. The hours of sunlight rise during the spring, are high in summer and become low again in winter. The rate of carbon dioxide supply is adjusted to correspond to the amount of heat and light being received by the plants.

Light intensity is higher when there is no cloud, so the rate of carbon dioxide supply is increased so that carbon dioxide is not a limiting factor under these conditions.

A good opening sentence that introduces the factors that affect the rate of photosynthesis.

The candidate introduces the idea of limiting factors.

The candidate links economics to limiting factors.

Examiner comment

An excellent answer that refers to each of the patterns in the data. The candidate has used the correct biological terminology throughout. There is a good account of limiting factors and the candidate has linked this to the cost of providing carbon dioxide.

B2 Understanding how organisms function

This section starts by introducing proteins and their functions, both inside and outside the cells of living organisms. It then focuses on enzymes, their role as biological catalysts and their mechanism of action. Examples of enzymes, both within the body and as used in the home and in industry, are explored.

Respiration is fundamental to life, and comparison of aerobic and anaerobic respiration is covered in terms of the chemicals used and produced by these processes, and their impact on the human body during exercise.

The two types of cell division, mitosis and meiosis, are then described in relation to the types of cells produced. This leads to a discussion of stem cells, what they are, how they are produced, and the social and ethical issues raised by techniques used to produce them. Genes and alleles are introduced, leading to opportunities to use and interpret genetic diagrams to explain inheritance. Further ethical issues are explored in relation to the inheritance and treatment of genetic disorders.

The final part of this section looks at the use of fossils as evidence for the theory of evolution, exploring why some species become extinct and how new species may form.

Test yourself

1 Explain the importance of proteins in a balanced diet.

2 Explain why body temperature is controlled in humans.

3 Explain the importance of exercise for keeping healthy.

4 State the purpose of respiration in cells.

5 Describe sexual reproduction in terms of gametes and genetic variation.

6 Describe the theory of evolution via natural selection.

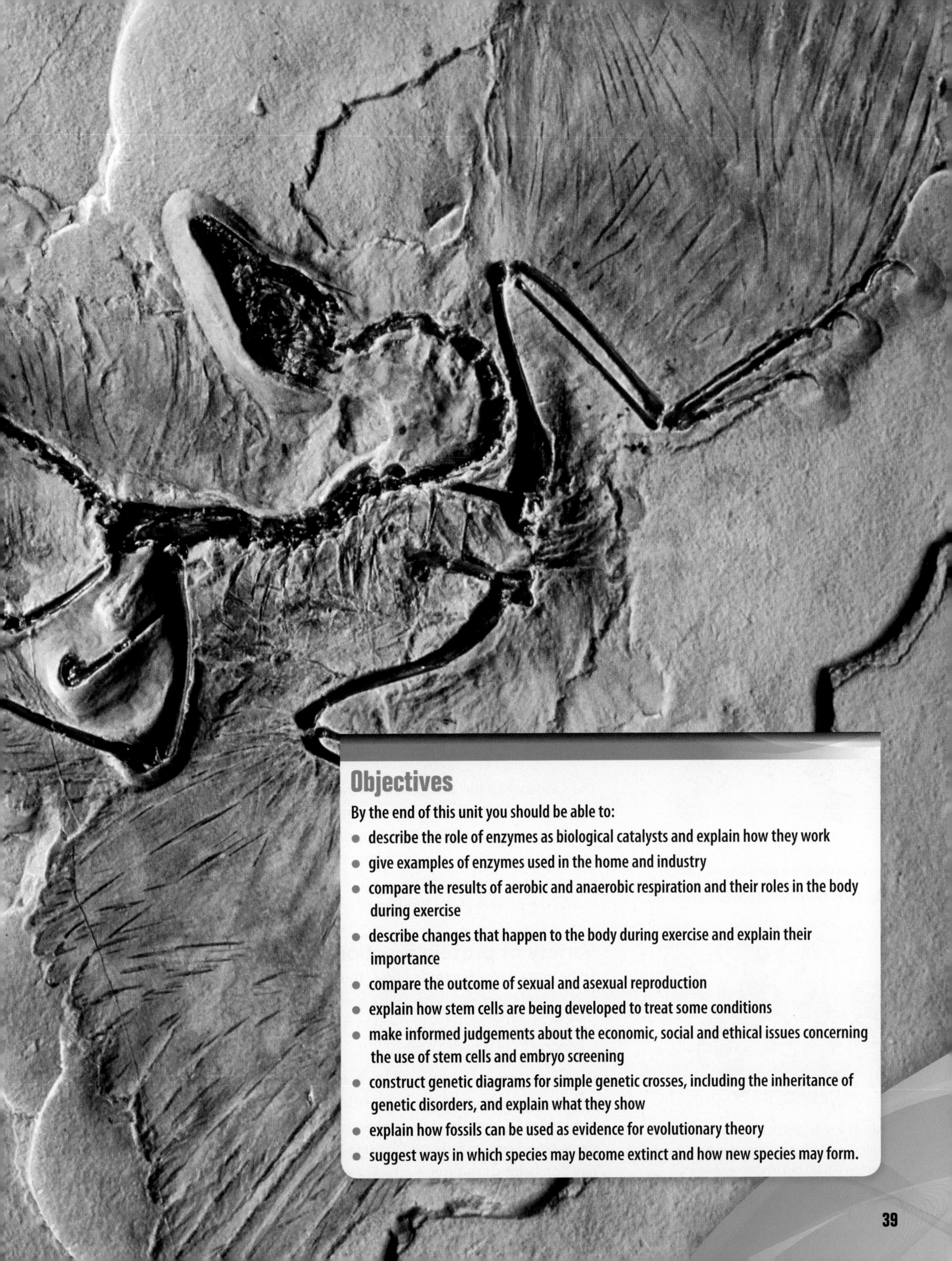

Objectives

By the end of this unit you should be able to:

- describe the role of enzymes as biological catalysts and explain how they work
- give examples of enzymes used in the home and industry
- compare the results of aerobic and anaerobic respiration and their roles in the body during exercise
- describe changes that happen to the body during exercise and explain their importance
- compare the outcome of sexual and asexual reproduction
- explain how stem cells are being developed to treat some conditions
- make informed judgements about the economic, social and ethical issues concerning the use of stem cells and embryo screening
- construct genetic diagrams for simple genetic crosses, including the inheritance of genetic disorders, and explain what they show
- explain how fossils can be used as evidence for evolutionary theory
- suggest ways in which species may become extinct and how new species may form.

B2 4.1 Protein structure, shapes and functions

Learning objectives

- describe the structure of protein molecules
- explain some functions of proteins in animals
- describe the role of catalysts in chemical reactions
- explain the functions of enzymes in living organisms.

The most diverse biological molecules

Nearly 20% of the fat-free mass of your body is made up of protein – water makes up about 72%. Your hair, skin and nails are made of protein. Protein molecules, in the form of haemoglobin, carry oxygen in your blood and help if you are injured by clotting blood. Throughout the living world proteins are extremely important molecules. They take part in most chemical reactions within cells and are part of the structure of most organelles.

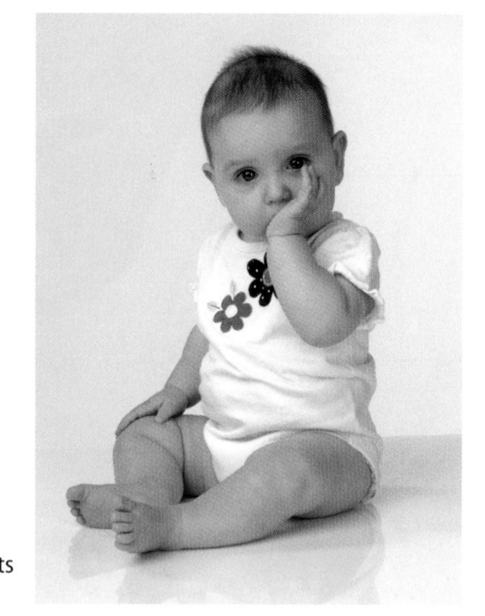

Proteins are structural components of muscle, hair and fingernails.

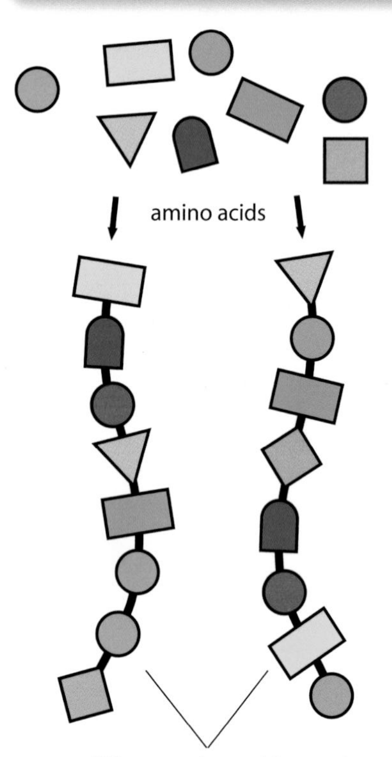

Figure 1 Two different polypeptide chains.

Structure of protein molecules

All proteins contain the same four elements: carbon, hydrogen, oxygen and nitrogen. Proteins are large molecules, known as macromolecules. They are made of a long chain of smaller, soluble molecules called **amino acids**. Twenty different amino acids are found in living organisms. These amino acids are joined together in a long chain, known as a **polypeptide**. Any number of amino acids can be joined together and any of the 20 types of amino acid can be used.

The sequence of the amino acids in the polypeptide chain is specific to every protein. These long chains of amino acids usually bend and fold extensively, forming a precise and specific three-dimensional shape. The shape is held together by chemical bonds. On the surface of the molecule there is often a depression or 'pocket' and this is known as the **binding site**. Other molecules can fit into the protein at the binding site.

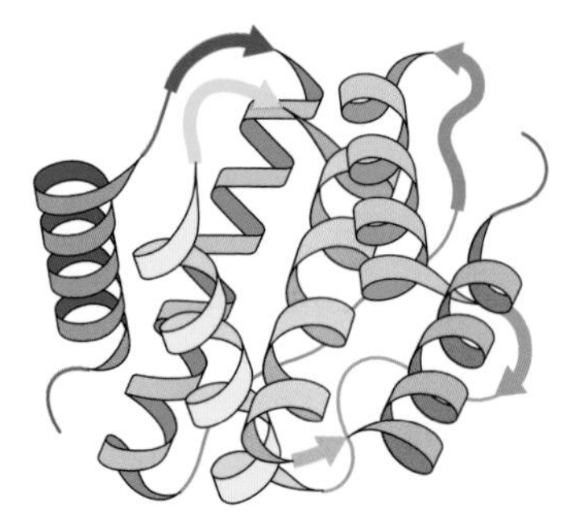

Figure 2 A folded protein contains helical (spiral) sections, and sections where the chain forms a flat zig zag (shown by arrows).

Variety of protein functions

Some proteins form minute fibres. These have very long chains of amino acids and a simple specific shape. These give a framework or structure to some tissues such as muscles. Muscle contraction relies on proteins acting as structural components of muscle tissue.

Hormones are proteins. Insulin is a hormone that your pancreas produces. It controls your blood sugar level within narrow limits no matter what you eat. Antibodies are protein molecules with a precise 3D shape. They are produced by white blood cells to fight off invading pathogens, such as bacteria. This immune reaction helps us to survive attacks from microorganisms.

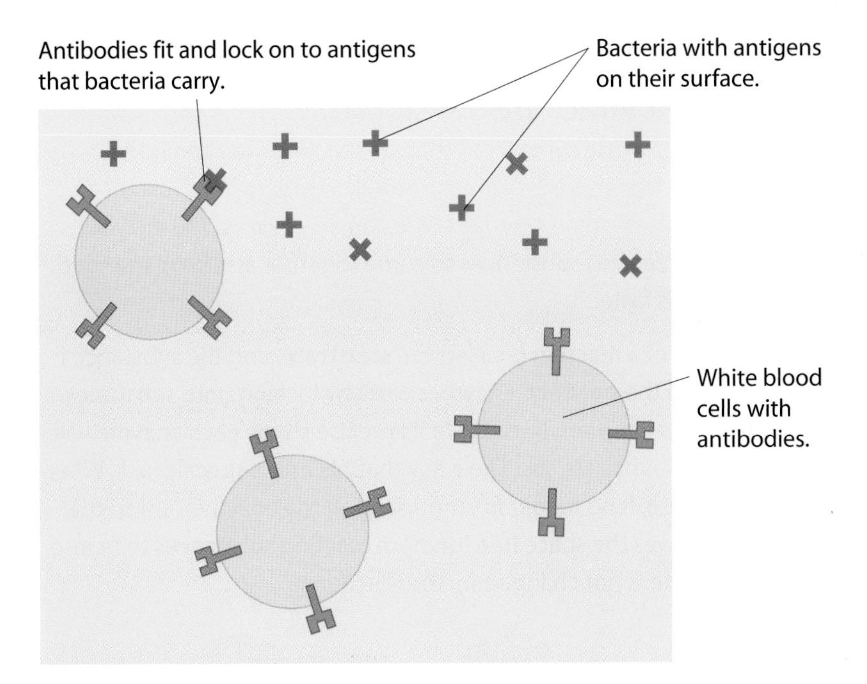

Figure 3 Proteins act as antibodies.

Biological catalysts are made of protein

Thousands of chemical reactions are taking place in the cells of animals and plants all the time. The rate of these chemical reactions is increased by the action of protein enzymes, which are **catalysts**. Catalysts are chemicals that speed up the rate of reactions, but are neither reactants nor products of the reaction. As enzymes speed up reactions in living organisms they are called biological catalysts. They catalyse processes such as respiration, growth, photosynthesis and protein synthesis.

Questions

1. Why are proteins such important molecules throughout the living world?
2. How are amino acids initially arranged in protein molecules?
3. Give three reasons why the amino acid composition of proteins is so varied.
4. Describe how each protein acquires its specific three-dimensional shape.
5. Give four functions of proteins.
6. Look at Figure 1. Using the same colours and shapes to represent amino acids, draw five different polypeptide chains each with eight amino acids.
7. A man cut his chin while shaving. List at least three ways in which proteins in his body are involved in this action and the consequences of it.
8. Explain what is meant by a catalyst. Explain where biological catalysts, or enzymes, act in living organisms, giving examples of two processes in which they are involved.

Examiner feedback

Remember 'CHON' to remind you of the four elements that are always present in proteins and amino acids (carbon, hydrogen, oxygen and nitrogen).

Science in action

A cell's proteome is the total amount of proteins present in it at one time. This produces a very large-scale set of data. Scientists use this set of data to construct hypotheses about the ancestry of modern organisms and they illustrate the fundamental importance of proteins in living organisms.

Route to A*

Denaturation of a protein is a change in its characteristic three-dimensional shape, making it no longer functional. It is an irreversible process.

Taking it further

In a polypeptide chain, the amino acids are linked together by chemical bonds called peptide bonds.

Find out what type of chemical reaction takes place between two amino acids when a peptide bond is formed.

Characteristics of enzymes

Learning objectives

- explain how enzyme shape is related to function
- describe the effect of temperature on enzyme action
- identify how pH affects enzyme action
- explain how the loss of 3D shape results in denatured enzymes
- interpret graphs to show the optimum conditions for the action of certain enzymes.

Molecules with a vital, special shape

During a chemical reaction the substances that are reacting are chemically rearranged as chemical bonds are broken or formed, to make new substances. We can speed up the rate of some reactions by using an enzyme. The enzyme makes it easier for the reacting substances to come together and be rearranged, so the reaction happens faster.

The starting substance of a reaction is called the **substrate**, and the substance it is converted to is called the **product**. Enzymes work by locking onto substrates. Figure 1 shows how this happens. Because of its precise shape each enzyme will only act on one type of substrate, just like a key that fits into a specific lock. When the substrate has reacted, it no longer fits the space on the enzyme, and so the products leave. This leaves the space free for more reacting substances to fit into the enzyme. The enzyme is not changed by the reaction.

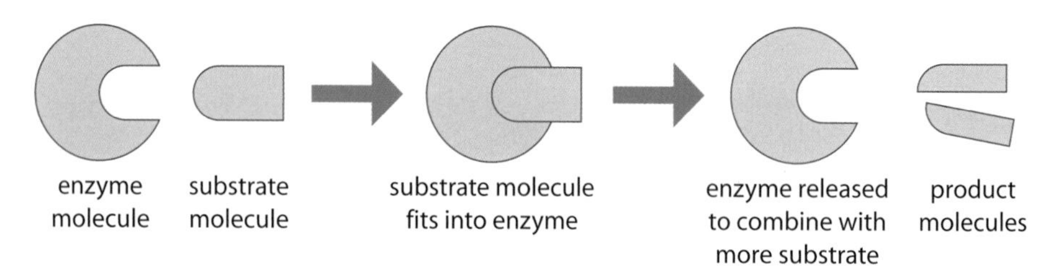

Figure 1 Some enzymes catalyse the breakdown of products.

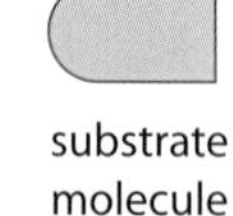

Figure 2 High temperatures change the shape of the enzyme.

Effect of temperature

Most chemical reactions are speeded up by an increase in temperature. Molecules move around more rapidly as the temperature rises. This causes more collisions to occur between enzymes and substrate molecules, and so increases the rate of reaction.

As the temperature continues to rise above a certain level, 37 °C in humans, the rate of enzyme-controlled reactions falls rapidly. High temperatures change the shape of enzymes: they denature them. A denatured enzyme has a different shape and so cannot lock onto the shape of the substrate. Therefore they no longer speed up the reaction.

Examiner feedback

Enzymes are molecules so they are *not* alive. When they stop working because their shape has changed they are denatured or destroyed, *not* dead and *not* killed.

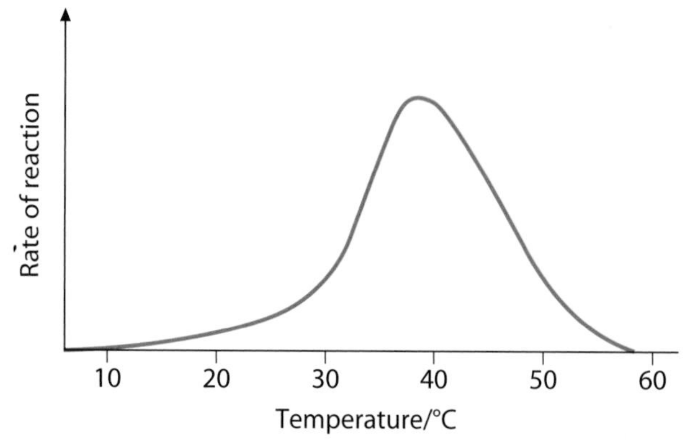

Figure 3 The effect of temperature on an enzyme-controlled reaction.

Effect of pH

Different enzymes work best at different pH values. The pH at which an enzyme works best is called the optimum pH. The optimum pH of an enzyme depends on the pH conditions where the enzyme works. For example, intestinal enzymes work best in alkaline conditions and have an optimum pH of 8. Stomach enzymes have an optimum pH of 2 because the stomach is acidic. Intestinal enzymes will not work at all in very acidic conditions.

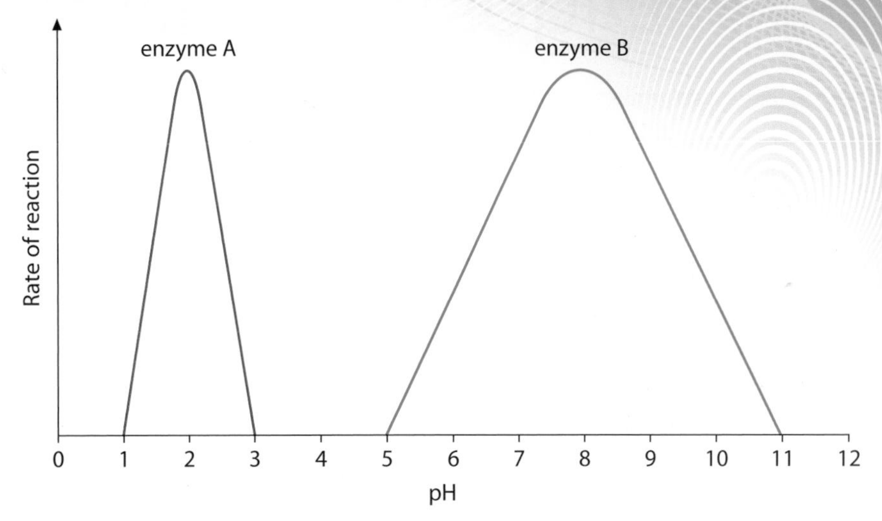

Figure 4 Different enzymes have different optimum pH values.

Science skills

Potato cells contain an enzyme called catalase. This enzyme speeds up the breakdown of hydrogen peroxide. As hydrogen peroxide breaks down, bubbles of oxygen are released forming a froth. A group of students investigated how changing temperature affects the action of catalase. They were provided with potato tissue and hydrogen peroxide and were told that catalase leaves the potato across its cut surfaces. The students decided to measure the rate of reaction by recording the height of froth formed in each test tube.

Table 1 The results of the investigation.

Temperature/°C	Height of froth/cm
15	2.5
25	4.2
35	4.5
45	4.1
55	3.5

a Suggest a suitable control for this investigation.

b What is the independent variable in this investigation?

c One group of students carefully cut the potato into small discs. They used a ruler to make sure the discs were all cut to the same size. Another group of students added the same mass of potato to each test tube.

- **i** Why is it necessary to add the same amount of potato to each tube?
- **ii** Which method is the more accurate – measuring the size or the mass? Give reasons for your answer.

Questions

1. Describe what would happen if we did not have enzymes in our bodies.
2. Explain why the shape of an enzyme affects the way it works in a reaction.
3. High temperatures destroy the shape an amino acid chain makes. **(a)** What effect would high temperatures have on an enzyme-controlled reaction? **(b)** Explain your answer.
4. How does pH affect the rate of enzyme action?
5. **(a)** Make a list of four reactions that you know are catalysed by enzymes. **(b)** Describe two processes in a plant that would be affected if it contained no enzymes.
6. Look at the results in Table 1. Explain the rate of reaction when the temperature increases from: **(a)** 15 to 35 °C **(b)** 35 to 55 °C.
7. Explain why a graph of temperature versus rate of reaction for two human digestive enzymes usually shows a single curve but a graph of pH versus reaction rate of two different digestive enzymes may have two separate curves.
8. Explain as fully as you can why enzymes are sometimes described as a lock that substrates, acting as a key, fit into.

Digestive enzymes

Learning objectives

- relate digestive enzymes to their function and where they work in the digestive system
- describe the breakdown products after catalysis by digestive enzymes
- describe the conditions that different digestive enzymes work in.

Using the food you eat

Your food contains proteins, starches and sugars, and fats and oils. The molecules of proteins, starch and fats are huge – much too large for you to absorb into your body. This means they have to be **digested**, broken into smaller molecules, in your gut. These digestion reactions need to be quick so that you can absorb what you need from your small intestine, before the remains pass out of your body. All of these digestive reactions are catalysed by enzymes to speed them up.

Science skills

Read the information below about food transit time in the gut.

After a meal the time taken for food to travel through your gut depends on many factors. Roughly, it takes 2.5–3 hours for 50% of stomach contents to empty into the intestines. Total emptying of the stomach takes 5–6 hours. Then 50% emptying of the small intestine takes 2.5–3 hours. Total emptying of the small intestine takes 5–6 hours. Finally, transit through the large intestine takes 30–40 hours.

a Tabulate the data in the paragraph above and include total transit time for each region of the gut. Then graphically display the data for total transit time.

Soon to be catalysed by enzymes.

Enzymes work outside cells that produce them

Your gut is simply a hollow tube of different diameters and wall types. Digestive enzymes are produced by specialised cells in glands and tissues lining the gut. They are made inside cells, but they move out of the cells into the gut where they work as they come in contact with food molecules. Some mix with the food in the gut, others remain attached to the outside of cells in the gut wall. They are **extracellular** enzymes.

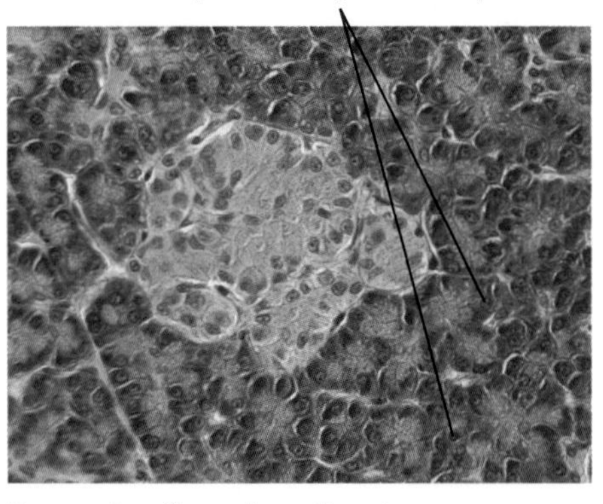

Pancreatic cells produce digestive enzymes.

The right tools for the job

Each type of food needs a particular enzyme to break it down into products that are useful to the body. Different enzymes work on different substances.

- **Amylase** enzymes catalyse the breakdown of starch to sugars.
- **Protease** enzymes catalyse the breakdown of protein into amino acids.
- **Lipase** enzymes catalyse the breakdown of lipids (fats and oils) into fatty acids and glycerol.

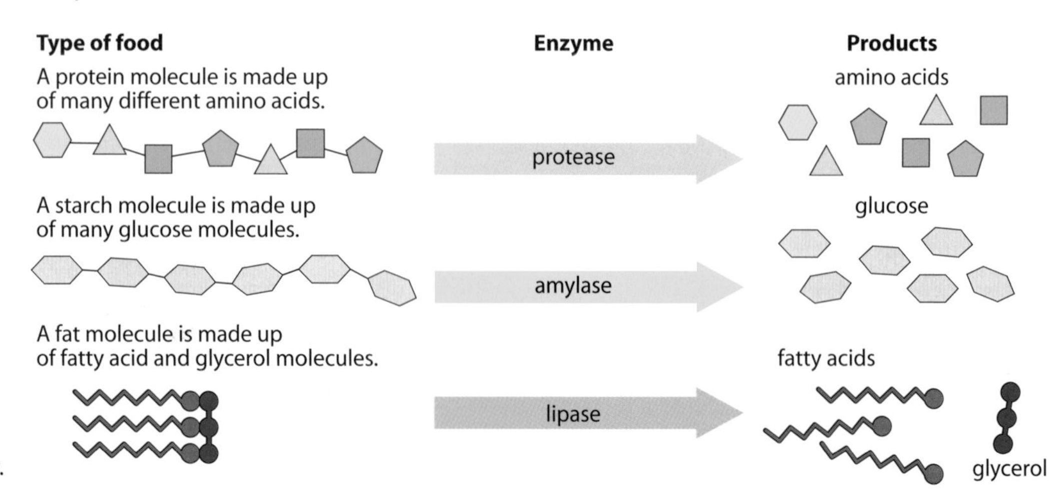

Figure 1 Specific enzymes break food down into products useful for your body.

Where are these digestive enzymes produced and where do they work?

Salivary glands in the mouth produce amylase enzymes.

The **stomach** produces protease enzymes. Hydrochloric acid is also produced. The enzymes in the stomach work best in acid conditions.

The **liver** produces bile, which is stored in the gall bladder. Bile neutralises the acid produced by the stomach and provides alkaline conditions.

gall bladder

The **pancreas** produces protease, amylase and lipase enzymes.

The **small intestine** produces protease, amylase and lipase enzymes. Enzymes in the small intestine work best in alkaline conditions.

Figure 2 Enzymes, hydrochloric acid and bile are released into the gut for digestion.

Examiner feedback

Protease, amylase and lipase enzymes are all produced in the pancreas and small intestine.

Bile is not an enzyme: it is a pH regulator.

Controlling gut pH

In each region of the gut other substances are released to control the pH. For example, the stomach produces hydrochloric acid because the protease enzymes there work best in an acid solution. The liver produces **bile**, which is stored in the gall bladder before being released into the small intestine. Bile neutralises the acid that was added to food in the stomach and provides alkaline conditions for enzymes in the small intestine. These are other protease, lipase and amylase enzymes.

Table 1 Each part of the digestive system releases different enzymes to digest food.

Type of enzyme	Where is it produced?	pH conditions for enzymes	Where does it work?
protease enzymes	stomach	acid	stomach
	pancreas small intestine	alkaline	small intestine
amylase enzymes	salivary glands	slightly alkaline	mouth
	pancreas small intestine	alkaline	small intestine
lipase enzymes	pancreas small intestine	alkaline	small intestine

Questions

1. There are many kinds of enzyme in your gut. Explain why.
2. What would happen if you didn't have enzymes in your gut?
3. Look at Figure 2. **(a)** In which two parts of the gut is amylase made? **(b)** Protease enzymes are made in the small intestine wall. Where else are they made? **(c)** In which part of the gut does lipid digestion take place?
4. Look at Figure 2. The stomach makes a chemical that is not an enzyme. **(a)** What is the chemical? **(b)** Suggest the conditions that enzymes in the stomach work best in.
5. You eat a cheese sandwich. Write down the stages of digestion of the starch and fat as they pass through your gut.
6. **(a)** What kind of tissue produces enzymes in general and what organs in the gut produce enzymes? **(b)** Why are digestive enzymes described as extracellular?
7. **(a)** Describe how bile reaches the small intestine.
   (b) What is one role of bile in digestion?
8. What is meant by digestion and why is it necessary in humans?

B2 4.4

Enzymes used in industry

Learning objectives

- explain why microorganisms are useful in enzyme technology
- describe how three types of enzyme are used in the food and drinks industry
- describe some industrial uses of enzymes
- evaluate the uses of enzymes in industry.

Enzymes from microorganisms

Vast amounts of microorganisms, such as bacteria, are grown in industry to supply us with useful enzymes. Many of the enzymes produced by microorganisms are passed out of the cell, which enables scientists to use the enzyme from the microorganism. Microorganisms can be grown relatively cheaply inside vats, known as fermenters, and no expensive equipment is needed. As they can multiply very rapidly, microorganisms produce large amounts of enzymes quickly. The fermenter is kept at 28 °C so energy costs for enzyme production are low.

Millions of bacteria can be grown quickly in these fermenters. Enzymes extracted from the liquid are used for many purposes.

Enzyme technology

For many years, inorganic catalysts (those not containing carbon) have been used in industrial reactions: for example, iron is used in the production of ammonia in the **Haber process**. Recently there have been many developments in the use of enzymes as industrial catalysts, a process known as **enzyme technology**. Enzymes are highly efficient catalysts: only a small amount of enzyme is needed to produce a large quantity of product. This is why they are more useful in industrial processes than inorganic catalysts.

Enzymes in the food and drinks industry

Enzymes are now used a lot in the catering industry. In some baby foods, proteases are used to help 'pre-digest' the proteins. This makes the food softer, or less fibrous, for babies to eat and easier for babies to digest.

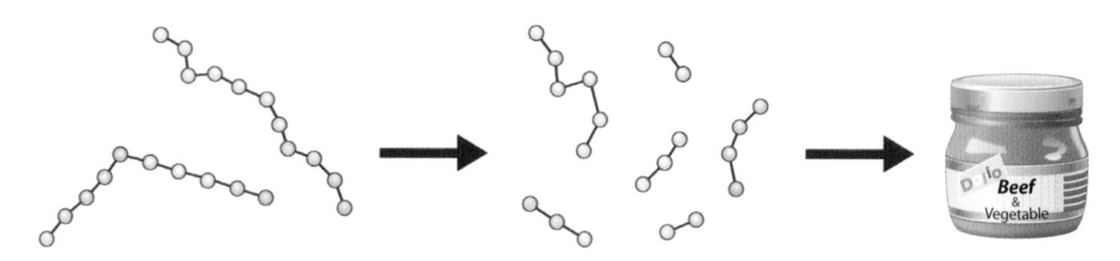

Figure 1 Protease enzymes are used to break down long-chain protein molecules into short chains that are easier for babies to digest.

Enzymes are also used in the production of sugar syrups used as sweeteners in the food and drinks industry. It is cheaper to get starch than sugars from plants to use in our food. Starch can be obtained from potatoes and cereals that are cheaper than sugar sources like sugarcane and sugar beet. Starch can be converted to sugar syrup using **carbohydrases** to catalyse the reaction. This means we can make products like sugary drinks, cakes and sweets more cheaply.

Fructose and glucose are both sugars with the same energy value. Isomerase is an enzyme used to convert glucose syrup into fructose syrup. Fructose is a much sweeter sugar than glucose, so less needs to be added to foods and drinks to make them taste sweeter. This is very useful in the production of slimming foods and low-calorie drinks.

Stage 1: production of glucose syrup

starch from maize grains

Carbohydrase enzymes are added to the starch to digest it into glucose.

↓

glucose syrup

Stage 2: conversion of glucose to fructose

glucose syrup

isomerase enzyme

↓

fructose syrup

Figure 2 Carbohydrase and isomerase enzymes are used in the production of sweeteners.

Science skills

The amount of fructose produced is affected by how fast the glucose syrup flows through the reactor containing immobilised enzymes. Figure 3 shows the result of increasing the rate of flow of glucose syrup into the reactor.

a What rate of flow should scientists use in the reactor? Explain your answer.

b How much more fructose is produced when the rate of flow is increased from 3 to 4 dm^3/min?

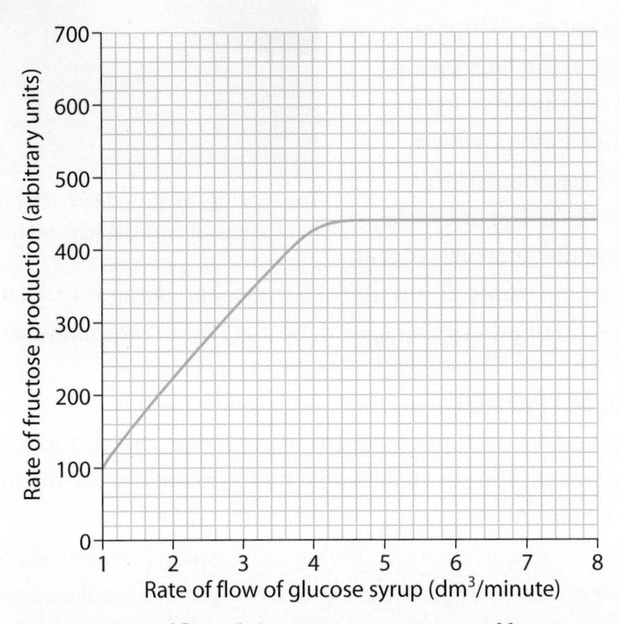

Figure 3 Rate of flow of glucose syrup versus rate of fructose production.

Science in action

When a reaction is complete, the enzyme and product are mixed up with each other. It is very expensive for an industry to keep producing enzymes and to keep separating products from enzymes. To avoid this expense scientists have found ways of fixing enzymes to the surface of small beads. This is called immobilising the enzyme. The diagram shows how isomerase enzymes fixed in this way can be used over and over again. It also means that there are no enzyme molecules mixed with the product, as they are all trapped in the beads.

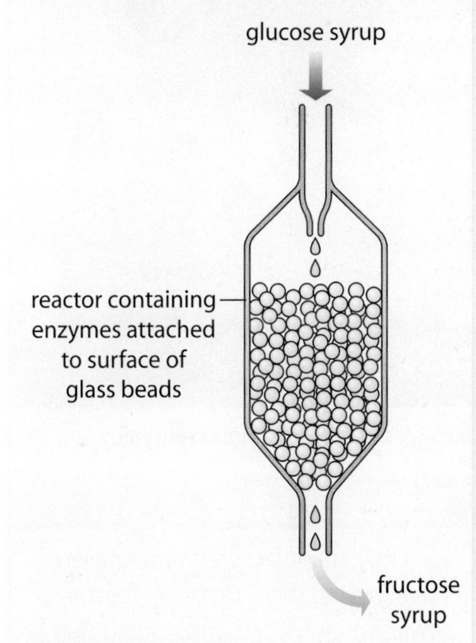

Figure 4 Immobilised enzymes can be used over and over again.

Questions

1. Which organisms produce the enzymes obtained in industry?
2. Where are these organisms grown and how are the enzymes obtained?
3. Give two advantages to industry of using enzymes.
4. What do the following enzymes digest: **(a)** protease **(b)** carbohydrase **(c)** lipase?
5. **(a)** Give three examples of the use of enzymes in the food industry.
   (b) For each example, name the type of food the enzyme works on and the reason for using an enzyme in the food production.
6. Evaluate the use of enzymes in the food industry.
7. Explain fully how protease enzymes in the food industry could be used over and over again to break down large protein molecules into smaller ones.
8. Evaluate the use of immobilised enzymes in the food industry.

A*

Taking it further

Isomers are molecules that contain the same types and numbers of atoms, but arranged in different ways.

B2 4.5

Home use of enzymes

Learning objectives

- explain why enzymes are used in biological detergents
- describe the types of enzyme biological detergents contain
- evaluate the use of enzymes in the home and in industry.

Biological detergents contain enzymes.

Science in action

Recently, manufacturers and retailers have tried to show they care for the environment by recommending that many garments are washed at 30 °C. However, Korean scientists have shown that washing at 30 °C is much worse at removing dust mites and pollen than washing at 60 °C. Allergy sufferers could have a problem with low-temperature washes.

Examiner feedback

An 'evaluate' question requires you to give some advantages and disadvantages and then to make a conclusion. Your conclusion should say whether or not you think enzymes (in this case) are useful, linked to a reason for your judgement.

Enzymes from bacteria

The enzymes produced by bacteria living in hot springs will work at high temperatures.

Bacteria help to clean our laundry. This is true if you use biological detergents containing enzymes. The first biological detergents that were produced would only work in warm water. However, proteases and lipases have now been produced that work at much higher temperatures. Most of the enzymes in washing powders are obtained from bacteria living in hot springs, which means the bacteria are adapted to live in water above 45 °C. The enzymes obtained from these bacteria will work at moderately high temperatures. This is useful because the detergents in washing powders, which get rid of greasy stains, work best at higher temperatures.

Biological detergents

The dirt that we get on our clothes comes from our bodies, our surroundings and from the food we eat. The substances that make up the dirt are mostly proteins, fats and sugars. All washing powders contain detergents to dissolve stains so that they can be washed away. Biological washing powders also contain protease and lipase enzymes. Protease enzymes catalyse the breakdown of proteins present in stains such as blood, grass and egg. Lipase enzymes catalyse the breakdown of lipids in stains such as fat, oil and grease. The protein and fat molecules in stains are broken down into smaller, soluble molecules that dissolve easily in water and can be washed away. In the home, biological detergents are more effective at low temperatures, such as 30 °C, than other types of detergents. For a wash at 60 °C or 90 °C biological detergents are not recommended because most enzymes are denatured at high temperatures and stop working. Some people still use non-biological detergents because they get an **allergic reaction**, such as a skin rash, to the enzymes.

Evaluating enzyme use in home and industry

It helps you to decide if using enzymes is useful or not by looking at their advantages and disadvantages.

Table 1 Advantages and disadvantages of enzymes.

Advantages	Disadvantages
Enzymes bring about reactions at normal temperatures and pressures, saving on energy.	Most enzymes are denatured at high temperatures.
Enzymes save on expensive equipment.	Many enzymes are expensive to produce.
Biological detergents are more effective at low temperatures than other types of detergent.	Allergy sufferers may be allergic to the enzymes and/or require a higher temperature wash to destroy allergens in bedding and clothes.

Science skills

A group of students carried out an investigation to find the conditions in which biological washing powders work best. The students used photographic film to demonstrate the action of the washing powders. The film contains black grains stuck on by a layer of gelatin. Gelatin is a protein. When the gelatin is broken down by the enzymes in the washing powder the film becomes clear as the black grains come away.

The students prepared a 1% solution of washing powder by dissolving 1 g of powder in 100 cm^3 of water. Figure 1 shows how the students designed the investigation.

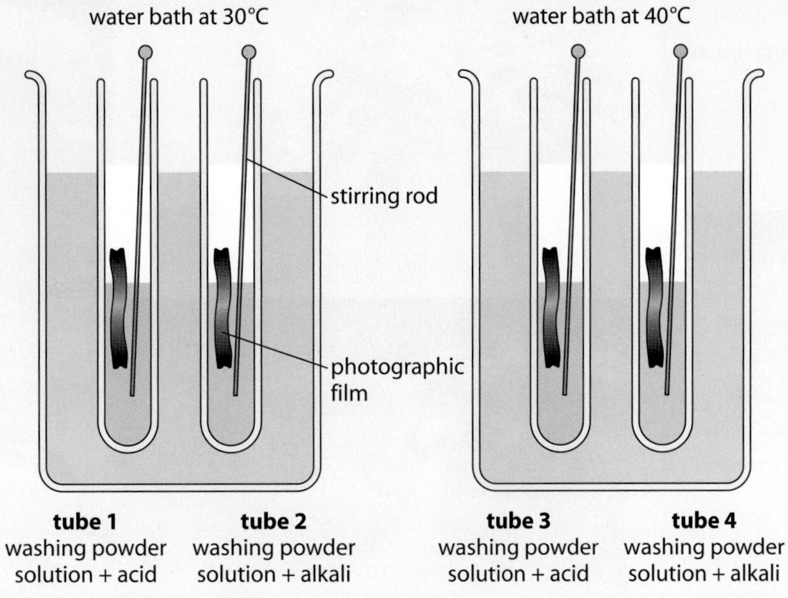

Figure 1 A biological washing powder can dissolve the protein on photographic film.

a Identify the two independent variables.

b What type of variable is the dependent variable?

c **i** Explain why a 1% solution of washing powder was used in all four test tubes.

ii Explain why a stirring rod was used.

d Suggest what the students should do to make their results more reliable.

Table 2 The students' results.

	Tube 1	Tube 2	Tube 3	Tube 4
Temperature/°C	30	30	40	40
pH	4	8	4	8
Time taken for film to go clear/min	not digested	25	40	10

Questions

1. Which organisms produce the enzymes found in biological detergents?
2. Which enzymes digest: **(a)** protein **(b)** fat?
3. Suggest two stains that: **(a)** protease **(b)** lipase would act on.
4. Suggest: **(a)** two advantages of using enzymes to help get clothes clean **(b)** one disadvantage of using enzymes in biological detergents.
5. Explain why it is easier for stains to leave clothes when they have been digested by enzymes.
6. A student washed his clothes using biological detergent on a very hot wash at 90 °C. They still came out with stains on. Explain to him the advantages of using enzymes in detergents and why his clothes would have been washed cleaner if the water temperature had been at 40 °C.
7. Suggest two reasons why hotels do not use biological detergents to wash their cotton sheets.
8. Use Table 1 to evaluate the use of enzymes in home and industry.

Aerobic respiration

Learning objectives

- describe how aerobic respiration in mitochondria uses glucose and oxygen to release energy
- summarise aerobic respiration with a word equation
- explain what cells use the energy released in respiration for.

Taking it further

The process of releasing energy from substances such as glucose takes place through a complex series of reactions. These reactions occur in different parts of the mitochondrion.

Respiration for energy

Respiration is carried out in plant cells and animal cells all the time to provide the energy that organisms use to stay alive and to do all that they need to do. Without respiration organisms would die.

All these activities need aerobic respiration.

Most animal and plant cells respire aerobically. **Aerobic respiration** literally means respiration using air. More accurately, it uses oxygen from the air. The oxygen is used to chemically break down 'fuel' molecules, usually a simple sugar called glucose. The breakdown of glucose releases energy and produces two waste products – carbon dioxide and water. The overall reaction is:

glucose + oxygen ⟶ carbon dioxide + water (energy given out)

This equation is only a summary. In respiration, glucose is broken down in many separate stages. Like other reactions inside cells, the reactions in respiration are controlled by enzymes.

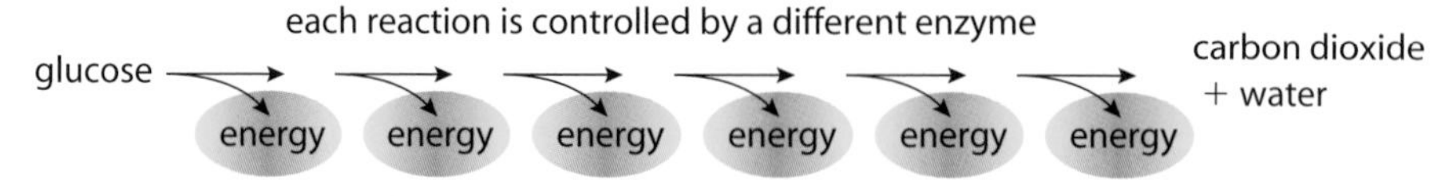

Figure 1 The breakdown of glucose in aerobic respiration involves many stages, each controlled by an enzyme.

Powerhouses of the cell

Most of the reactions in aerobic respiration occur inside small structures called mitochondria, which are found inside all plant and animal cells. Cells that use a lot of oxygen, such as muscle cells, tend to have many more mitochondria than others. In animals, the glucose comes from the breakdown of food molecules in the gut, or from the breakdown of food stores in the body. In plants, the glucose is a product of photosynthesis.

Practical

We can use this apparatus to prove that animals produce carbon dioxide during respiration. With adaptation, it can also show that plants produce carbon dioxide during respiration.

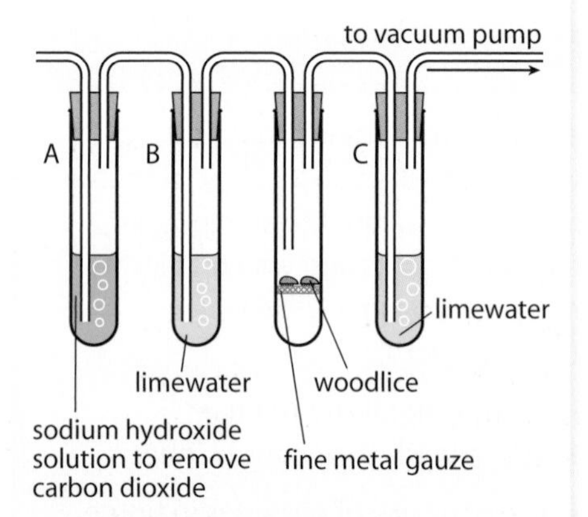

Figure 2 Apparatus to demonstrate carbon dioxide production during respiration.

In most animals, oxygen and glucose are carried to all cells in the blood. The waste products of respiration (carbon dioxide and any water the cell doesn't need) are carried away in the blood.

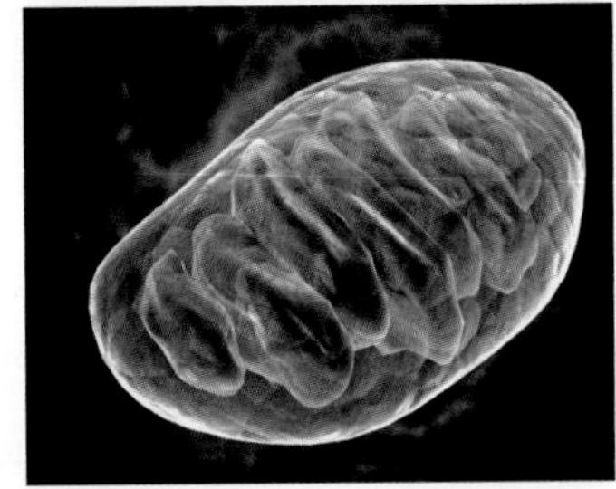

Mitochondria are sometimes called the powerhouses of cells because this is where energy is released in respiration.

Using the energy from respiration

The energy from respiration is used for many processes in a living organism. It can be used to make large, complex molecules from smaller subunits – for example animals build proteins from amino acids or fats from fatty acids. Plants start with even smaller molecules, for example they make amino acids from sugars and nitrogen-containing substances from the soil.

Energy is also needed to break down large molecules into smaller ones, for example during digestion in animals. The smaller molecules can more easily be moved around the body to where they are needed. Animals also need energy to make muscles work and to pump the blood round the body. Mammals and birds also use energy from respiration to keep their body temperature at a constant level above the temperature of their surroundings.

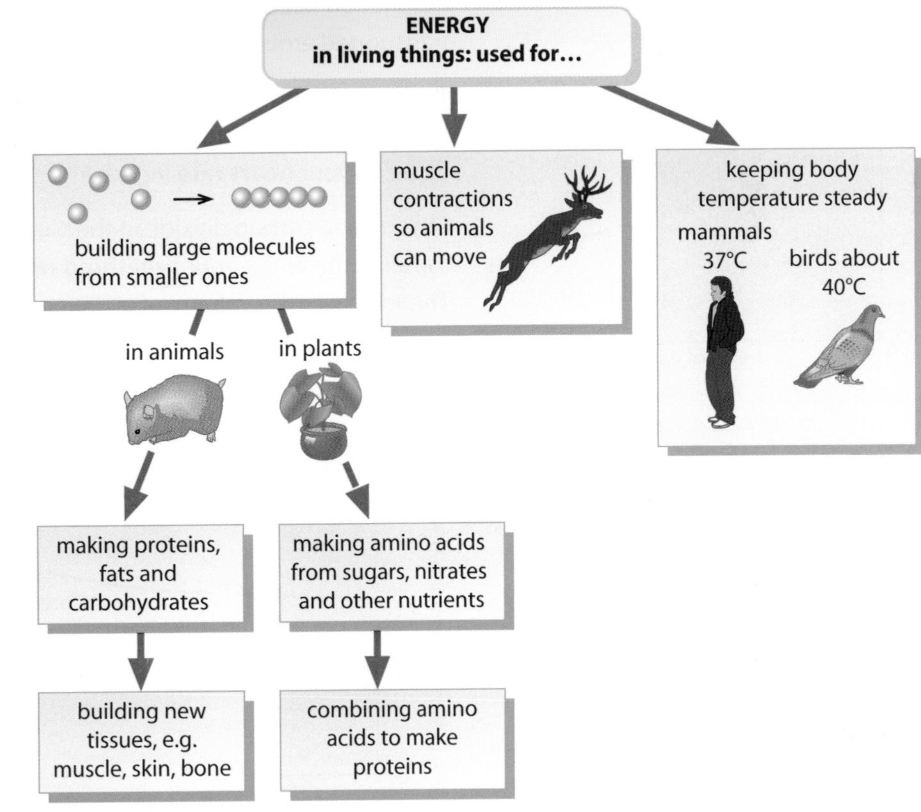

Figure 3 Living things use the energy they get from respiration in many ways.

Science skills

Reactions that are controlled by enzymes are usually affected by temperature. The rate of a reaction usually increases as temperature increases up to around 40 °C, after which it decreases.

a Identify the dependent and independent variables in an experiment to show the effect of temperature on the enzymes that control the reactions of respiration.

Questions

1. Make a list of 10 things that you use energy for. Try to include some that allow you to stay alive.
2. **(a)** Where does the glucose for respiration come from in animals? **(b)** Explain how a plant gets the glucose it needs for respiration.
3. A muscle cell has more mitochondria than a brain cell. Suggest a reason for this.
4. Give as many reasons as you can that could explain why one person would be breathing faster than another.
5. The rate of respiration of a mouse, a plant and a lizard are measured as the volume of carbon dioxide used in 1 hour. **(a)** Describe how the plant experiment should be set up to make sure the results are reliable. Explain your answer. **(b)** Explain what must be done to all the results so that they can be compared fairly. **(c)** Put the organisms into order to show how quickly they respired, slowest first. Explain your order.
6. Explain fully the role of the circulatory system in enabling respiration to take place in the cells.
7. Cyanide blocks an enzyme that controls one of the reactions in respiration. Explain fully why cyanide is a very effective poison.

Changes during exercise

Learning objectives

- describe some changes that take place in the body during exercise
- interpret data on the effects of exercise on the human body
- explain that the effect of changes during exercise is to supply sugar and oxygen faster to muscles and remove carbon dioxide more rapidly.

Energy for exercise

Your muscles need energy to contract. This energy comes from respiration. When you exercise you need more energy so that your muscles can contract more frequently and for longer, so the rate of respiration increases the harder you exercise.

When you begin exercising, glucose moves into the mitochondria of muscle cells. As the level of activity increases, more glucose moves out of the blood and into muscle cells. However, if there isn't enough glucose in the blood for the increased level of respiration, then stores of **glycogen** in muscle and liver cells are converted to glucose to supply what is needed.

Breathing rate and heart rate

Respiration needs oxygen and produces carbon dioxide. These gases are also transported around the body in the blood.

To supply all the extra oxygen and sugar that is needed when you exercise, and to remove all the extra carbon dioxide from cells, your blood needs to circulate faster, so your **heart rate** increases.

Oxygen and carbon dioxide in the blood also need to be exchanged faster with the air in the lungs. Your **breathing rate** increases and you breathe more deeply. This increases the volume of air being moved into and out of the lungs.

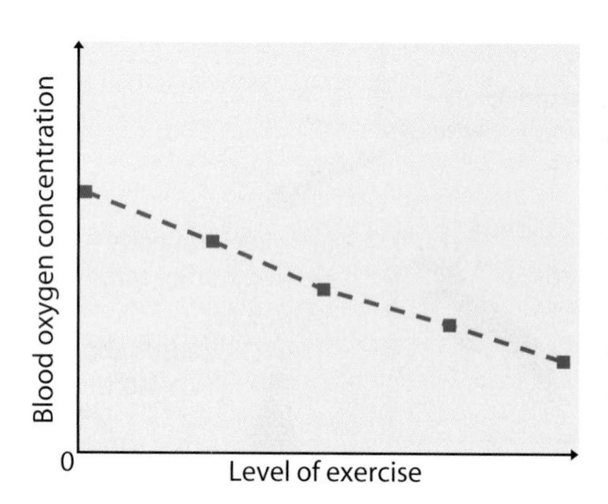

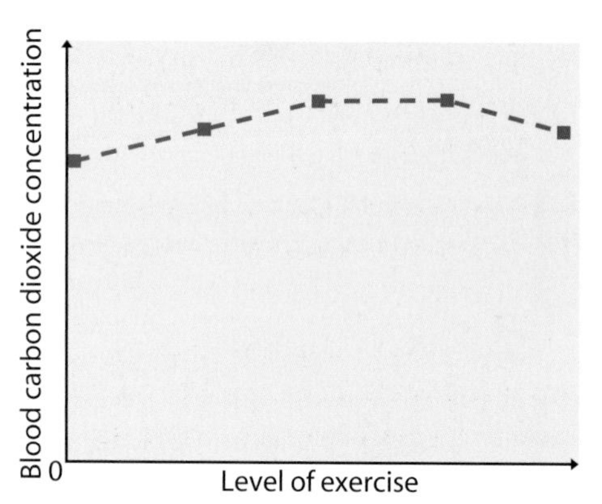

Figure 1 Concentrations of oxygen and carbon dioxide in the blood change with increasing level of exercise.

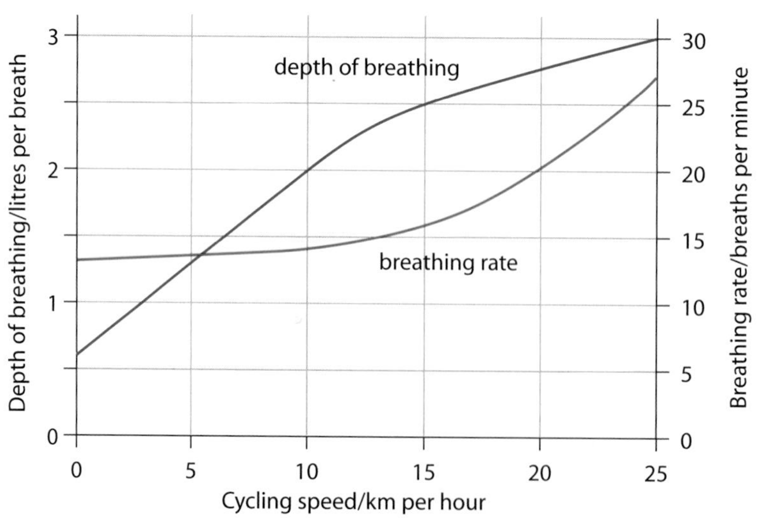

Figure 2 Changes in breathing rate and depth of breathing with exercise.

Science skills

Table 1 Data showing the heart rate of someone of average fitness after 1 minute of different levels of exercise.

Exercise level	Heart rate/ beats per min
resting	69
gentle	84
moderate	123
vigorous	162

a Explain why heart rate was measured after 1 minute at each level.

b Graph the data and explain the shape of the curve.

c Suggest how the curve might differ for a highly fit athlete and for an unfit person.

Examiner feedback

During exercise both breathing rate and heart rate increase, so more oxygen and glucose are transported to the actively respiring cells.

If you exercise regularly your body gets fitter and better able to provide the increased blood supply that muscles need during activity. Our bodies are adapted for regular and frequent activity as a result of human evolution. In the UK seven out of 10 adults do not get enough exercise. This is leading to an increase in health problems such as high blood pressure and heart disease.

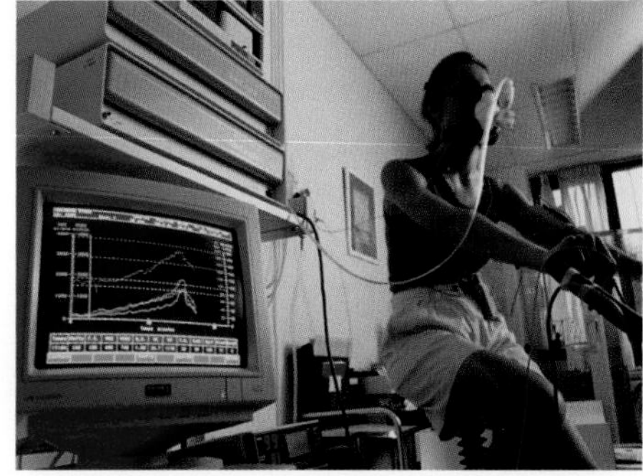

Heart rate, breathing rate and breathing depth can all be measured while a person exercises.

Science skills

Figure 3 suggests that being unfit is more closely linked to risk of death than being overweight.

d How would you decide if these results are reliable?

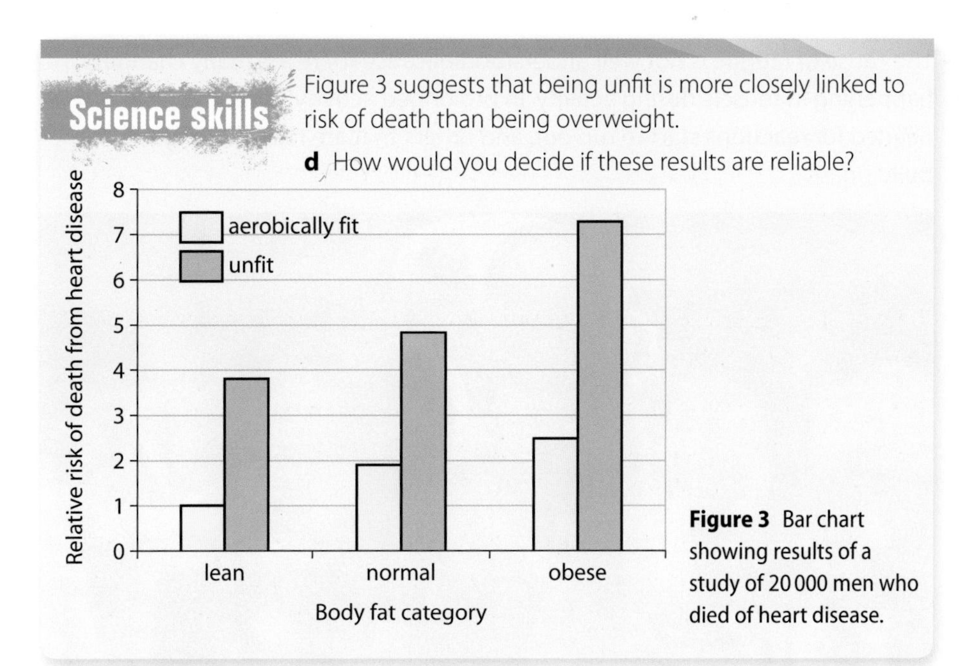

Figure 3 Bar chart showing results of a study of 20 000 men who died of heart disease.

Questions

1. Describe what effect an increasing level of activity has on: **(a)** the oxygen concentration in blood near cells **(b)** the carbon dioxide concentration in blood near cells.
2. The oxygen and carbon dioxide concentrations shown in Figure 1 were measured in venous blood. Explain why these measurements were taken from the veins, not the arteries.
3. Look at Figure 2. **(a)** How many breaths per minute were taken when the person was cycling at 10 km/hour? **(b)** What was the volume of each breath when the person was cycling at 15 km/hour? **(c)** Calculate the total volume of air breathed in and out in 10 minutes when the person was cycling at 20 km/hour. Show your working.
4. **Table 2** Blood supply to different parts of the body. (Skeletal muscles include those of the arms and legs.)

Part of body	Blood flow at rest/cm^3 per min	Blood flow during exercise/cm^3 per min
brain	750	750
heart muscle	300	1200
gut and liver	3000	1500
skeletal muscles	1000	1600

Compare the values at rest and during exercise and explain any changes, or lack of change.

5. Draw a concept map to show how your body responds to increased exercise. Add notes to your map to explain why those changes happen.
6. **(a)** Which has the higher concentration of oxygen, the air in the lungs or the blood coming to the lungs from the tissues? Explain your answer. **(b)** Which has the higher concentration of carbon dioxide, the air in the lungs or the blood coming to the lungs from the tissues? Explain your answer. **(c)** The rate of diffusion of a gas increases if there is a greater difference in concentration. Explain how breathing faster and deeper increase the rate of exchange of oxygen and carbon dioxide between the blood and air in the lungs.
7. For most of human evolution we lived as hunter–gatherers, moving around to find food. Explain in terms of respiration and human evolution why this still affects the way our bodies work, and why this might be the cause of health problems due to being unfit.

Anaerobic respiration

Learning objectives

- explain how anaerobic respiration supplies energy to muscle cells when insufficient oxygen is available
- describe anaerobic respiration as the incomplete breakdown of glucose
- explain why anaerobic respiration results in an oxygen debt that has to be repaid
- explain why muscles become fatigued after long periods of activity.

Taking it further

You will find that advanced texts talk about lactate rather than lactic acid. This is because when lactic acid is in solution in water, it dissociates into negative lactate ions and positive hydrogen ions. However, the two terms mean the same thing in this instance.

Examiner feedback

A different form of anaerobic respiration occurs in many microorganisms, including yeast. It is commonly called fermentation and its end-product is ethanol (ethyl alcohol), not lactic acid as in animal cells. It is important to distinguish between the two processes because of their different end-products.

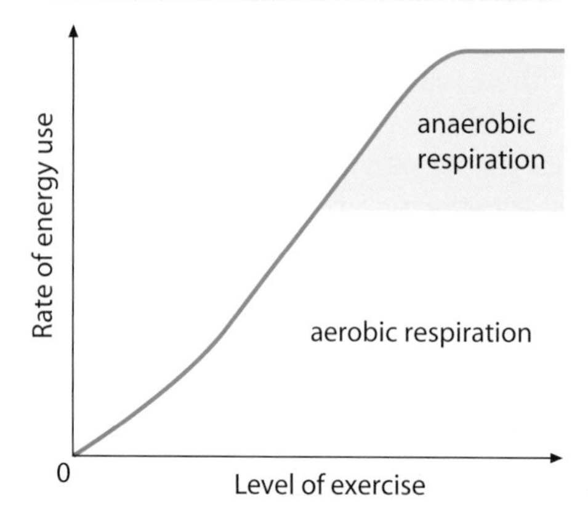

Figure 1 This graph shows the contribution of aerobic and anaerobic respiration to energy production at different levels of activity.

Running out of oxygen

If you exercise for a long time, your muscles start to **fatigue**. This means that they don't contract as strongly as they normally do, and cannot do as much work. You feel an increasing weakness and pain or cramps in the muscles.

The cause of fatigue is not well understood because there are many changes happening in muscle during activity. In prolonged activity some chemicals needed for reactions start to run out, and others that are made during activity build up.

This athlete has muscle fatigue after running a marathon.

Another source of energy

Most of the time muscles get the energy to contract from aerobic respiration. However, if you suddenly start exercising vigorously, or if you exercise vigorously for some time, your muscle cells may not be able to get enough oxygen to keep contracting hard.

Fortunately, if oxygen levels in muscle cells are low, the cells can also use **anaerobic respiration**. This process releases energy without the need for oxygen to break down glucose. Anaerobic respiration does not replace aerobic respiration. It provides muscles with extra energy beyond what they can get from aerobic respiration.

Comparing the two types of respiration

Anaerobic respiration also breaks down glucose, but it does not make the same products as aerobic respiration. The equation for anaerobic respiration is:

glucose ⟶ **lactic acid** (energy given out)

Anaerobic respiration produces much less energy per glucose molecule than aerobic respiration. This is because the glucose is only partly broken down and there is still a lot of energy locked in the bonds of the lactic acid molecules. However, the breakdown of glucose to lactic acid is much faster than the breakdown of glucose to carbon dioxide and water, so anaerobic respiration can supply energy quickly.

Science skills

Studies of human athletes show that different sports depend on different combinations of aerobic and anaerobic respiration.

Table 1 Types of respiration for different activities.

Activity	Type of respiration
short-distance sprint	mostly anaerobic
middle distance, e.g. 400 m run	anaerobic and aerobic
long distance, e.g. marathon	mostly aerobic

a Explain why different kinds of athletes need to train differently to improve the efficiency of their muscle cells to manage aerobic or anaerobic respiration.

Whales and seals keep lactic acid inside their muscles and out of their blood during diving because they dive for so long and the amount of lactic acid produced could damage other organs.

The oxygen debt

For a while after exercise, we continue to breathe deeply even though our muscles have stopped working as hard. The extra oxygen that our bodies need after exercise is called the **oxygen debt**.

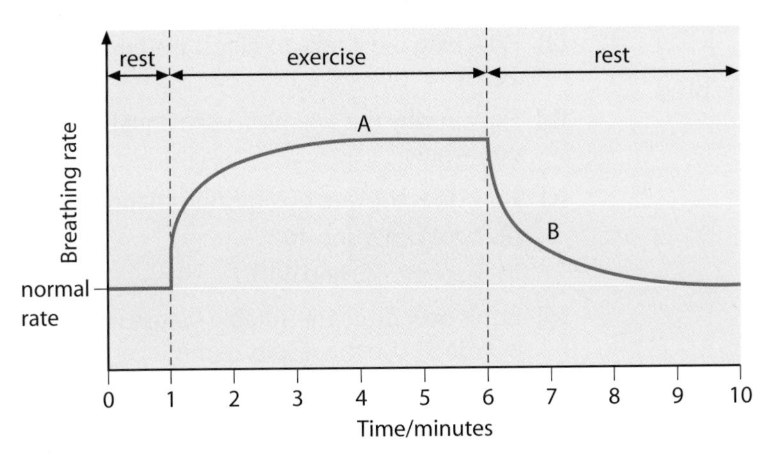

Figure 2 Breathing rate during and after exercise.

Some of the extra oxygen taken in at this time is used to return the body to its resting state. However, if anaerobic respiration has occurred, the lactic acid must also be removed from the muscle cells and recycled. It is transported in the blood to the liver where it is oxidised so that it can be used for aerobic respiration another time, when it is broken down to carbon dioxide and water.

Science in action

It used to be thought that lactic acid caused muscle fatigue, but scientists now realise that it plays an essential part in keeping muscles working when they are being overstimulated during vigorous exercise. Endurance athletes are now trained to develop both their aerobic and anaerobic capacity.

Examiner feedback

Make sure you know what anaerobic respiration is, how it can contribute during exercise and why it results in an 'oxygen debt'.

Questions

1. Explain what we mean by 'muscle fatigue'.
2. State one similarity and one difference between aerobic and anaerobic respiration.
3. Explain why the oxygen concentration in muscle cells may be low during vigorous activity.
4. Give one disadvantage and one advantage of anaerobic respiration compared with aerobic respiration.
5. Look at Figure 2. Explain the breathing rate at points A and B.
6. Why is it an advantage for an endurance athlete to develop both aerobic and anaerobic capacity?
7. Explain fully why whales and seals need a special adaptation to cope with lactic acid.
8. Explain the different proportions of aerobic and anaerobic respiration shown in Table 1.

Assess yourself questions

1 Proteins are the most diverse biological molecules. Write three or four sentences about proteins. Include the following words in your sentences: polypeptide, specific, amino acids, bonds, enzymes, twenty, three-dimensional.

In this question you will be assessed on using good English, organising information clearly and using specialist terms where appropriate. *(6 marks)*

2 An investigation was carried out to find out the effect of bile on the action of the enzyme lipase.

Table 1 Four test tubes and their contents.

Test tube	Contents
1	milk, lipase, pH indicator, bile
2	milk, lipase, pH indicator
3	milk, boiled lipase, pH indicator, bile
4	milk, lipase, pH indicator, boiled bile

Bile is alkaline. The pH indicator is yellow when the pH is 7 or less and red when the pH is over 7.

Table 2 Time taken for the indicator to change colour.

Test tube	Time taken for the pH indicator to change colour/min
1	15
2	38
3	no change
4	15

(a) What colour was the indicator in test tube 1 at the start of the investigation? Explain your answer. *(2 marks)*

(b) Explain why the action of lipase caused the indicator to change colour in test tube 1. *(3 marks)*

(c) Explain why there was no colour change in test tube 3. *(2 marks)*

(d) What do the results from test tubes 1 and 2 tell you about the effect of bile on the reaction? *(1 mark)*

(e) One student concluded that bile contains enzymes that digest fats. Which of the results shows that this conclusion is incorrect? Explain your answer. *(2 marks)*

3 Scientists working for a washing powder manufacturer carried out tests on a new protease enzyme that removes protein stains, such as egg and blood. They wanted to find out if the protease was suitable for use in washing powders.

They placed equal-sized cubes of egg white into test tubes. The test tubes were placed into water baths and kept at different temperatures from 0 to 60 °C. The same volume of the enzyme was added to each tube. The scientists recorded the time taken for the egg white to be digested. Figure 1 shows their results.

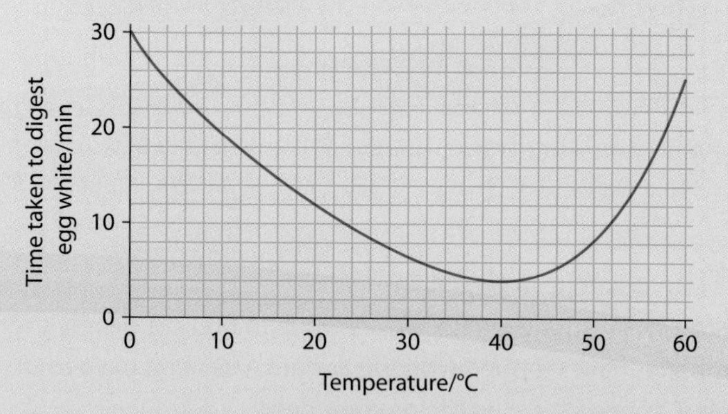

Figure 1 Results of the investigation.

(a) How long did it take to digest the egg white at 20 °C? *(1 mark)*

(b) Explain why the scientists used equal-sized cubes of egg white. *(1 mark)*

(c) How does the rate of digestion change:

(i) between 5 and 40 °C *(1 mark)*

(ii) between 40 and 60 °C? *(1 mark)*

(d) Is this new protease suitable for use in washing powders? Use the results of this investigation to explain your answer. *(2 marks)*

4 The graph below shows the action of an enzyme measured at different temperatures.

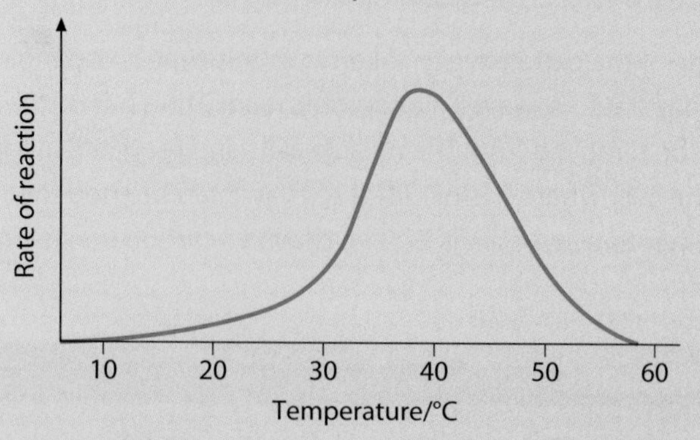

Figure 2 Enzyme action at different temperatures.

(a) Which variable is the dependent variable? *(1 mark)*

(b) What kind of variable is the independent variable? *(1 mark)*

(c) How could you show the graph was accurate? *(1 mark)*

(d) Explain the rate of enzyme action between 10 and 50 °C. *(3 marks)*

5 **Table 1** Some industrial uses of enzymes.

Use	Enzyme involved	Explanation
tenderising meat	protease	**(i)**
making sugar syrups	**(ii)**	starch is broken down into sweet sugars
(iii)	isomerase	fructose is a very sweet sugar so less of it is needed in the production of slimming foods

(a) From where are these industrial enzymes obtained? *(1 mark)*

(b) What would be the consequence of the absence of enzymes from one of these chemical reactions that uses them? *(1 mark)*

(c) Complete **(i)**, **(ii)** and **(iii)** in the table above. *(3 marks)*

(d) Explain what immobilised enzymes are, and how using them in reactions saves money for the food industry. *(3 marks)*

6 Respiration is a set of chemical reactions that use catalysts.

(a) Where in a cell does respiration take place? *(1 mark)*

(b) Copy and complete the word equation that summarises the reactions in respiration:

glucose + ______ ⟶ ______ + ______ + energy given out *(3 marks)*

(c) To what group of chemicals do the catalysts belong and why are they important? *(2 marks)*

(d) Give *two* ways in which the energy released during respiration is used in animals. *(2 marks)*

(e) Give *two* ways in which the energy released in respiration is used in plants. *(2 marks)*

7 A student measured her heart rate immediately after sitting, walking or jogging for 2 minutes. She carried out each test three times, and rested for 2 minutes between each test. Her heart rate is given in beats per minute.

Table 2 The effect of exercise on the student's heart rate.

Exercise	Sitting			Walking			Jogging		
	1	2	3	1	2	3	1	2	3
Heart rate/ bpm	71	68	72	91	85	84	113	110	119

(a) Explain why the student took three measurements for each level of exercise. *(1 mark)*

(b) Calculate the mean for each level of exercise. *(1 mark)*

(c) Explain what the student's results show. *(1 mark)*

(d) Another student's results for the same investigation were sitting 62, walking 84, and jogging 108. Give *one* reason for the difference between the two students' results. Explain your answer. *(1 mark)*

(e) Do the second student's results support your conclusion in part **(c)**? Explain your answer. *(1 mark)*

(f) Explain fully why heart rate changes with level of exercise. *(3 marks)*

8 The graph in Figure 3 shows the different sources of energy used by muscle cells as the effort put into exercise increases.

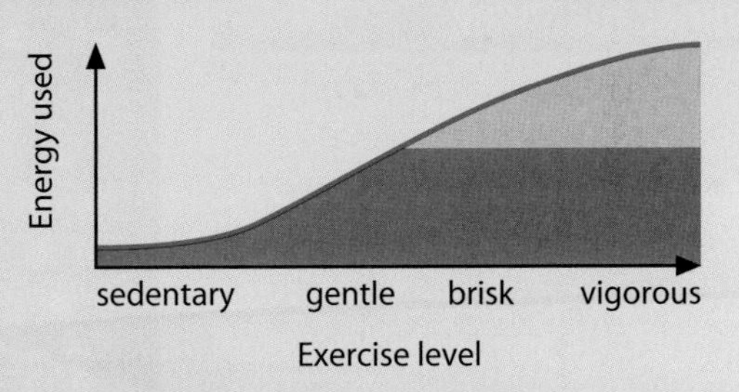

Figure 3 The contribution of aerobic and anaerobic respiration in supplying energy to muscles during activity.

(a) Explain why the energy from aerobic respiration levels off as effort increases. *(1 mark)*

(b) Write a word equation for anaerobic respiration. *(2 marks)*

(c) Anaerobic respiration produces less energy per glucose molecule than aerobic respiration. Explain why. *(1 mark)*

(d) Describe what happens to the product of anaerobic respiration after exercise has ended. *(2 marks)*

(e) Explain what is meant by the term 'oxygen debt'. *(2 marks)*

(f) Explain fully why lactic acid concentration increases in muscles during vigorous activity and why it is difficult to prove that lactic acid is the cause of pain in muscles. *(4 marks)*

9 Describe fully the role of aerobic and anaerobic respiration during exercise in humans.

In this question you will be assessed on using good English, organising information clearly and using specialist terms where appropriate. *(6 marks)*

Cell division

Learning objectives

- describe how new body cells are produced by mitosis
- explain that new cells are needed for growth or repair, or for asexual reproduction
- describe how gametes (sex cells) are produced by meiosis
- describe the main differences between mitosis and meiosis
- explain how fertilisation produces a new cell with two sets of chromosomes that then divide by mitosis.

Examiner feedback

Chromosomes don't usually stay in pairs or sets, so you won't normally see them like this. However, you need to remember that body cells have two sets of chromosomes, while gametes (sex cells) in many organisms only have one set.

Practical

The cells at the tip of a root or shoot in a plant are dividing rapidly for growth. We can investigate the stages of division in plant cells by making a **root tip squash**.

Taking it further

The division of cells during mitosis and meiosis is considerably more complex than shown here. Each type of division goes through several different phases which are named and described separately. However, it is important to remember that division is one continuous process and these stages run into one another. So, when looking at slides of division, it can be difficult to identify exactly at which stage each cell is.

Chromosome sets

In the body cells of most organisms, there are two sets of genetic information, or **chromosomes**, because one set comes from each parent. For example, humans have 46 chromosomes in almost every body cell, which can be arranged in 23 pairs (two sets of 23).

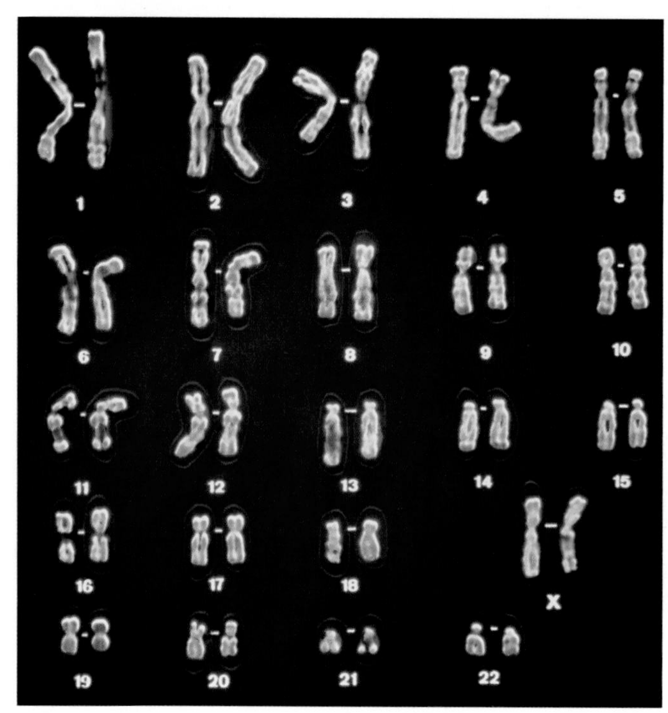

A human karyotype shows all the chromosomes from one cell arranged in pairs, from biggest to smallest.

Mitosis

During growth, or for replacement of damaged cells, a body needs to make new cells. To do this, existing body cells divide in a type of division that is called **mitosis**.

Before a cell divides, every chromosome is copied. During division, one copy of each chromosome moves to one side of the cell and the other copy moves to the other side. The cell then splits in two, creating two new cells with the same number of chromosomes as the original. Since body cells start with two sets of chromosomes, mitosis will produce more cells with two sets of chromosomes.

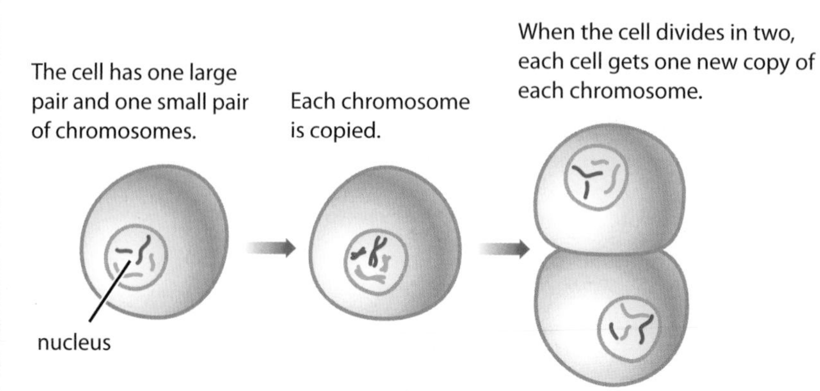

Figure 1 Mitosis of a body cell with four chromosomes (two pairs). The chromosomes are drawn short here, and coloured, so it is easier to see what is happening. They don't really look like this.

Mitosis is also the kind of division that occurs during **asexual reproduction**, when new offspring are produced by the division of parent cells. The cells of the offspring receive copies of every chromosome in the parent cell. So the offspring carry the same genes as their one parent, and are called **clones**.

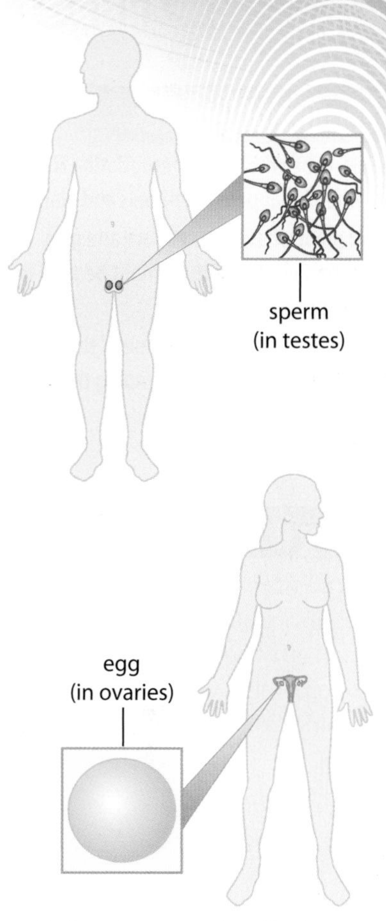

Figure 2 Gametes are produced in the reproductive organs.

Meiosis

The production of **gametes** (sex cells) in the reproductive organs uses a different kind of cell division, called **meiosis**. In humans, meiosis occurs in the **testes** in men and the **ovaries** in women. During meiosis, each new cell only receives one set of chromosomes, so human gametes only contain 23 chromosomes.

Meiosis begins with a cell that contains two sets of chromosomes. All the chromosomes are copied, as at the start of mitosis. However, in meiosis the original cell divides twice so at the end of the division there are four cells that each contain only one set of chromosomes. In men the four cells then develop into four sperm. In women, one of the four cells becomes an egg.

The cell has one large pair and one small pair of chromosomes.

Each chromosome is copied.

The cell divides in two and then in two again. Each new cell gets a copy of one chromosome from each pair.

nucleus

These cells are not identical.

Figure 3 Meiosis of a cell with four chromosomes.

After fertilisation

During **fertilisation**, when an egg cell fuses with a sperm cell, the chromosomes from each gamete come together in the new nucleus. So the fertilised cell contains two sets of chromosomes again. The fertilised cell divides by mitosis, so that every body cell of the new individual will have two sets of chromosomes.

Examiner feedback

Often in science exams if you misspell a word but it looks like the correct word then you will be awarded the mark. You cannot afford to spell either mitosis or meiosis incorrectly as they look too similar to each other.

Questions

1. Which kind of division happens in the cells of a root tip squash? Explain your answer.
2. Explain the purpose of mitosis in most organisms.
3. Draw up a table to compare the similarities and differences between mitosis and meiosis.
4. Explain why an offspring of sexual reproduction will be different from both its parents but an offspring of asexual reproduction will be identical to its parent.
5. Describe the role of mitosis and meiosis in the life cycle of a human.
6. Explain why meiosis is needed before fertilisation can take place.
7. Komodo dragons live on a range of islands in Indonesia, some of which are very small and can only support a few of these large lizards. Scientists assumed they only reproduced sexually, but females in zoos have produced eggs through asexual reproduction. Explain as fully as you can why Komodo dragons may use both forms of reproduction.

Differentiated cells

Learning objectives

- describe what cell differentiation is and when it happens in animals and plants
- explain what stem cells are and give examples of how they might be used to treat some conditions
- identify and evaluate some of the social and ethical issues surrounding the use of embryonic stem cells.

Taking it further

The process of differentiation (specialisation) involves the switching on and off of genes as the cell develops. This is why all cells have the same genes, but don't all look the same.

Differentiated cells

Your body contains many different kinds of cells, and almost all of them are **differentiated** (specialised) to do different jobs. When an animal egg cell is fertilised it starts dividing to make an **embryo**. The cells in the early stages of an animal embryo can differentiate to form almost any kind of cell. As cell division continues, the cells become increasingly specialised and the range of types of cell they can develop into decreases. By the time you are born, almost all your cells are differentiated.

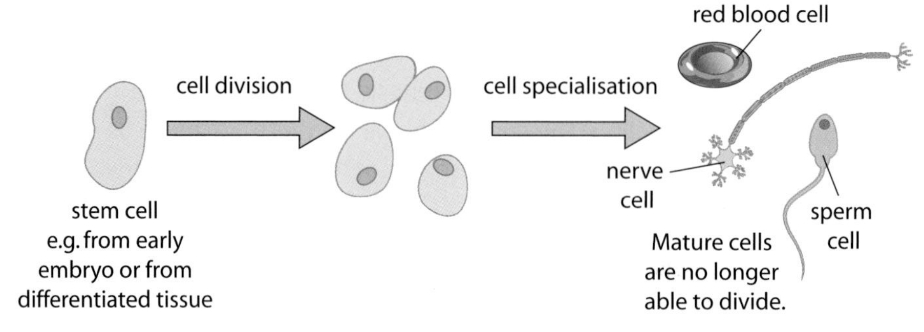

Figure 1 Differentiation of animal cells.

In plants, it is different. Many cells in a fully grown plant keep the ability to differentiate into any kind of cell. We can take cuttings from plant parts or use tissue culture to grow a whole new plant, but we cannot do this in animals.

1 Cut shoot from a plant.

2 Dip cut end into hormone rooting powder.

3 Plant in pot of soil as soon as possible after cutting.

Figure 2 In a plant cutting, cells at the cut end divide and differentiate into root cells to form a new root system.

Science in action

Scientists are currently researching how specialised cells can be returned to the stem cell state and then directed to produce new, different specialised cells. The success of this process will change the scientific problems that need tackling to produce effective treatments for people with disorders caused by defective cells, as well as change the ethical issues raised. Society needs to be aware of any changes, and the questions that politicians and the general public will need to consider.

Stem cells

Cells that can differentiate into a range of other cells are called **stem cells**. The cells in an early embryo (**embryonic stem cells**) have the ability to differentiate into any kind of body cell.

In differentiated body tissues there are a few **adult stem cells**, which divide when needed for growth or repair to damaged tissue. However, these cells can normally only differentiate into a limited range of other cells. Stem cells in **bone marrow**, for example, normally only produce different kinds of blood cell.

Using stem cells for treatment

Stem cells are already being used to treat some human disorders, for example bone marrow cells are used to treat leukaemia, a type of cancer in blood cells. New treatments are being developed for many other disorders where cells are damaged, such as to replace damaged nerve cells in people who have been paralysed after their spinal cord was broken in an accident. The advantage of using embryonic stem cells is that they are easier to work with, but a practical disadvantage is that putting cells from one body into another means lots of medication for the rest of the patient's life. Using stem cells created from a patient avoids this problem.

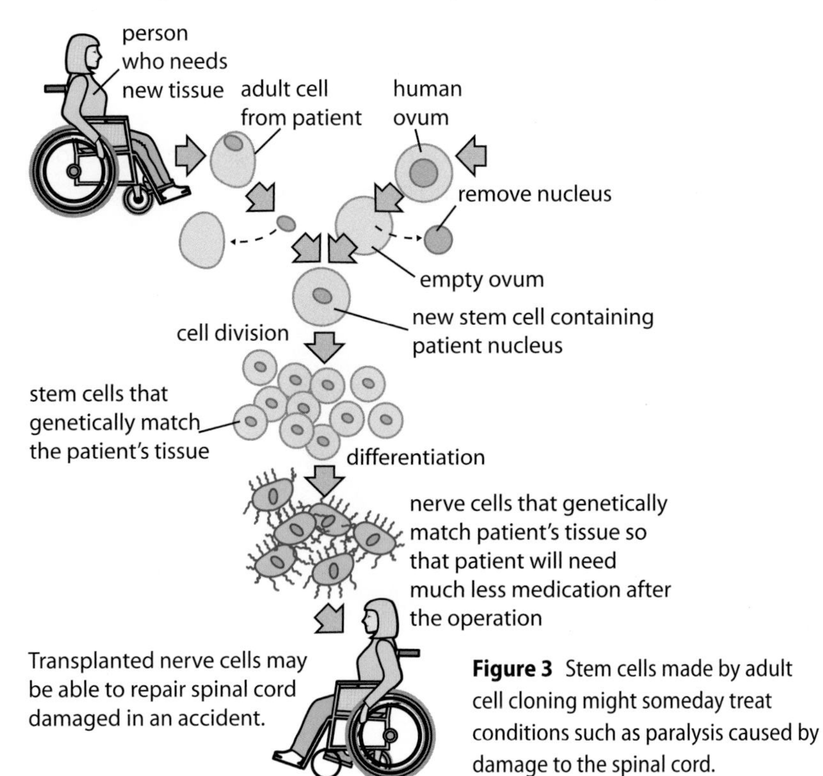

Figure 3 Stem cells made by adult cell cloning might someday treat conditions such as paralysis caused by damage to the spinal cord.

Is it ethical?

Many countries ban the use of embryonic stem cells for treatment. This is partly because research is still trying to solve some of the practical problems they cause, such as the need for lifetime medication. However, there are also **ethical questions**. Some people say that using such embryos is destroying a new life. However, most of the embryos used are those unwanted after fertility treatment and would be destroyed anyway. Research into embryonic stem cells is allowed in many countries because they are easier to extract from tissues than adult stem cells.

Science in action

Scientists are researching many other ways to encourage adult stem cells to differentiate into a greater range of specialised cells. For example, in March 2010 a boy was given a replacement windpipe with stem cells taken from his own bone marrow. The stem cells were treated to grow into all the kinds of cells needed in the new windpipe. Doctors will monitor the success of this new treatment.

Examiner feedback

You will not be expected to remember the details of Figure 3, but it will help you to understand the link between techniques for adult cell cloning and gene therapy. It also shows why people who are against the use of embryos in research and treatment don't like this technique, because an embryo is created and then destroyed.

Questions

1 Explain why we can grow a whole new plant from a leaf but not a complete human from a leg.

2 Define the term 'stem cell' in your own words.

3 List the similarities and differences between embryonic and adult stem cells.

4 Explain how the adult cell cloning technique makes it possible to use the patient's own cells to treat a disorder.

5 Explain why there are ethical problems with using embryonic stem cells.

6 Bone marrow transplant is a form of stem cell therapy that has been used since the 1960s to treat patients with leukaemia. Suggest why this kind of stem cell treatment has been available for so long, while other stem cell treatments are still in the development phase.

7 If a couple have a child that is paralysed due to nerve damage, it is possible to mix their sperm and eggs in the lab to create an embryo from which stem cells could be taken and used to cure the child's paralysis. **(a)** What are the advantages to the child of doing this? **(b)** What are the advantages to the parents? **(c)** Should the law allow this kind of treatment? Explain your answer.

8 Stem cells can be taken from the umbilical cord of a baby just after birth. A parent could decide to do this for a baby and have the cells stored in case they are needed later in life to cure a disorder or paralysis. Storage costs money. Evaluate the advantages and disadvantages of this idea.

Genes and alleles

Learning objectives

- define the terms *chromosome*, *DNA*, *gene* and *allele*
- explain why different alleles of a gene produce different forms of a characteristic
- explain why sexual reproduction produces variation in the offspring.

Chromosomes and DNA

Chromosomes are immensely long molecules of DNA (deoxyribonucleic acid) that are found in the nucleus of almost every cell in your body. The DNA molecule is made of millions of subunits of just four types. The order of these subunits forms a code, which can be translated into genetic instructions for building the cell and for maintaining its functions.

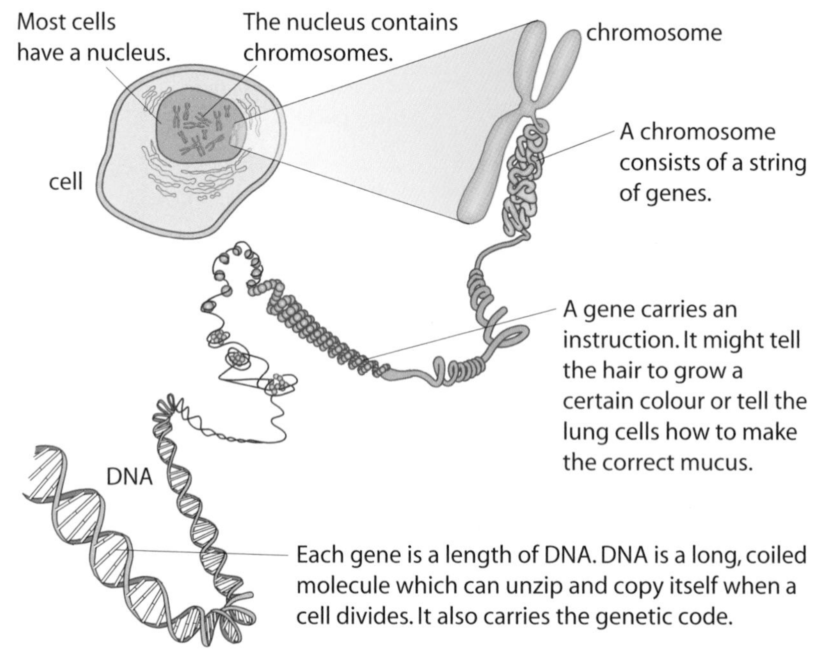

Figure 1 DNA has a **double helix** shape, like a twisted ladder.

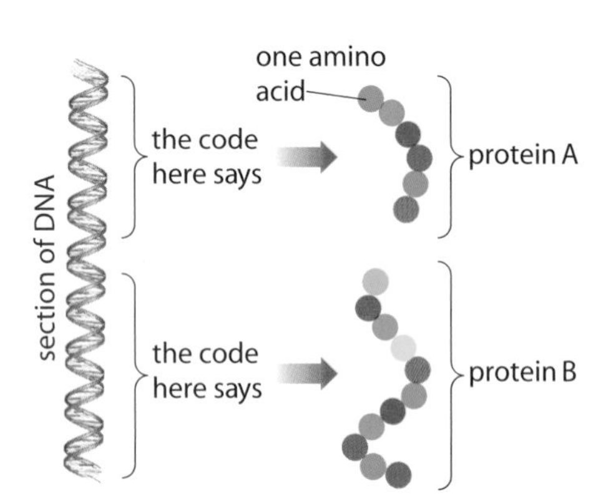

Figure 2 Building proteins from the code in DNA.

Genes and proteins

Along a chromosome there are many genes. Each gene is a small section of DNA. A gene codes for a particular sequence of amino acids, and therefore a particular protein. Proteins play many important roles in our bodies.

- Large structures, such as hair and nails, are made from proteins.
- Proteins form the basis for many tissues, such as bone, brain and muscle.
- Within cells, proteins form a large part of structures such as mitochondria.
- Proteins control what can enter and leave the cell.
- Proteins produce the colour of your eyes, hair and skin.
- Enzymes are proteins – they control the rate at which different chemical reactions take place in a cell.

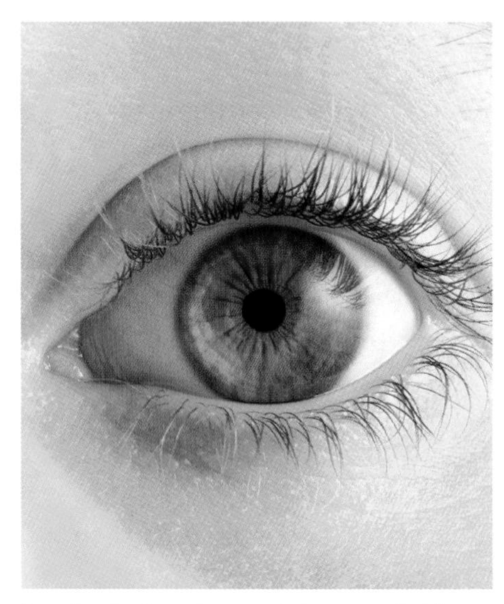

Very few human characteristics are coded for by a single gene. Eye colour is the result of interactions between several genes.

Variation in the code

Variation between individuals occurs because each gene occurs in slightly different forms, called **alleles**. For example, people with blue eyes and people with brown eyes have different alleles of the genes that produce eye colour.

Remember that, as a result of sexual reproduction, you have two sets of chromosomes in your cells, one set from your mother and one from your father. The *genes* on each chromosome of almost every pair are the same – however the *allele* for each gene may be different. The characteristic that you have will depend on how those alleles of the gene on each chromosome pair interact.

We can see the variation in the genetic code between individuals not only in their characteristics, but also when we produce an image of their DNA. This is known as **DNA profiling** (also called **DNA fingerprinting**).

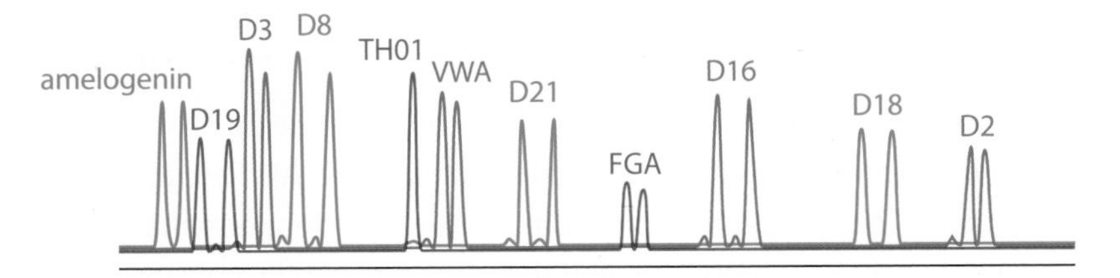

Figure 3 Part of a DNA profile showing a few identified genes. This profile was made using **gas chromatography** (older methods used **gel electrophoresis**).

Only identical twins will have identical DNA profiles, although members of the same family will show similarities. Children have profiles that match half of each parent's profile. Children in the same family may have some or many similarities in their profiles.

Questions

1. A couple have two children: one child has blue eyes, the other has brown eyes. Explain how this is possible.
2. Figure 4 shows three DNA profiles made by gel electrophoresis. The three profiles are from two parents and a child. Identify the child and explain your choice.

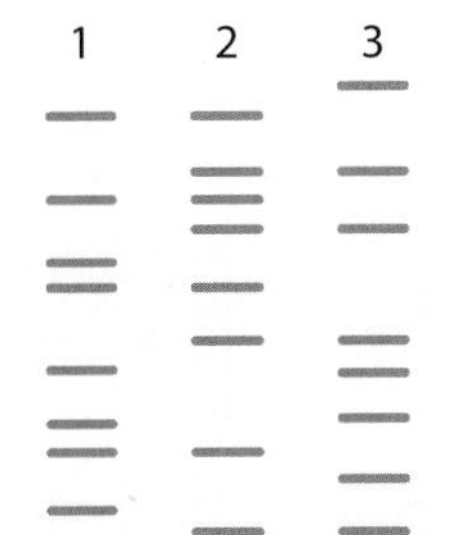

Figure 4 DNA profiles.

3. Gas chromatography is a newer technique for making DNA profiles than gel electrophoresis. What important feature do the techniques have in common?
4. Some people have argued that the police should have a database of DNA profiles from everyone in the UK. Explain an advantage of having a DNA profile database of everyone.
5. Give one social, one economic and one ethical disadvantage of having a DNA profile database for everyone.
6. Do you think police should be allowed to hold DNA profiles for everyone? Explain your answer.
7. Explain fully how genes produce different characteristics, such as dark or fair hair.
8. Explain fully how two children of the same parents could have DNA profiles that are either 100% identical, or 0% identical, and the implications this has for matching DNA profiles with family members in a criminal case.

A*

Science in action

DNA profiling is a key tool in **forensic science**. Tiny samples of tissue such as blood, semen or hair can be used to produce a DNA profile that can help to identify a person who was at the scene of a crime, either by comparing it with a police DNA database or with DNA taken from suspects. However, it cannot specifically identify who carried out the crime.

Practical

DNA can be extracted from any cell with a nucleus, but it is easier to extract from plant cells such as from kiwi fruit.

Science in action

DNA analysis is being used in many areas of science to identify individuals and species, not just in human forensic science. For example, samples taken from whale meat for sale in Japanese markets can show whether or not the meat comes from protected species or from species that are legal to hunt. Comparing the DNA of different species also helps to show evolutionary relationships. For example, evidence from DNA samples taken from fossils of Neanderthal people show that in Europe, where they lived at the same time as *Homo sapiens* (our own species), the two species probably interbred quite frequently.

B2 6.4 Inheriting characteristics

Learning objectives

- explain how sex is inherited in humans
- draw genetic diagrams to explain the effect of inheriting dominant and recessive alleles
- evaluate the work of Mendel on inheritance in peas.

Inheriting sex chromosomes

Of the 23 pairs of chromosomes in humans, only one pair differs between males and females. These are the **sex chromosomes** that determine whether you are male or female. In women, both chromosomes in the pair are the same size: they are X chromosomes (see Figure 1 in lesson B2 6.1). In men, one of the sex chromosomes (the Y chromosome) is shorter than the other (an X chromosome).

Gametes contain only one chromosome from each chromosome pair (see lesson B2 6.1), so only one sex chromosome ends up in each gamete. All the eggs that a woman produces will contain an X chromosome, while half the sperm that a man produces will contain an X chromosome and the other half a Y chromosome.

We can explain the inheritance of sex chromosomes using a **genetic diagram** that shows the chromosomes in the body cells, the gametes of the parents, and the full range of possible chromosome combinations in the children.

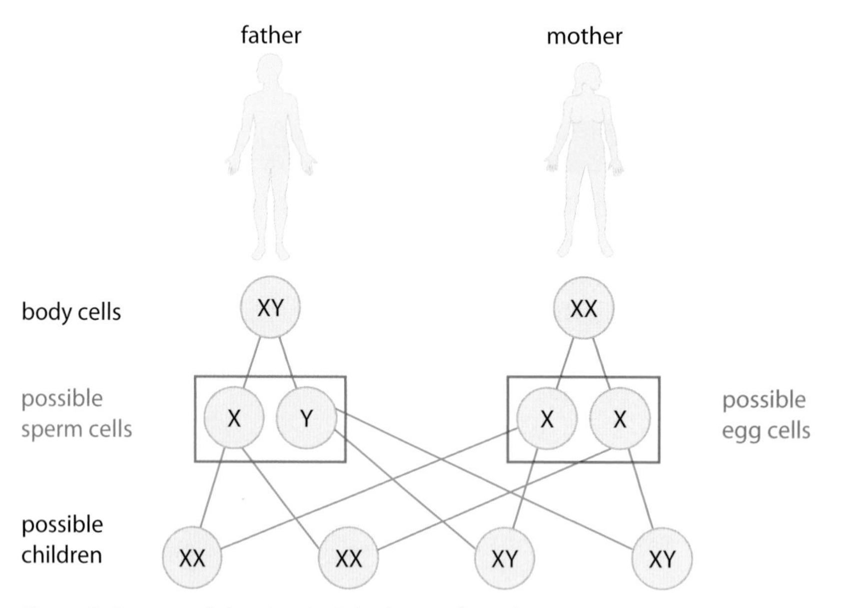

Figure 1 One way of showing the inheritance of sex chromosomes.

		mother	mother
		X	X
father	X	XX	XX
father	Y	XY	XY

Figure 2 A Punnett square showing the inheritance of sex chromosomes.

Another way of showing the outcome of this uses a different kind of genetic diagram called a **Punnett square**.

Examiner feedback

It is better to draw a Punnett square in an exam rather than a line diagram. This is because the examiner sometimes can't see where the lines are going unless the diagram has been drawn very well.

Inheriting alleles

We can use Punnett squares like the one in Figure 2 to investigate the inheritance of alleles. In this case, we show the alleles on the chromosome pair for the particular gene we are investigating. Each gamete has only one allele, because the gametes have only single chromosomes rather than pairs. However, the fertilised eggs have two alleles again, one from each parent.

Different pairs of alleles can produce different characteristics. This is because some alleles are **dominant**, where the characteristic they code for always shows whether the individual has one or two copies of the allele. Other alleles are **recessive**, where the characteristic they code for only shows when both chromosomes of the pair have that allele.

Examiner feedback

When choosing letters to represent alleles, use something that looks different in capitals and lower case, so it is easy to see the difference. The capital stands for the dominant allele. Always define your use of letters before you answer the question.

In pea plants, the allele for purple flowers is dominant to the allele for white flowers. So when you cross a purple-flowered plant that has two alleles for purple flowers with a white-flowered plant, all the offspring have purple flowers. If you then cross two of the offspring, you can prove that they both contained one allele for purple and one for white flowers because some of the offspring have white flowers. Figure 3 shows that the chance of having purple flowers compared with white is 75%, or a ratio of 3:1.

Mendel's work on inheritance

Gregor Mendel (1822–1884) spent many years studying the inheritance of characteristics in pea plants. He was trying to explain why some characteristics (such as white flower colour) seem to disappear in one generation but reappear in the next. Mendel interpreted his results in terms of separate 'inherited factors', which we now call genes. Although Mendel published his work in 1865, it was largely ignored until about 1900.

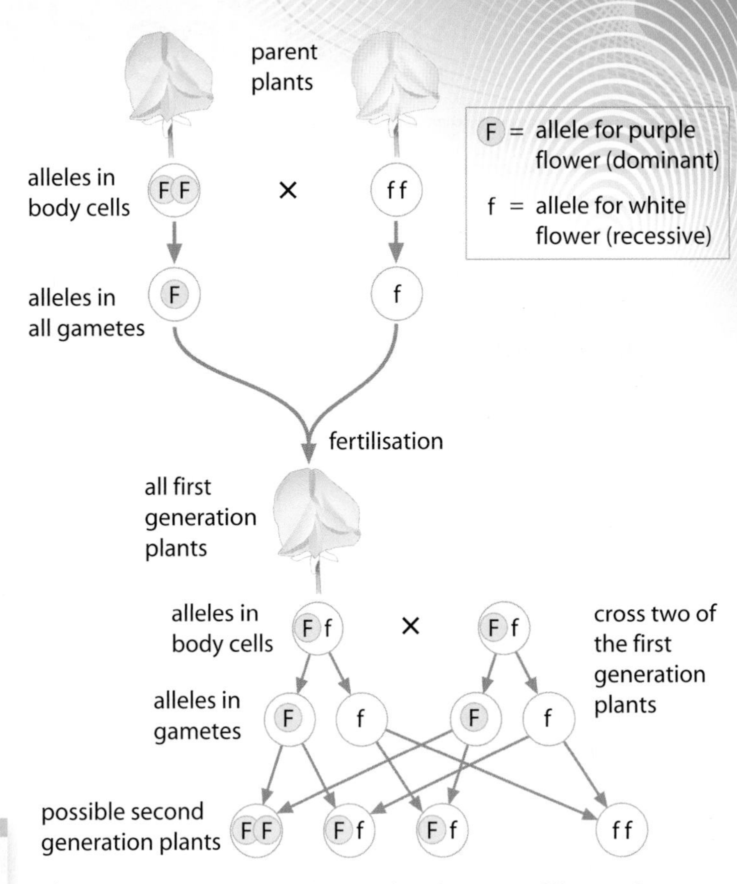

Figure 3 A genetic diagram showing the inheritance of flower colour in pea plants over two generations.

Science skills

Table 1 shows some of Mendel's results. For each cross:

a State what the first generation plants would have looked like.

b Draw a genetic diagram to work out the theoretical result at the second generation.

c Compare the theoretical result from your diagram with the actual result and suggest why there are any differences.

Table 1

Parent varieties	Results at second generation
round seed × wrinkled seed	336 round : 101 wrinkled
tall stem × dwarf stem	787 tall : 277 dwarf

Science in action

Mendel was very careful about how he carried out his experiments. He spent two years creating **pure-breeding** plants (ones that produced only that character when crossed with each other) for each characteristic before he started his crosses. He then transferred pollen from one plant to another using a brush, to make sure he knew which cross had been made. And he repeated each cross many times.

Questions

1 Describe the difference between dominant and recessive alleles.

2 Look at Figure 1. What is the average chance of a couple having a baby boy? Explain your answer.

3 A couple already have three boys. What is the chance that the next child they have is a girl? Explain your answer.

4 Draw a Punnett square for a cross between a purple-flowered plant with two alleles for purple with a white-flowered plant. Describe the proportion of flower colour in the offspring.

5 Draw a Punnett square for a cross between a purple-flowered plant with one allele for purple and one for white crossed with a white-flowered plant. Describe the proportion of flower colour in the offspring.

6 Before Mendel's experiments, many scientists thought that inherited characteristics were blended in the offspring, e.g. a cross between red- and white-flowered plants would give plants with flower colour somewhere between white and red. Explain how Mendel's experiments proved this wrong.

7 Evaluate Mendel's method and explain why it helped him get reliable results.

B2 6.5

Inheriting disorders

Learning objectives

- interpret family trees that show the inheritance of a disorder
- construct genetic diagrams to explain the inheritance of some disorders
- explain the outcome of crosses between individuals with different combinations of dominant and recessive alleles of the same gene.

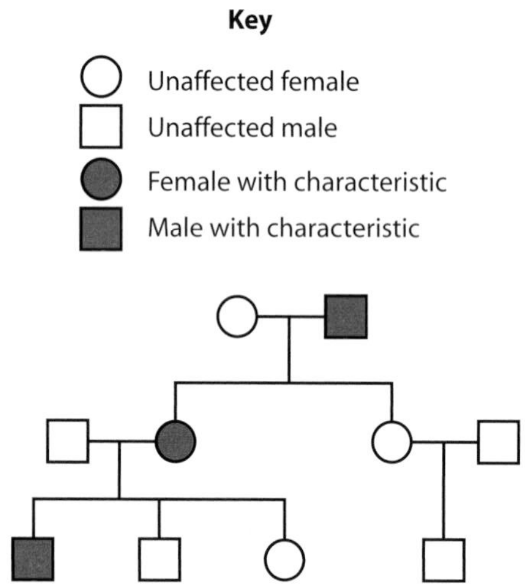

Family trees

We sometimes talk of characteristics 'running in families', which means that the characteristic occurs more frequently within the members of a family than within the general population. We can see this in a **family tree**. Characteristics that run in families are inherited in the genes.

Science in action

As we learn more about how characteristics develop, we are finding that many disorders and diseases, such as heart attack and cancer, are produced by a combination of genes and the influence of the environment. Using family trees can help us identify people who are more likely to have a faulty gene, and help them reduce the risk of becoming ill by avoiding the environmental factors linked to the disease.

Key

- Unaffected female
- Unaffected male
- Female with characteristic
- Male with characteristic

Figure 1 This family tree shows which members of one family have a particular characteristic.

Inheriting dominant disorders

A family tree can tell us who has a disorder, but not whether a child will inherit the disorder. For this we need to construct a genetic diagram. **Polydactyly**, having more fingers, thumbs or toes than usual, is an inherited disorder that occurs in 1 in 1000 people. One type of polydactyly is caused by a dominant allele, which means that a person only needs one polydactyly allele for the disorder to show. A genetic diagram can show us that the likelihood of the children inheriting the disorder, if one of their parents has one copy of the allele, is 50% (sometimes quoted as 1:1 or one in two chances).

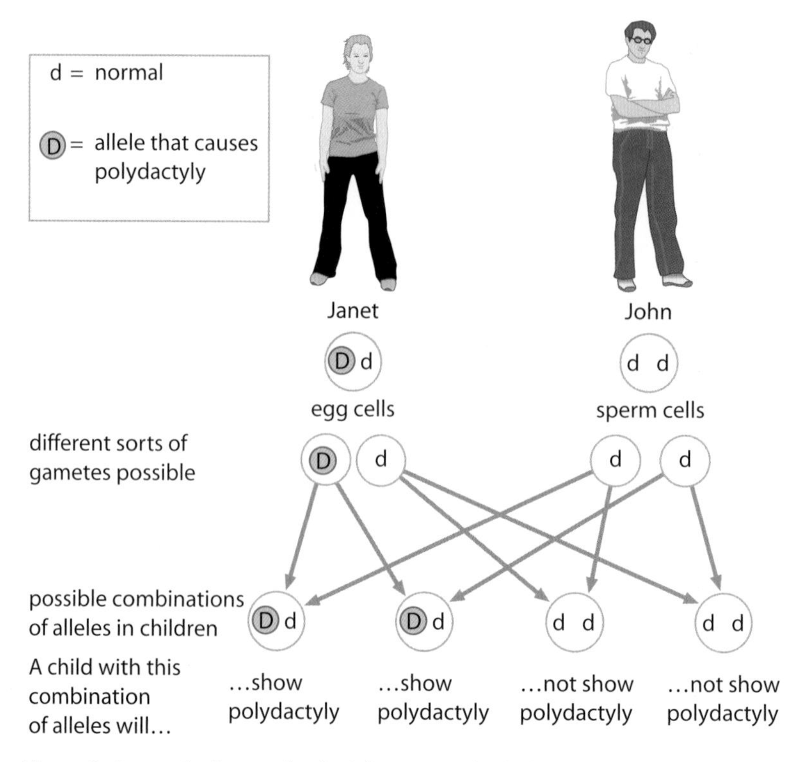

Figure 2 A genetic diagram for the inheritance of polydactyly.

Examiner feedback

Practise drawing Punnett squares for crosses involving dominant and recessive alleles, such as those in Figures 2 and 3. This will help you to become familiar with the layout and expected results from each type of cross.

Inheriting recessive disorders

Some inherited disorders are caused by recessive alleles. This means a person will only show the disorder if they carry two alleles for it. If they only have one allele for the disorder they will not show the disorder themselves but will be able to pass the allele on to their children. They are called **carriers** for the disorder.

Disorders caused by recessive alleles may seem to 'disappear' in a family tree for one or more generations before appearing again, like white flower colour in pea plants.

Cystic fibrosis is a disorder of cell membranes that affects the lungs and other parts of the body. It is caused by a recessive allele, so it can be inherited even if neither parent suffers from the disorder. Figure 3 shows that the chance of two carriers for the disorder having a child with cystic fibrosis is 25% (or one in four, or 1:3).

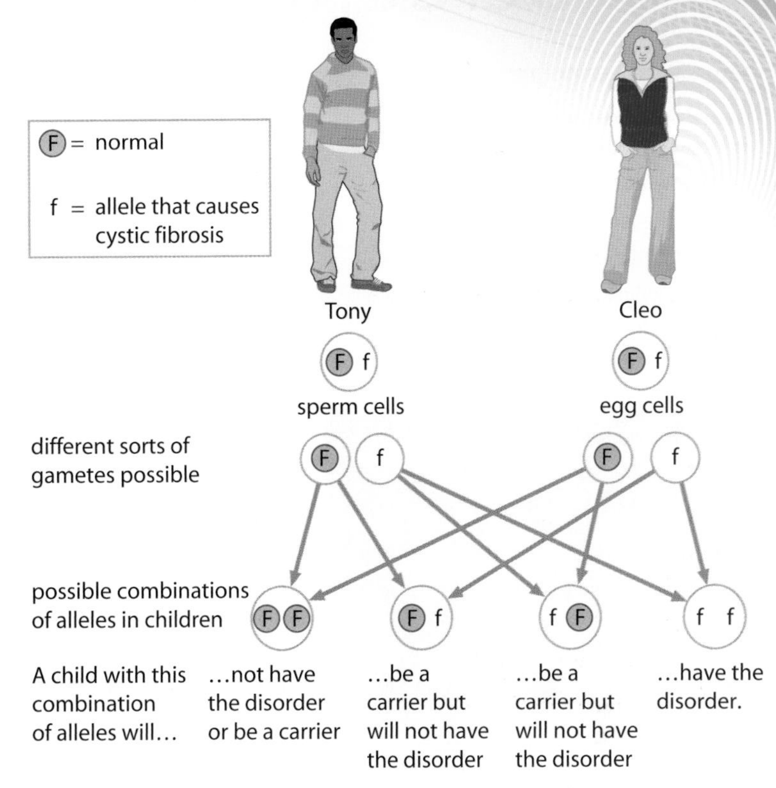

Figure 3 A genetic diagram showing the inheritance of cystic fibrosis.

Questions

1 Look at Figure 1. **(a)** How many generations of the family does it show? **(b)** How many children did the oldest couple have, and what sex were they? **(c)** How many people in this family show the characteristic? **(d)** Some disorders are commonly only found in men (sex-related characteristics). Is the characteristic shown in this family tree sex-related? Explain your answer.

2 Draw genetic diagrams for polydactyly in which one parent has one allele for the disorder and the other parent has none. Use your diagram to predict the chance that a child of the couple will have an additional finger or thumb.

3 Figure 4 shows a family tree for a characteristic in humans called dry earwax. **(a)** Use what you know about the inheritance of characteristics to decide whether this is a dominant or recessive characteristic. **(b)** Identify the possible alleles that each individual in the tree might have, making clear how certain you are about each choice.

Key

Unaffected female
Unaffected male
Female with dry earwax
Male with dry earwax

1 2 3 4 5 6 7 8 9 10 11

Figure 4 Family tree showing inheritance of dry earwax.

4 Explain why a disorder caused by a recessive allele may seem to 'disappear' in one generation, only to reappear again in the next. Use Punnett squares to support your answer.

5 Compare the use of family trees and genetic diagrams to describe the inheritance of genetic disorders, identifying their advantages and disadvantages.

Screening for disorders

Learning objectives

- explain that embryos can be screened for genetic disorders
- describe some of the advantages and disadvantages of screening for disorders
- make judgements on some of the issues related to embryo screening.

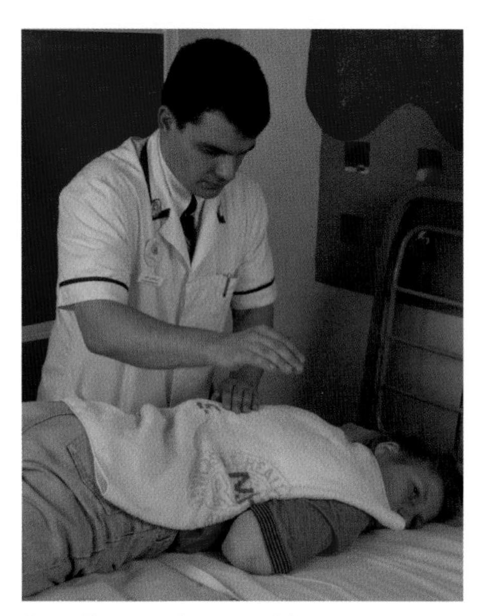

Cystic fibrosis sufferers need firm patting on the back or chest every day to loosen the sticky mucus so they can cough it out.

The problems with genetic disorders

Some genetic disorders can seriously affect the sufferer, making it difficult for them to live a normal life. For example, someone who suffers from cystic fibrosis needs **physiotherapy** every day to remove mucus that is clogging their air passages and lungs. They also need frequent treatment with antibiotics to clear infections caused by microorganisms getting trapped in the mucus.

People with genetic disorders may need extra support and care from their families, and they may need additional medicine and healthcare. Some disorders can even affect how long the person is likely to live.

There are also problems with knowing whether children will inherit a disorder. For example, people with a disorder caused by a faulty dominant allele that doesn't show **symptoms** for many years might pass the allele on to their children before they know they have it. Also, people who are carriers for a recessive disorder may never know they have that allele until their child shows symptoms.

Making decisions

People who know they have faulty alleles have to make difficult decisions about whether to have children. A couple may have to decide between taking the risk of having an affected child or adopting a child, or even not having children at all.

One way to make these decisions easier is **embryo screening**. Eggs are taken from the woman's ovaries and fertilised by sperm from her partner in a dish in the laboratory. This is called **IVF** (*in vitro* fertilisation). The embryos are tested to see if they carry the alleles for the disorder. Any embryos that do not have the faulty gene can be placed in the womb of the woman to develop into a baby.

Science in action

Embryos conceived naturally can also be tested for genetic disorders, either by removing a sample of fluid from the womb in which the unborn baby is developing, or by taking cells from the unborn baby itself. Both techniques carry a risk of causing the pregnancy to abort, so scientists are looking for another technique that is less invasive, such as testing the mother's blood. All these techniques raise difficult decisions for parents about whether or not to abort a pregnancy if the baby is shown to have a disorder.

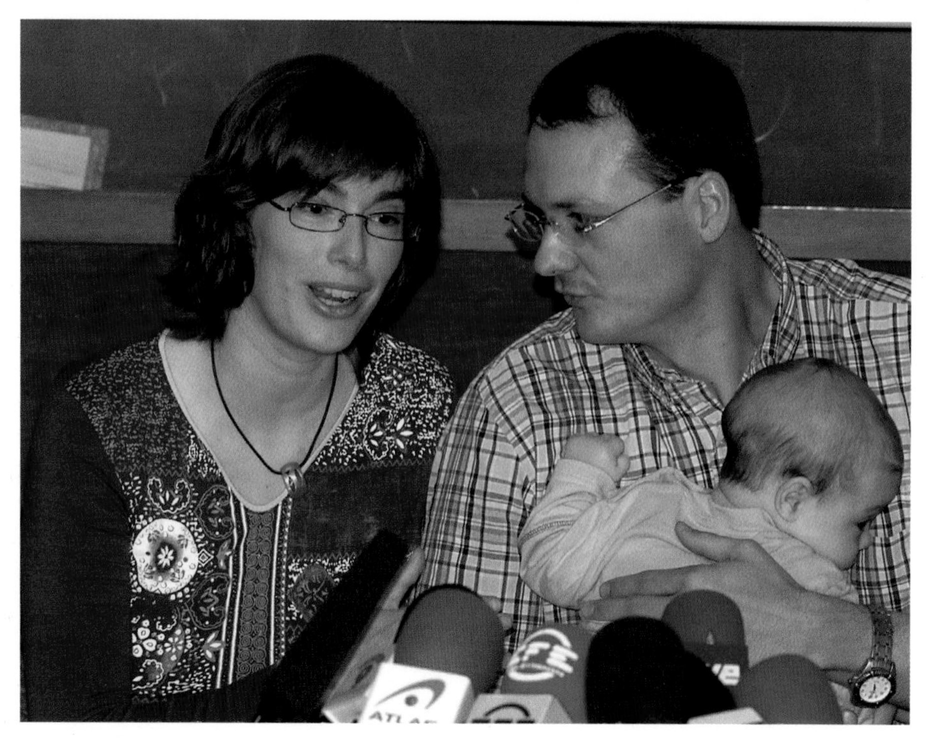

Baby Roger Farre's embryo was screened to make sure he didn't have the allele for a disorder caused by a dominant allele.

Some people think that embryo screening is a good thing, because it means parents can make decisions about whether or not to raise a child with a genetic disorder and face the problems that might cause. Other people feel that avoiding producing children with a genetic disorder changes how everyone thinks about people with such disorders, and that we should value everyone equally no matter what their genetic make-up.

Many people who are unhappy with the idea of embryo screening are concerned that it could be used to test for all kinds of alleles. Parents might wish to screen embryos for characteristics such as a particular hair colour, or (if scientists can identify them) for characteristics such as intelligence or sporting ability. Deciding when it is and is not acceptable to carry out embryo screening is a decision for society, not scientists. Table 1 shows different kinds of issues we have to think about.

Table 1 Issues around embryo screening.

Type of issue	Explanation
ethical	whether we think something is right or wrong
social	how something affects people, both individuals and all of society
economic	where money is involved

Examiner feedback

You will be given information on unfamiliar examples in the exam and expected to make judgements on them. To get the best marks, remember to consider all the different types of issues, and to suggest how different groups of people may respond differently on those issues. For example, some religious and ethical groups have very strong views on the rights of unborn children; politicians will have views that may be closely linked to popular opinion or to any financial costs such as for medical treatment, as these are supported by money that goes to the government in taxes.

Questions

1 Explain what embryo screening is.

2 Explain how embryo screening could mean that no people in future will be born with genetic disorders that damage health.

3 Explain why it is possible for a parent with an allele for a genetic disorder to pass it on to their children before realising they carry the allele.

4 Give one social and one economic advantage of there being no people with genetic disorders.

5 Is it ethical to avoid producing babies with genetic disorders?
Explain the reasons behind your view.

6 Two couples are planning to start families. Jill and Carl have just discovered that Jill's sister has a baby who suffers from cystic fibrosis. Paul and Kim are worried because Paul's father died when he was 42 from a genetic disorder caused by a dominant allele. **(a)** Explain fully the chances that Paul and Kim might have a child with the same disorder that Paul's father had. **(b)** Explain fully the chances that Jill and Carl might have a child with cystic fibrosis. **(c)** Embryo screening is an expensive technique. Should both couples be offered embryo screening to help them have a baby without a genetic disorder? Explain your answer.

7 Some genetic disorders cause a lot of pain and early death, but others, such as some kinds of cancer and blindness, may only cause problems after years of 'normal' health and can be treated by medicine. Should all genetic disorders be treated the same? Present a range of examples to support your argument.

B2 7.1 Fossil evidence

Learning objectives

- describe ways in which fossils form
- interpret fossil evidence in relation to evolutionary theory
- explain why scientists are not certain about how life began on Earth.

Examiner feedback

It is important to understand why, in relation to all the organisms that have lived, very few organisms have been found as fossils. You will be expected to explain this in terms of the following:

- Few organisms die in places where good fossilisation can occur.
- Many fossils are in rocks deep below the surface and so have not been discovered.
- Fossils are also destroyed by geological processes.

You will also be expected to understand and explain the impact of this on the challenges of interpreting evolution from fossil evidence.

Homo sapiens (modern day human) — c.200 000 years to present
Homo erectus — 1.8–<0.5 million years ago
Homo habilis ("Handy man") — 2.3–1.4 million years ago
Australopithecus afarensis — 3.8–2.9 million years ago

Australopithecus → *Homo habilis* bigger brain case, smaller teeth and jaws
Homo habilis → *Homo erectus* flatter jaw, high forehead, intermediate brain case
Homo erectus → *Homo sapiens* very large brain case, flat face

Figure 1 Evidence from fossil skulls shows how the human head has evolved over time.

Fossil formation

Fossils are the remains or traces left of organisms that once lived. They can form in several different ways.

The fossil of a skeleton.

Mineral replacement

Minerals in the groundwater replace minerals in the hard parts of the skeleton, such as in bone or shell. Or, if soft tissue doesn't decay, the minerals can fill the spaces in cells and show what soft parts of the body looked like.

A mummified body preserved in a bog.

Mummification

If the growth of decay organisms is prevented by conditions in the ground (e.g. too cold or too acid), the soft tissues are preserved as a **mummy**.

A fossil ammonite.

Moulds and casts

Sometimes a dead organism is pressed into soft sediment. As the organism decays, the shape left in the sediment is filled with other minerals, making a **cast**. The shape in the sediment is a **mould**.

The footprint of a hadrosaur.

Trace fossils

Trace fossils are things made by organisms, such as footprints, burrows or root shapes. These can make casts in the right conditions.

A fragmentary record

Only a very small proportion of dead organisms end up in conditions where they can become fossilised; in particular, soft-bodied organisms may decay completely before they can form a fossil. Many fossils that do form are destroyed by geological activity. So fossils tell only a fragment of the story of life on Earth.

We can find out how old fossils are using **radioactive dating** methods. This makes it possible to date the rocks in which the fossils are found, and so date the fossils. Fossils provide key evidence for the **theory of evolution** (see lesson B1 7.1) because we can use them to show how organisms have changed over time.

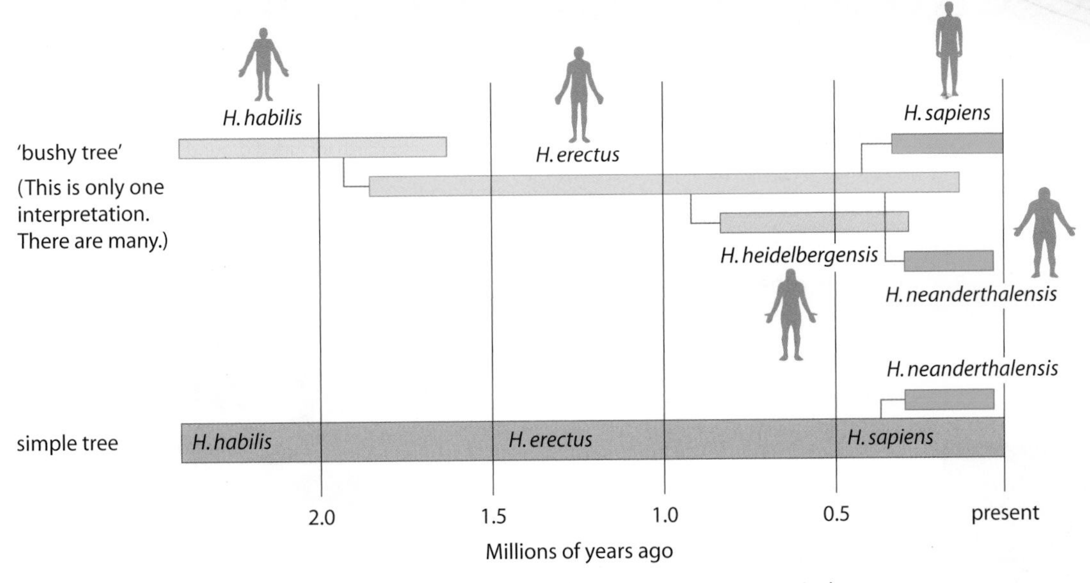

Figure 2 Even when we have many related fossils, as we do for humans, scientists can't always agree how to arrange them in an **evolutionary tree**. Some scientists separate slightly different fossils as individual species, while others group them together into one species.

The earliest life on Earth

We are reasonably certain that the Earth formed about 4600 million years ago, from evidence in the Solar System. Evidence from the earliest rocks suggests that there was life on Earth at least 3500 million years ago. Until around 2000 million years ago the only forms of life on Earth were bacteria. Bacteria are microscopic, so finding fossils of them is difficult. Also, some chemical processes make shapes that look like fossil bacteria. To separate the non-living shapes from real fossil bacteria, scientists analyse the rocks for substances such as proteins and nucleic acids.

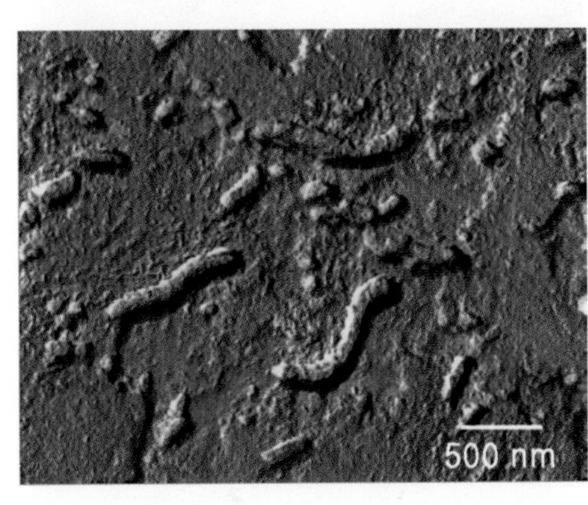

Bacteria-like shapes in this Martian meteorite rock suggest that there may once have been life on Mars. However, many scientists think the shapes are not bacteria.

Questions

1. Look at Figure 1. Describe how skull shape has evolved in humans.
2. What evidence can a mummified fossil give that a mineralised fossil cannot?
3. Give as many reasons as you can why we have so few fossils of all the organisms that once lived.
4. Sometimes geological processes are so great that they fold rocks over and older rocks are found above younger rocks. What evidence could you use to prove this had happened?
5. On a visit to a museum with a 9-year-old child, how would you explain that the 'bones' she sees in a fossil dinosaur are not bone at all?
6. Use evidence from variation in modern humans to explain why there is more than one evolutionary tree proposed for humans.
7. Explain why scientists cannot be certain how life began on Earth.
8. Write an argument to justify how the discovery of feathered dinosaur fossils strengthens support for the theory of evolution.

A*

Science in action

One argument against the theory of evolution in Darwin's time was the problem of explaining something complex, such as a bird's wings. The theory requires that at every stage of evolution any feature must be the best adaptation to the environment so that it is selected for. Recent fossil discoveries have shown that feathers evolved before wings strong enough for flight, suggesting that feathers evolved first for another purpose, such as insulation or display. Flight later became possible as wings got larger.

The causes of extinction

Learning objectives

- describe how changes to the physical environment can cause extinction
- describe how other organisms may cause the extinction of species
- evaluate evidence for the causes of mass extinctions.

Many frogs and other amphibians, especially tropical species, are threatened with extinction because of a deadly fungal disease.

Extinction of species

A species evolves as its characteristics change over time. This is driven by **natural selection**, as a result of changes in the physical environment or in the other organisms that live in the same place. If the individuals cannot evolve to survive these changes, the species will become **extinct** (see lesson B1 7.2).

The fossil record shows us that species become extinct, but it cannot tell us why. We need to look for evidence of changes that we think might have caused the extinction. For example, evidence from rocks shows that sea level and temperature have varied greatly over Earth's history. Large or rapid changes in either temperature or sea level could lead to the extinction of species.

Figure 1 The green animals were South American species that spread north, and the blue animals were North American species that spread south when the two continents joined.

The different species in an area affect each other, for example through **predation** or **competition**. About 3 million years ago, North and South America joined together for the first time. Before then the two continents had been separate and many species of plants and animals that lived on them were very different. When they joined, some species moved from one continent to the other.

After this exchange, many species quickly became extinct. Some died out because other species moved into their area and competed successfully for resources of food and space. In other cases new predator species came into an area and killed off whole populations of prey species. New diseases that came with the invaders could also have caused some extinction of local species.

Mass extinctions

Sometimes conditions change so rapidly that many species die out at the same time. Such events are known as **mass extinctions**. The worst was about 251 million years ago, when it is thought that over 90% of life on land and in the oceans died out. Another mass extinction about 65 million years ago killed off the last of the dinosaurs.

Most mass extinctions can be linked to rapid changes in sea level or temperature. Some, like the ones 251 million and 65 million years ago, are also linked to volcanic eruptions that lasted over a million years. These would have changed the Earth's climate, affecting both plants and animals. There is some evidence that the extinction 65 million years ago was due to a large asteroid impact.

A series of major volcanic eruptions could produce huge clouds of ash and dust that would block the sunlight and cause a 'global winter'.

Science skills

a The five worst mass extinctions were 65 million, 205 million, 251 million, c.370 million and c.445 million years ago. Look at the graphs of temperature and sea level change in Figure 2 and suggest what part these factors may have played in these extinctions.

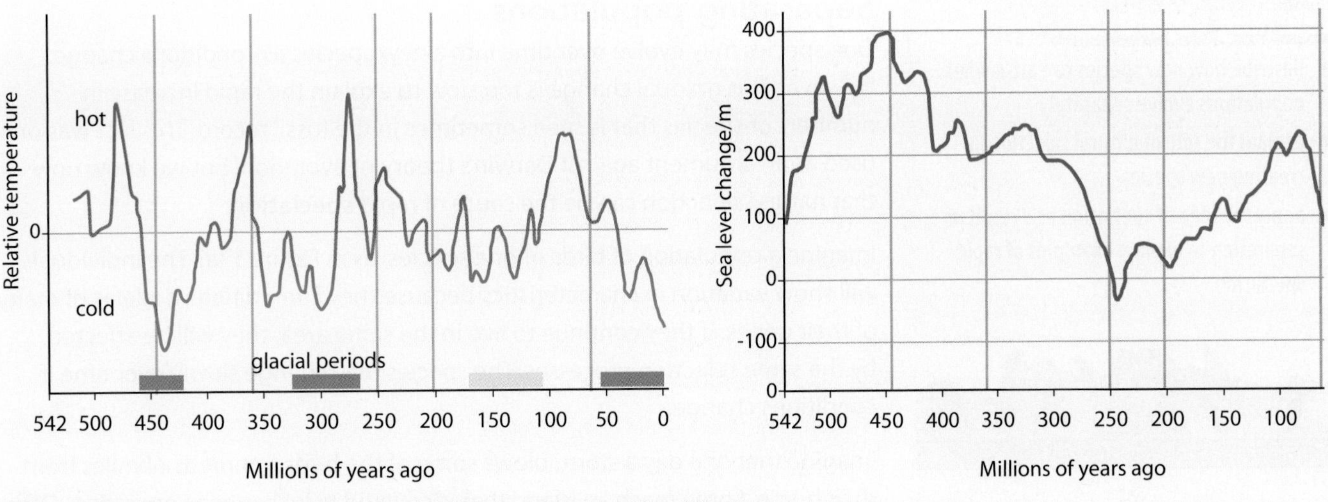

Figure 2 Variation in world temperature and sea level over the past 540 million years.

A changed species

Sometimes the fossil record shows one species disappearing and a similar one taking its place. However, this doesn't necessarily mean that an older species died out and the newer species came from somewhere else. It could be that the original species has evolved so much that it is classified as a new species. Sometimes it is not clear which interpretation is true. This is the case for human evolution (see lesson B2 7.1).

Questions

1. Describe the factors that can cause the extinction of a species.
2. It has been estimated that the volcanic eruptions that occurred about 251 and 65 million years ago both lasted between 1 and 2 million years. Explain how they could cause a mass extinction.
3. In the mass extinction 65 million years ago, about 75% of life on Earth became extinct. Suggest what survived.
4. Woolly mammoths became extinct about 8000 years ago. What evidence would you look for to decide whether the extinction of mammoths was the result of climate change or hunting by humans?
5. When scientists looked at a sample of *Australopithecus* fossil bones, some thought there were two species, one larger than the other, while other scientists thought there was just one species. Suggest why.
6. Before South America was connected to North America, many of the large animals were marsupials. Today, only a few marsupials are found there, while many marsupial species live in Australia. Suggest an explanation for this difference.
7. Much of our understanding about how organisms become extinct has come from looking at the impact of humans on the environment today. Using your answer to question 1, explain how this evidence is limited when we try to apply it to examples of extinction in the fossil record.
8. Explain how a large asteroid impact could cause the extinction of life.

B2 7.3

The development of new species

Learning objectives

- describe how new species can arise when populations evolve separately
- explain the role of natural selection in creating new species
- apply the idea of speciation as a result of separation to explain examples of rapid speciation.

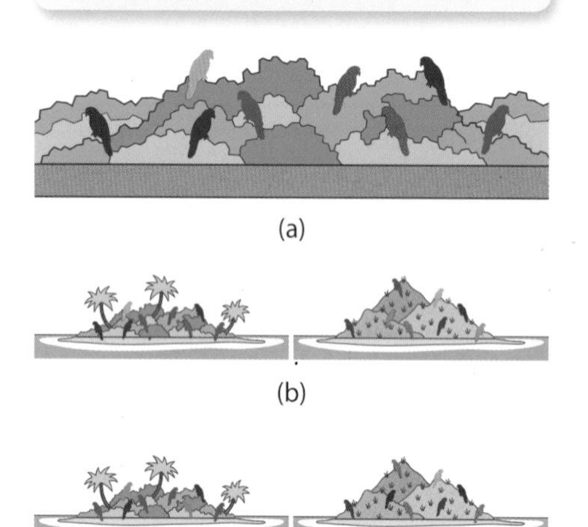

Figure 1 Rapid speciation. A population of birds (a) is blown by a storm to two different islands (b). The two groups adapt differently to their new environments (c).

These closely related fish are adapted to very different environments in Lake Malawi.

Separating populations

One species may evolve over time into a new species as conditions change. However, this gradual change is too slow to explain the rapid increase in numbers of species that is seen sometimes in the fossil record. This fact was once used as an argument against Darwin's theory of evolution, but we know now that natural selection can be the cause of rapid **speciation**.

Imagine a population of birds of one species, as in Figure 1 (a). The individuals will show variation in characteristics because they carry different alleles of many of their genes. If they continue to live in the same area, they will be affected by the same selection processes. The species may change slowly over time if conditions change.

Imagine that one day a storm blows some of the birds hundreds of miles from their home. Some reach an island that is covered in lush green vegetation. Others land on an island covered in rough brown grass. On both islands, the birds begin to adapt to their new environment.

Both islands have predators that eat birds, so camouflage is an important adaptation. Over time, the birds on the green island will be selected for green colour and those on the grass island for brown colour. There may also be other changes in each island population.

Eventually the two populations could become so different that, even if they came back together, they would not be able to breed with each other. They have become different species, as in Figure 1(c).

An example of rapid speciation

Lake Malawi in Africa has dried up many times. The lake we know is only about 14 000 years old. There are several hundred species of one type of fish living in the lake. Some of these species are predators, some are herbivores, and some sift food from the mud on the lake floor. They come in many different colours and sizes. However, classification has shown that they all evolved from a few, possibly just one, species of river fish.

Scientists suggest that when the lake started to form, different parts of the lake provided slightly different environments for the river fish, such as open water, shallow muddy bays and deep areas of weed. Adaptations that were better suited to each of these places were selected, so that different forms of the fish evolved in different parts of the lake. Female fish chose to mate with males that came from the same area, so many new species evolved at the same time.

Science in action

In the past, species were classified by looking at their structure. However, DNA evidence is now being used to help identify relationships between species. Some of the results have been surprising. For example, mole species live on most continents: bone evidence suggests they are related, but DNA evidence shows that each species evolved from a completely different animal.

The European mole and the golden mole from South Africa look similar, but they are not closely related. They both spend much of their time underground.

Taking it further

There are different ways in which speciation could be caused. For GCSE, you only need to know how it can result from geographical separation, but you will find other mechanisms suggested. These include genetic mutation, which is common in plants, behavioural separation, particularly where females choose males to mate with based on visible characteristics, and temporal separation, where different groups within the same population breed at different times. All mechanisms of speciation depend on restricting mating (and so exchange of genes) between the individuals of the original population, so that natural selection can act differently on each group.

Questions

1. Draw a flowchart to show how speciation can occur when populations of a species are separated.
2. Explain why separation of populations can lead to a rapid increase in the number of species.
3. **(a)** Suggest why moles from different continents were once classified as one group. **(b)** Suggest how mole species could have evolved to look similar even though they have different ancestors.
4. Suggest why rapid speciation was once an argument *against* the theory of evolution.
5. Explain how female choice of males for breeding supports the process of speciation.
6. Explain the importance of using DNA evidence to identify relationships in speciation.
7. Different species of *Anolis* lizards that live in different environments on one Caribbean island are more closely related to each other than to *Anolis* lizards on other islands. Suggest why.
8. Pollution and a new predator in Lake Victoria have reduced the number of fish species. However, a new species has been discovered that has much larger gills. This enables the new species to live in polluted muddy water where other species, and the predator, cannot live. Outline how this new species could have evolved.

Route to A*

To get the best marks on questions about speciation, you must demonstrate a clear understanding of how separation of individuals, which are then subjected to different selection pressures, can lead to the evolution of individuals that are sufficiently different that they are no longer capable of interbreeding.

ISA practice: carbohydrase enzymes

Carbohydrase enzymes are used in industry to break down starch into sugar syrup. A manufacturer of sugar syrup has asked some students to investigate the effect of temperature on the time it takes for carbohydrase to break down starch into syrup.

Section 1

1 Write a hypothesis about how you think temperature affects the rate of enzyme action. Use information from your knowledge of rates of reaction to explain why you made this hypothesis. *(3 marks)*

2 Describe how you could carry out an investigation into this factor.

You should include:

- the equipment that you would use
- how you would use the equipment
- the measurements that you would make
- a risk assessment
- how you would make it a fair test.

You may include a labelled diagram to help you to explain the method.

In this question you will be assessed on using good English, organising information clearly and using specialist terms where appropriate. *(9 marks)*

3 Design a table that you could use to record all the data you would obtain during the planned investigation. *(2 marks)*

Total for Section 1: 14 marks

Section 2

Two groups of students, Study Groups 1 and 2, investigated the effect of temperature on the breakdown of starch by amylase. Figures 1 and 2 show their results.

Temperature 10°C

Starch present after 0, 2, 4, 6, 8, 10, 12, 14, 16 minutes. No starch after 18 and 20 minutes.

Temperature 20°C

Starch present after 0, 2, 4, 6, 8, 10 and 12 minutes.

No starch 14, 16, 18 and 20 minutes.

Temperature 40°C

Starch present after 0, 2, 4 and 6 minutes.

No starch 8, 10, 12, 14, 16, 18 and 20 minutes.

Temperature 60°C

Starch present after 0, 2, 4, 6, 8, 10 and 14 minutes.

No starch 12, 16, 18 and 20 minutes.

Temperature 80°C

Starch present after 0, 2, 4, 6, 8, 10, 12, 14, 16, 18 and 20 minutes.

Figure 1 Study Group 1's results.

4 **(a)** Plot a graph of these results. *(4 marks)*

(b) What conclusion can you draw from the investigation about a link between the temperature and the rate of enzyme action? You should use any pattern that you can see in the results to support your conclusion. *(3 marks)*

(c) Look at your hypothesis, the answer to question 1. Do the results support your hypothesis? Explain your answer. You should quote some figures from the data in your explanation. *(3 marks)*

Here are the results of three more studies.

Figure 2 shows the results from another two students, Study Group 2.

Temperature 10°C
Starch present after 0, 2, 4, 6, 8, 10, 12, 14, 16 minutes. No starch after 18 and 20 minutes.

Temperature 20°C
Starch present after 0, 2, 4, 6 and 8 minutes.
No starch 10, 12, 14, 16, 18 and 20 minutes.

Temperature 40°C
Starch present after 0, 2 and 4 minutes.
No starch 6, 8, 10, 12, 14, 16, 18 and 20 minutes.

Temperature 60°C
Starch present after 0, 2, 4, 6 and 8 minutes.
No starch 10, 12, 14, 16, 18 and 20 minutes.

Temperature 80°C
Starch present after 0, 2, 4, 6, 8, 10, 12, 14, 16 and 18. No starch 20 minutes.

Figure 2 Study Group 2's results.

Figure 3 is a graph drawn from the results of Study Group 3, who carried out another investigation into the effect of a factor on the rate of enzyme action.

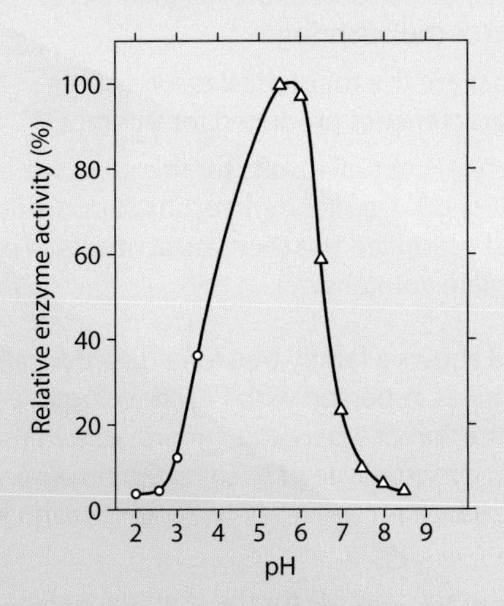

Figure 3 Graph of Study Group 3's results.

Study Group 4 was a group of researchers, who looked on the internet and found Figure 4: a graph showing the effect of temperature on the rate of reaction of a carbohydrase obtained from bacteria.

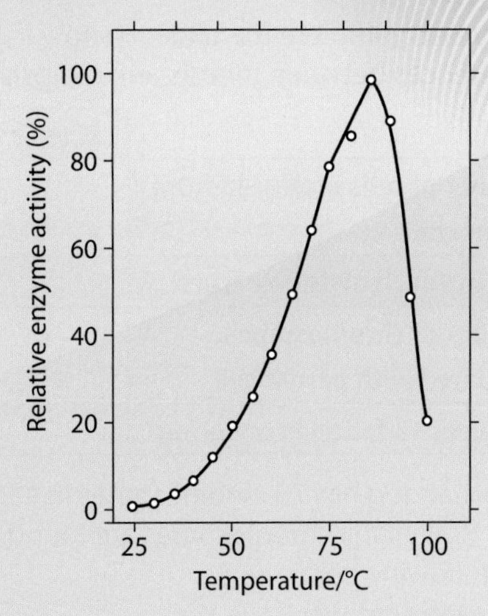

Figure 4 Graph of enzyme activity for a bacterial enzyme.

5 **(a)** Draw a sketch graph of the results from Study Group 2. *(3 marks)*

(b) Look at the results from Study Groups 2 and 3. Does the data support the conclusion you drew about the investigation in answer to question 5(a)? Give reasons for your answer. *(3 marks)*

(c) The data contain only a limited amount of information. What other information or data would you need in order to be more certain whether the hypothesis is correct or not?
Explain the reason for your answer. *(3 marks)*

(d) Look at the results from Study Group 4. Compare them with the results from Study Group 1. Explain how far the data shown supports or does not support your answer to question 5(b). You should use examples from Study Group 4 and Study Group 1. *(3 marks)*

6 **(a)** Compare the results of Study Group 1 with Study Group 2. Do you think that the results for Study Group 1 are *reproducible*?
Explain the reason for your answer. *(3 marks)*

(b) Explain how Study Group 1 could use results from other groups in the class to obtain a more *accurate* answer. *(3 marks)*

7 Applying the results of the investigation to a context.
Suggest how ideas from the original investigation and the other studies could be used by the manufacturers to decide on the best temperature at which to use the carbohydrase enzyme to break down starch, and the best source for this enzyme. *(3 marks)*

Total for Section 2: 31 marks

Total for the ISA: 45 marks

Assess yourself questions

1 **(a)** Copy and complete the table to show the differences between mitosis and meiosis. *(8 marks)*

	Mitosis	Meiosis
Number of cells produced from parent cell		
Cells in which division occurs		
Number of chromosomes compared with parent cell		
Produces variation in offspring?		

(b) An adult dog has 78 chromosomes in each body cell. How many chromosomes are found in each of the following cells? *(3 marks)*

(i) dog sperm cell

(ii) dog liver cell

(iii) fertilised dog egg cell.

(c) Explain why asexual reproduction produces clones. *(2 marks)*

(d) Explain fully the importance of meiosis in sexual reproduction. *(4 marks)*

2 **(a)** Define the following words. *(3 marks)*

(i) chromosome

(ii) gene

(iii) allele

(b) In birds, the sex chromosomes of a male are ZZ and those of a female are WZ.

Draw a Punnett square to show how sex is inherited in birds, identifying the sex of the offspring. *(4 marks)*

(c) Figure 1 shows DNA profiles of a mother and her baby, and those of two men who could have been the baby's father.

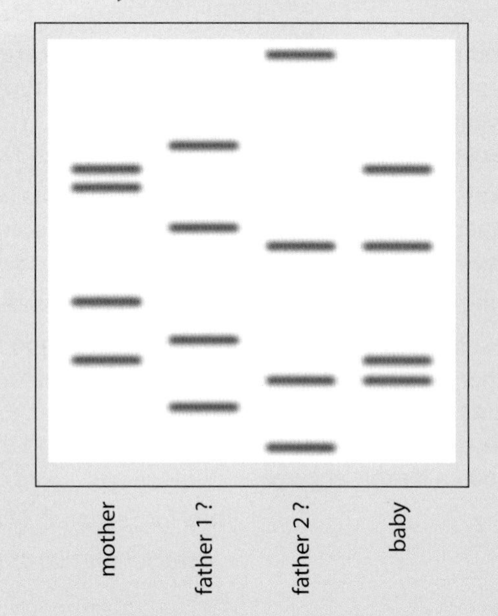

Figure 1 DNA profiles.

(i) Explain why a DNA profile can be used to identify individuals. *(1 mark)*

(ii) Explain the similarities and differences between the mother's and baby's profiles. *(2 marks)*

(iii) Explain fully whether either profile of the two possible fathers proves he is the father. *(3 marks)*

3 Mendel crossed pure-breeding tall pea plants with pure-breeding short pea plants. Tall is dominant over short. See Figure 2.

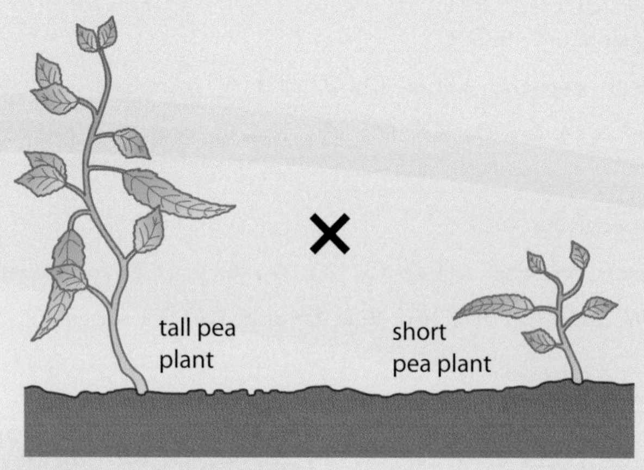

Figure 2 A cross between pea plants with different alleles for height.

(a) What does pure-breeding mean? *(1 mark)*

(b) Why was it important for Mendel to start with pure breeding parents? *(3 marks)*

(c) Draw a genetic diagram to show the results of this first cross, and then the result of a cross between two of the offspring. *(5 marks)*

(d) What are the theoretical results of the characteristics produced for this cross? *(1 mark)*

(e) Mendel's actual results for this cross were: 787 tall, 277 short. Do Mendel's results support the model that produced the theoretical results in part **(d)**? Explain your answer. *(3 marks)*

4 Figure 3 shows a family tree for a disorder called PKU in two families. A person with PKU develops a very high concentration of a particular amino acid in their blood. This can eventually lead to severe brain damage. PKU is caused by a recessive allele. A person who is a carrier shows no ill effects.

(a) Using the letter D for the dominant allele and d for the recessive allele, what is the genotype of person 1? *(1 mark)*

(b) What is meant by the term 'carrier'? *(2 marks)*

(c) Draw a genetic diagram to explain how person 11 inherited PKU when neither of his parents had the disease. *(3 marks)*

(d) Person 7 is pregnant with her third child. What is the chance that this child will have PKU? Explain your answer. *(2 marks)*

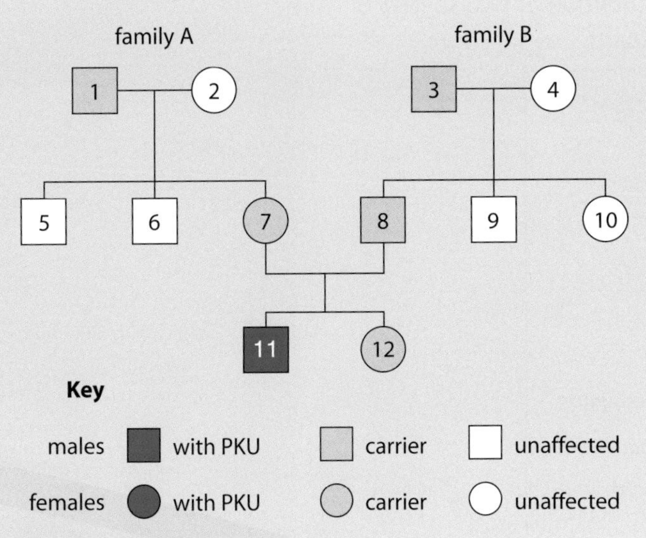

Figure 3 A family tree.

5 Multiple sclerosis (MS) is a disorder of the nervous system caused by the gradual breakdown of insulating tissue around nerve cells. In May 2010 scientists in Bristol tested the effect of stem cells on six patients with MS. For each patient, they took stem cells from their bone marrow and injected the cells into the patient's blood.

(a) What are stem cells? *(1 mark)*

(b) Where are adult stem cells and embryonic stem cells found? *(2 marks)*

(c) What were the doctors trying to achieve with this treatment? *(1 mark)*

(d) Give *one scientific* advantage of using bone marrow cells from the patient rather than embryonic stem cells. *(2 marks)*

(e) Give *one ethical* advantage of using bone marrow cells from the patient rather than embryonic stem cells. Explain your answer fully. *(3 marks)*

6 Ammonites lived on Earth from 200 million to about 65 million years ago.

A fossil ammonite.

(a) Explain how this fossil formed. *(3 marks)*

(b) We have many ammonite fossils but very few fossils of worms. Does this mean that there were many more ammonites than worms? Explain your answer. *(2 marks)*

(c) What other evidence could we use to help judge whether there were many or only a few worms? *(1 mark)*

(d) No ammonites survived the mass extinction that occurred 65 million years ago. Give *two* examples of causes of mass extinctions. *(2 marks)*

(e) Different ammonite species can be classified by their shell forms. This shows that in some periods many new ammonite species formed. Suggest *one* way that these new species could have developed. *(4 marks)*

(f) Some ammonite species are not found in the fossil record after about 145 million years ago. Explain as fully as you can *two* reasons why this is. *(4 marks)*

7 A couple who are planning to have a child are offered embryo screening.

(a) Explain what we mean by embryo screening. *(1 mark)*

(b) Describe one situation in which embryo screening might be offered to a couple, and explain why it is offered. *(2 marks)*

(c) Explain as fully as you can the ethical, social and economic issues raised by embryo screening. *(3 marks)*

8 Cockatiels are birds that are commonly kept as pets. Breeders need to understand the genetics of their birds so they are certain which characteristics any offspring will have, because some characteristics are worth more than others. Dominant Silver is a characteristic that produces a silver colour and is dominant to other colours, including Normal Grey.

(a) Using an appropriate diagram, show the genotypic and phenotypic results expected from a cross between a heterozygous Dominant Silver male and a heterozygous Dominant Silver female. *(3 marks)*

(b) Explain fully why the actual results from this cross might all have a Normal Grey phenotype. *(1 mark)*

(c) A breeder who has a bird that shows Dominant Silver colouring wants to know if the bird is heterozygous or homozygous for the characteristic. Using genetic diagrams, explain how she could do this. *(4 marks)*

GradeStudio Route to A*

Here are three students' answers to the following question:

Lemmings are mouse-like rodents. The diagram shows the distribution of two species of lemming in North America: *Dicrostonyx torquatus and D. hudsonius.*

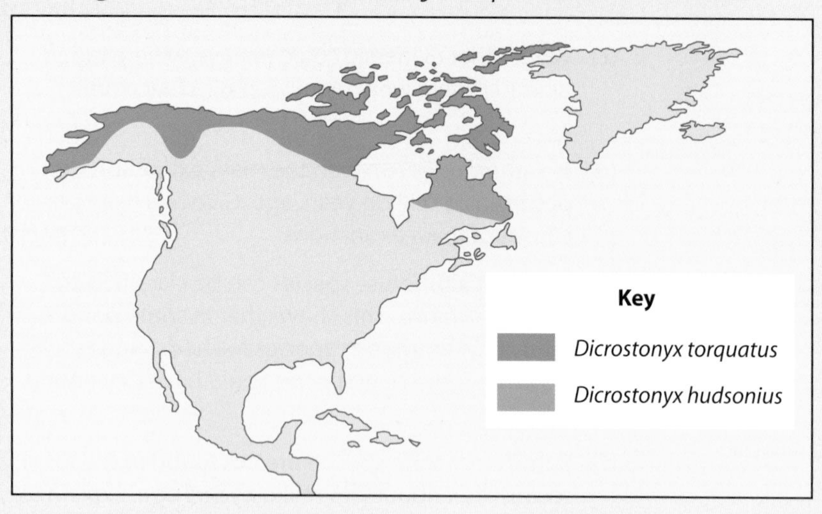

Figure 1 Distribution of Canadian lemmings.

Hudson Bay is a large ocean bay in northern Canada. *D. hudsonius* inhabits the eastern side of the bay and *D. torquatus* inhabits the western side. Before Hudson Bay was formed there was only one species of lemming present in the area.

Explain how the two species of lemming evolved from the original species.

In this question you will be assessed on using good English, organising information clearly and using specialist terms where appropriate. *(6 marks)*

Read the answers together with the examiner comments. Then check what you have learnt and try putting it into practice in any further questions you answer.

B Grade answer

Student 1

After the bay was formed the lemmings on the east and west sides of the bay did not mix because they were too small to swim across. Because they did not mix they evolved separately. Eventually they could not breed together so they became separate species. Natural selection has occurred.

It would be better to use the correct term, which is 'were geographically isolated'.

Separate evolution is mentioned in the question – no marks are gained for repeating this information.

The candidate has the argument the wrong way round – when they can no longer breed they are separate species.

Examiner comment

The candidate shows some understanding of speciation including an idea of isolation and an idea that different species do not interbreed. But there is very little use of biological terminology. There is no reference to variation. The reference to natural selection needs qualification to gain credit.

A Grade answer

Student 2

The first stage in the formation of two different species is geographical isolation. Here the two groups of lemmings became separated when Hudson Bay was formed. Conditions on the two sides of the bay were different and the two groups adapted to their different surroundings. This led to natural selection. Eventually the two groups became so different that they could not interbreed. They were now two different species.

It would be better to point out that it was the environmental conditions that were different.

It is not clear what 'This' refers to. The reference to natural selection is weak.

The candidate seems to be stating that the lemmings intended to adapt to their surroundings.

Examiner comment

A fairly good account that involves most of the main stages, but the account lacks the link between variation and natural selection, which together account for the eventual differences that led to speciation.

A* Grade answer

Student 3

The two groups of lemmings became separated geographically when Hudson Bay formed. Environmental conditions, such as temperature, may have been different on the two sides of the bay. There is natural variation in a lemming population. This is because there is a wide range of alleles in a population. Some variations enabled lemmings to survive in the different conditions. These lemmings would breed and pass their alleles onto the next generation. Survival of the fittest to breed is known as natural selection. In natural selection alleles that enable an organism to survive are selected. Due to natural selection the two populations of lemmings became different. The differences meant that they were unable to interbreed. They had become separate species.

Good use of appropriate biological terminology.

A good description of variation.

A good description of natural selection.

A good description of how the two populations became different.

Examiner comment

An excellent explanation that gives the main principles underlying speciation: isolation; variation; natural selection; no interbreeding. The principles were described in the correct sequence. Correct biological terminology was used throughout.

Read the whole question carefully.

- Before you begin to answer a question that involves a sequence of events, plan out your answer by writing a key word from each stage in rough. Then sequence the stages before you start your explanation.
- Use the correct biological terminology throughout.
- Make sure you refer to all the information given to you, but do not simply copy information without qualifying it.

Examination-style questions

1 The diagram shows part of a root hair cell.

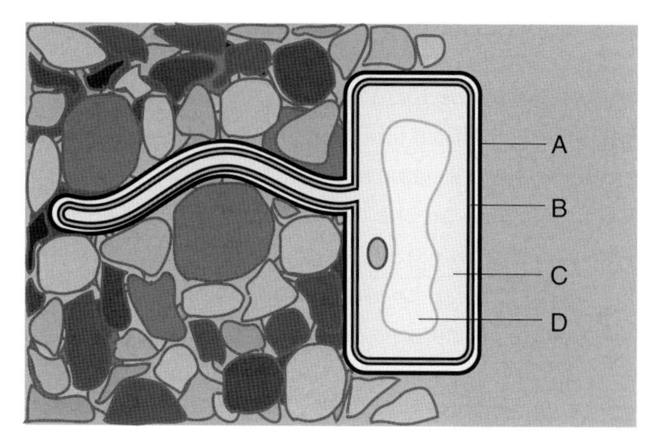

(a) Name the structures labelled A, B, C and D. *(4 marks)*

(b) The diagram below shows four ways in which molecules may move into and out of a cell. The dots show the concentration of molecules.

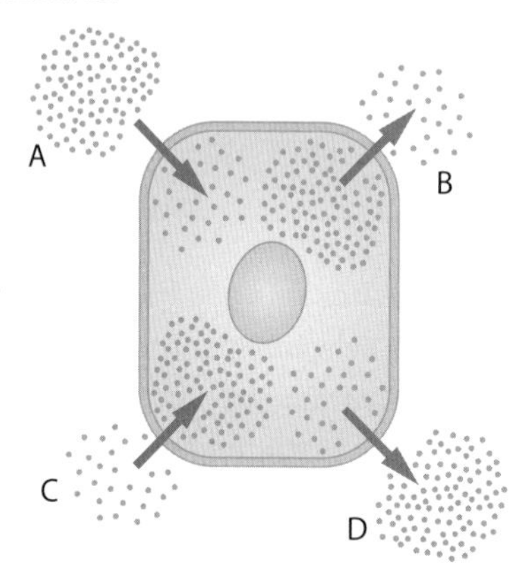

Which arrow – A, B, C or D – represents:

(i) Movement of oxygen molecules during respiration?

(ii) Movement of oxygen molecules during photosynthesis? *(2 marks)*

(c) The digestive system is adapted for the digestion and absorption of food. Describe, in terms of cells, tissues and organs, how the digestive system is adapted for these functions.

In this question you will be assessed on using good English, organising information clearly and using specialist terms where appropriate. *(6 marks)*

2 The graph shows the effect of light on the rate of photosynthesis.

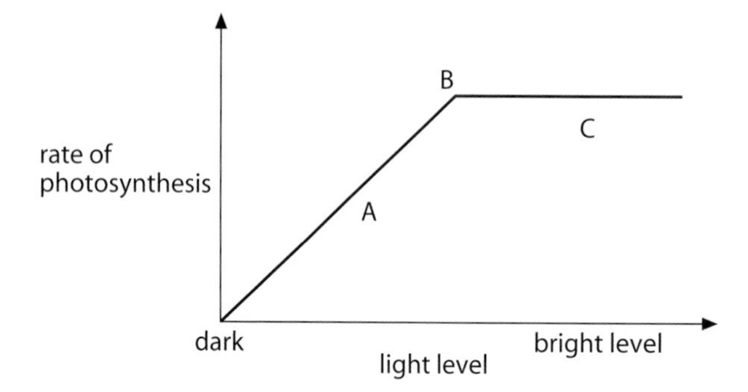

(a) Using the letters A, B and C on the graph, explain the shape of the graph. *(6 marks)*

(b) Give two ways in which the rate of photosynthesis at C could be increased. *(2 marks)*

3 Students estimated the number of dandelions on their school's field. They used quadrats, each with an area of 1 m^2.

The school playing field was rectangular in shape, with dimensions 90 m × 50 m.

The students counted the number of dandelions in 10 quadrats.

(a) Describe *one* way in which the students could have ensured that the quadrats were randomly distributed. *(2 marks)*

(b) The table shows the students' results.

(i) Calculate the mean number of dandelions per quadrat. *(2 marks)*

(ii) Use the mean you have calculated in part (i) to estimate the number of dandelions in the field. *(2 marks)*

Quadrat number	Number of dandelions
1	3
2	3
3	6
4	2
5	1
6	2
7	0
8	3
9	2
10	0

(c) The diagram shows the distribution of plants in a lake.

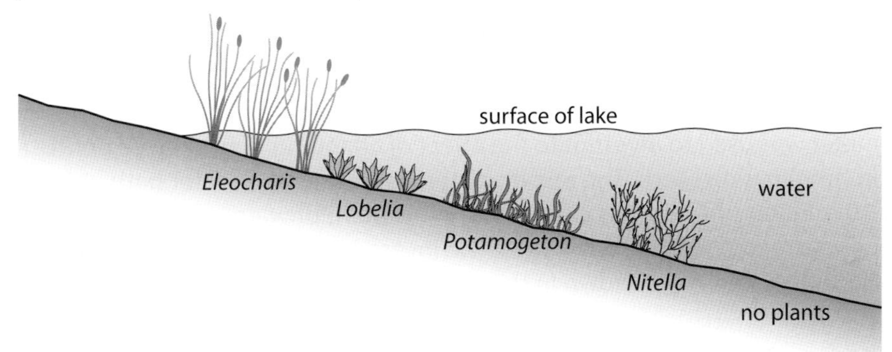

Suggest an explanation for the distribution of:

(i) *Eleocharis* *(2 marks)*

(ii) *Nitella.* *(2 marks)*

4 The graph shows the effect of pH on two enzymes from the human digestive system.

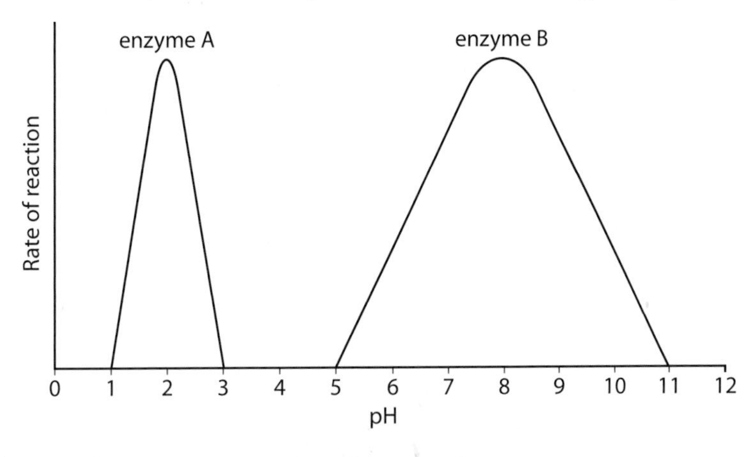

(a) Give two differences between enzyme A and enzyme B. *(2 marks)*

(b) What type of enzyme is enzyme A? Explain the reason for your answer. *(3 marks)*

5 The rate of respiration of an organism can be investigated using a respirometer. A respirometer measures the amount of oxygen used in respiration.

The diagram shows a respirometer containing germinating seeds. The amount of oxygen used during respiration is measured by the movement of the coloured liquid in the capillary tube.

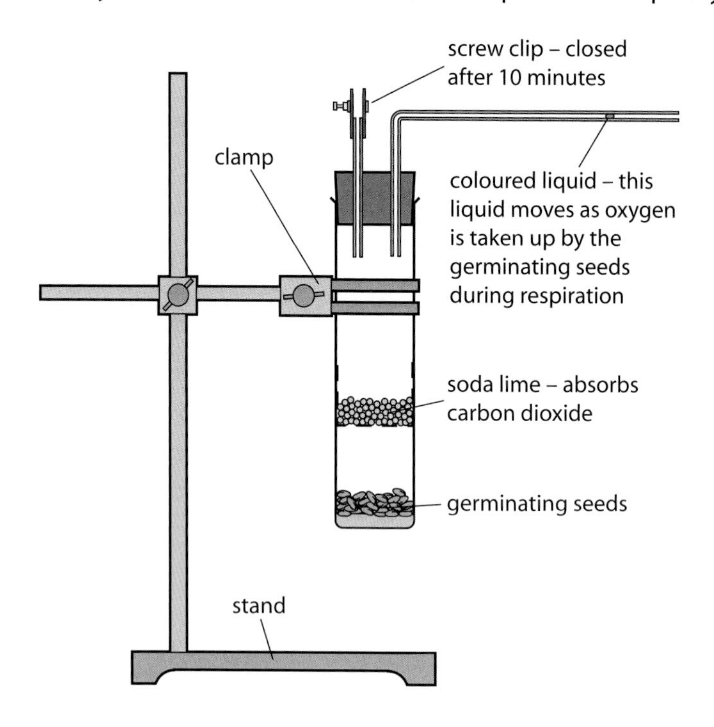

The respirometer was used to investigate the effect of changing temperature on the rate of respiration in germinating seeds. The respirometer was placed in a water bath at 20 °C with the clip open. After 10 minutes, the clip was closed. From this point onwards, the position of the liquid in the capillary tube was recorded every 5 minutes. The investigation was then repeated with the water bath set at 30 °C. The results are shown in the graph.

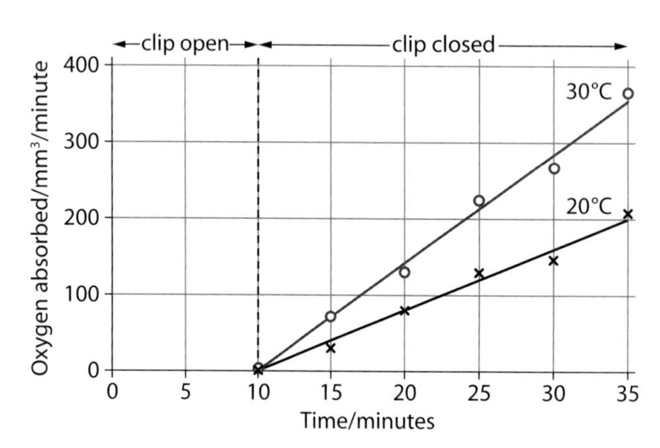

(a) Explain why the respirometer was left in the water bath for 10 minutes before closing the clip. *(2 marks)*

(b) Explain why the liquid moved after the clip was closed. *(2 marks)*

(c) Use the graph to calculate how much oxygen was used per minute at 20 °C. Show your working. *(2 marks)*

(d) Use the results to explain the effect of temperature on the rate of respiration in the germinating seeds. *(2 marks)*

6 The table shows the results of an investigation carried out to find out how breathing changes during exercise.

Activity/step-ups per minute	Mean volume of each breath/cm^3	Breathing rate/ breaths per minute	Total amount of air inhaled per minute/cm^3
20	500	18	
30	750	25	
40	1100	32	

(a) **(i)** How many breaths did the person take per minute when exercising at 30 step-ups per minute? *(1 mark)*

(ii) By how much did the volume of each breath increase between 20 and 40 steps per minute? *(1 mark)*

(b) Make a copy of the table. Calculate values for the fourth column and add them to the table. *(3 marks)*

7 The diagram shows the result of one cell division. The parent cell has two pairs of chromosomes.

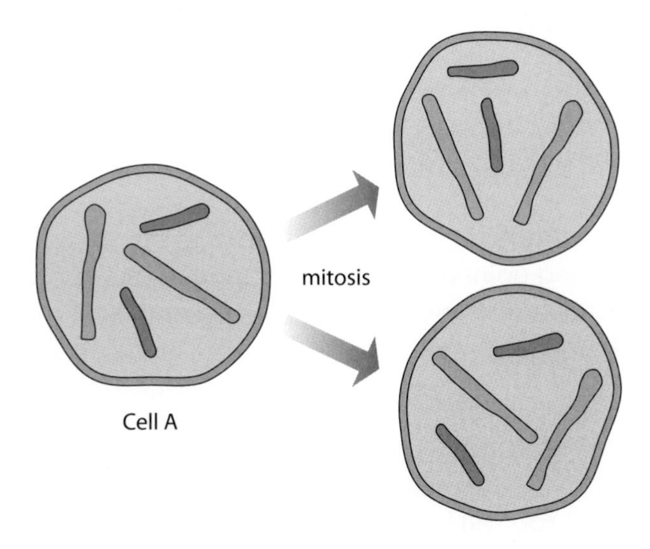

(a) This cell division is an example of mitosis. Explain why mitosis is important to living organisms. *(2 marks)*

(b) Draw the four cells that would be produced if the parent cell divided by meiosis. *(2 marks)*

8 A mother underwent a new medical procedure as she gave birth. Stem cells were collected from her baby's umbilical cord and stored for possible use in the future. The mother said that using stem cells in this way could save her child's life in the future. She said it was like having an insurance policy for her child.

'Stem cells are undifferentiated cells that act like the master cells of the body,' her doctor said. 'These cells can be used to treat heart problems and various forms of cancer.' Stem cells obtained from embryos produced by IVF can also be used to treat certain diseases.

(a) Explain what is meant by 'stem cells are undifferentiated cells'. *(2 marks)*

(b) Evaluate the use of stem cells to cure human diseases.

In this question you will be assessed on using good English, organising information clearly and using specialist terms where appropriate. *(6 marks)*

C2

The building blocks of chemistry

In this section you will look more deeply into the nature of matter, finding out how the arrangement and number of subatomic particles vary from element to element. In the first chapter you will see how the number and arrangement of electrons in an atom control the chemistry of the elements and the reactions they take part in. You will discover how non-metals can combine to form molecules, and how metals and non-metals combine to form salts and other ionic compounds.

In the second chapter you will learn how these chemical reactions at atomic level lead to the properties of the major groups of materials. Why do metals conduct electricity and why are they strong yet easy to shape? Why do salts have high melting points, yet can be brittle? Why are some non-metals, such as sulfur, soft and easy to melt, while others, such as diamond, are very hard with very high melting points? Why do different polymers have different properties? All these questions can be explained once you understand the processes of chemical bonding. You will also be introduced to nanotechnology.

Chapter 3 then looks at chemistry quantitatively. Why do different atoms have different masses? How can you use this knowledge to work out the formula of a chemical compound? You will learn how to calculate the yield of a product from a chemical reaction, once you know its balanced chemical equation. You will also learn about how substances are analysed using instruments such as mass spectrometers or techniques such as chromatography.

Test yourself

1. What controls the number of electrons in an atom and how are the electrons arranged around the nucleus?
2. What does the group number of an element in the periodic table tell you about its electron pattern?
3. How do metals and non-metals form compounds together?
4. How do non-metals form molecules?
5. How does a balanced chemical equation explain why the mass of products must equal the mass of reactants in a chemical reaction?

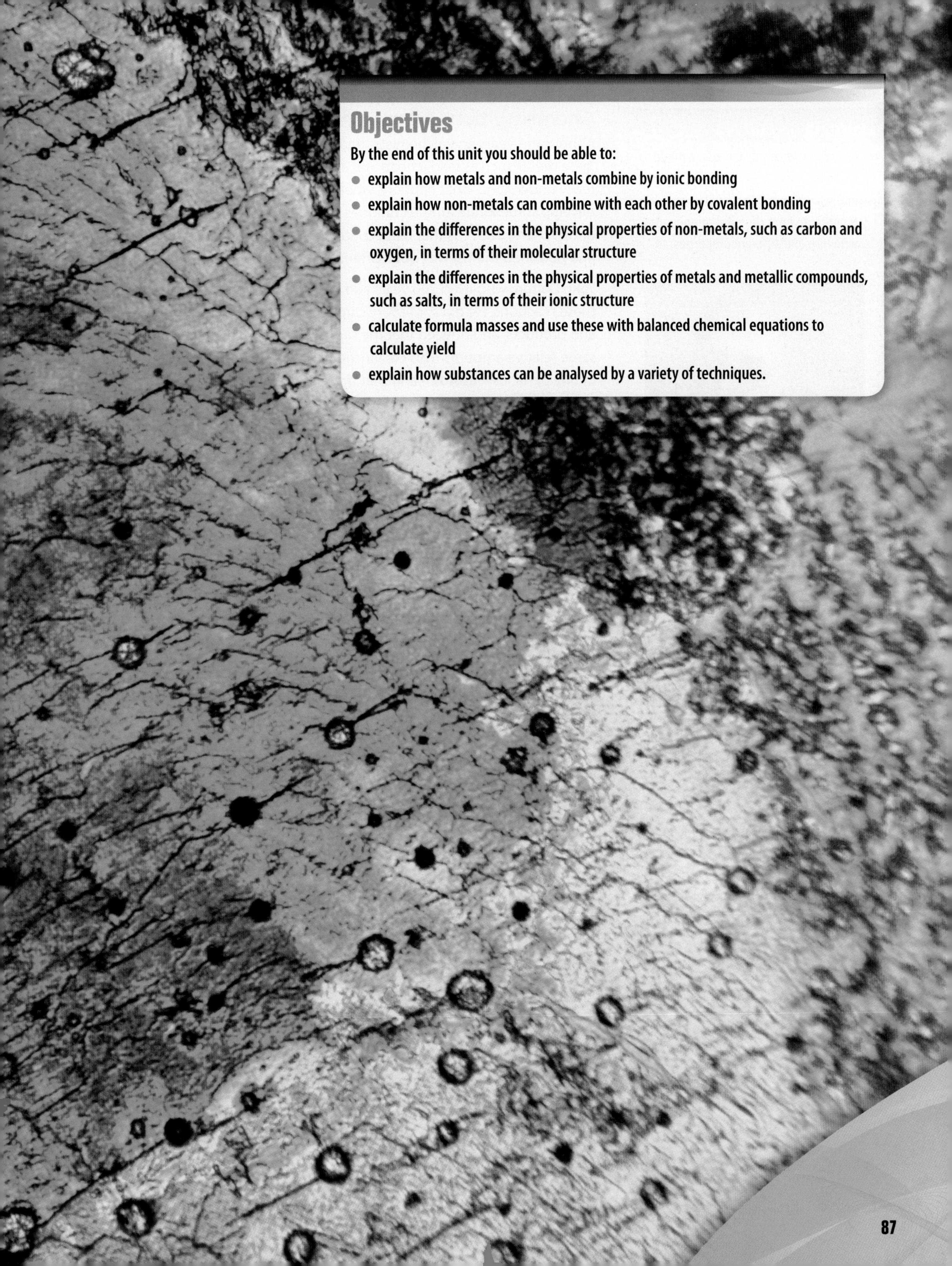

Objectives

By the end of this unit you should be able to:

- explain how metals and non-metals combine by ionic bonding
- explain how non-metals can combine with each other by covalent bonding
- explain the differences in the physical properties of non-metals, such as carbon and oxygen, in terms of their molecular structure
- explain the differences in the physical properties of metals and metallic compounds, such as salts, in terms of their ionic structure
- calculate formula masses and use these with balanced chemical equations to calculate yield
- explain how substances can be analysed by a variety of techniques.

C2 1.1

Understanding compounds

Learning objectives

- describe how all substances are made up of atoms
- understand how different atoms can join together to form chemical compounds with new and different properties.

Chemistry born from alchemy

A thousand years ago alchemists tried to make gold (rare and expensive) from 'base metal' such as lead (cheap and available). They failed, because they didn't really understand what happens when chemicals react. However, they did develop some interesting techniques such as distillation, and they spotted one or two useful patterns that helped later chemists. They also blew themselves up and poisoned themselves quite often.

John Dalton, who lived 200 years ago, was determined to make sense of all these reactions. He did some very careful experiments, weighing the reactants and products, and noticed that the same chemicals always reacted in the same relative amounts. This convinced him that everything must be made up of tiny particles – **atoms** – and that these atoms always combine in the same proportions for a given compound. In particular he concluded that:

- every **element** is made of its own distinctive atoms of a particular mass
- other chemicals are made from atoms that have joined together in some way.

Dalton gave his elements special symbols and drew pictures of some simple **compounds**. He may not have got everything quite right, but he certainly got chemistry moving in the right direction.

John Dalton – a great experimenter and a great chemist.

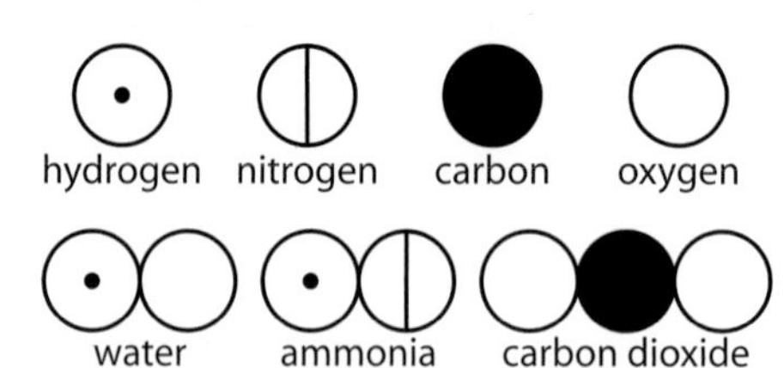

Figure 1 Some of Dalton's elements and compounds.

Elements, mixtures and compounds

We now know that there are just over 100 different elements. An element is a substance made from only one type of atom. It cannot be broken down into simpler substances. All the elements are listed in the periodic table (see lesson C2 1.5).

Hydrogen and oxygen are both elements. Hydrogen is a colourless gas that is flammable. Oxygen is a colourless gas in which other substances burn well. If you mix hydrogen and oxygen, then you still have the same substances: hydrogen and oxygen.

If a flame is put near the mixture of hydrogen and oxygen, a chemical reaction takes place producing the compound water. This is very different to both hydrogen and oxygen. It is a colourless liquid that can put fires out.

In a mixture of hydrogen and oxygen, atoms of oxygen are not joined to atoms of hydrogen; they are just mixed together. In the compound water, the hydrogen and oxygen atoms are joined to each other.

The equation for this chemical reaction is:

$$\text{hydrogen} + \text{oxygen} \longrightarrow \text{water}$$
$$2H_2 + O_2 \longrightarrow 2H_2O$$

Science in action

Hydrogen and oxygen react with a hot flame. Oxy–hydrogen torches are used to melt the surface of glass objects as a way of polishing off tiny scratches.

hydrogen (H_2) + oxygen (O_2) → mixing → mixture of oxygen (O_2) and hydrogen (H_2) → chemical reaction → water (H_2O)

Figure 2 The reaction between hydrogen and oxygen.

More about compounds

A compound is a substance made from different elements chemically joined together. Although there are only just over 100 elements, there are millions of known compounds.

The properties of each substance in a mixture are the same as before they were mixed. However, the properties of a compound are different to those of the elements from which it is made. For example, sodium chloride (common salt) is made from sodium and chlorine. Sodium and chlorine are both very reactive, dangerous elements. Sodium chloride is completely different to both sodium and chlorine. Sodium chloride is stable and safe enough to eat.

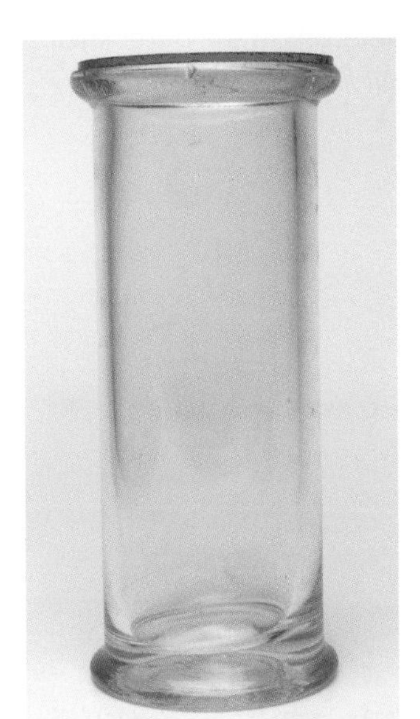
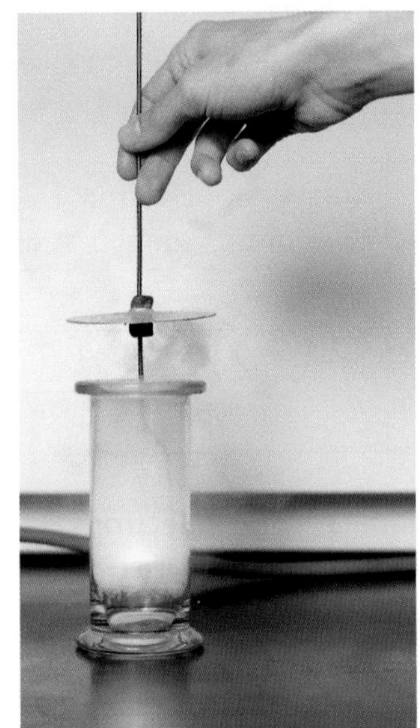

The reaction between sodium and chlorine.

The equation for this chemical reaction is:

$$\text{sodium} + \text{chlorine} \longrightarrow \text{sodium chloride}$$
$$2Na + Cl_2 \longrightarrow 2NaCl$$

Compounds can be broken down (decomposed) into simpler substances. This can often be achieved by heating (thermal decomposition) or using electricity (electrolysis). For example, passing an electric current through molten aluminium oxide breaks it down into aluminium and oxygen.

Questions

1 **(a)** What was Dalton's big (or small?) idea about the nature of materials? **(b)** How did he check to see if his ideas were correct?

2 How can the elements in the compound aluminium oxide be separated?

3 Which element do all of these acids have in common: sulfuric acid (H_2SO_4), phosphoric acid (H_3PO_4), nitric acid (HNO_3), and hydrochloric acid (HCl)?

4 What is the difference between a mixture of hydrogen and oxygen and the compound water? Explain in terms of how the atoms are arranged.

5 Here is a list of substances. Which are elements and which are compounds?
Mg, SO_2, Co, CO, Br_2, KBr, $CaCO_3$, K_2O.

6 Look at the diagrams in Figure 3 and say for each whether it shows an element, a compound or a mixture.

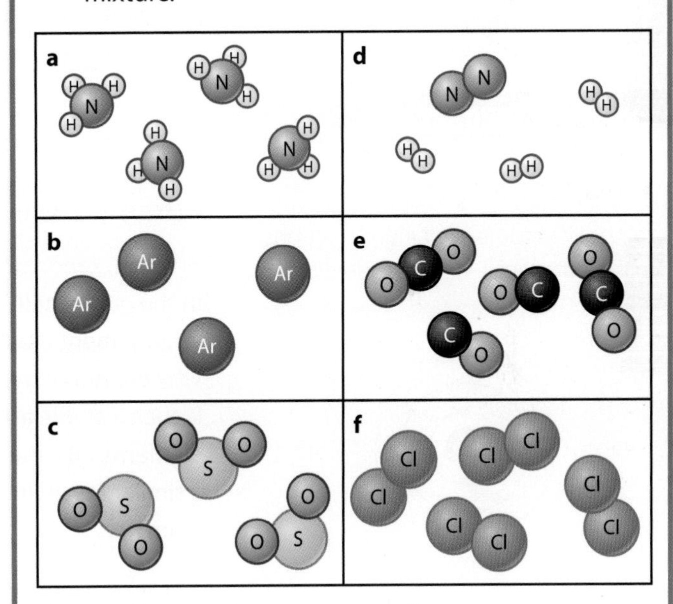

Figure 3 Element, mixture or compound?

7 Magnesium burns in oxygen to form magnesium oxide. What pattern does this table of masses show?

Mass of magnesium/g	Mass of oxygen reacting/g
0.3	0.2
0.36	0.24
0.24	0.16

8 Carbon dioxide is a compound made from the elements carbon and oxygen. Use these three substances to explain the difference between elements and compounds.

Ionic bonding

Learning objectives

- describe how electrons stack up in their energy levels (electron shells)
- explain why the noble gas electron structure is very stable
- understand why atoms with just a few outer electrons try to lose them
- understand why atoms just short of the noble gas structure try to gain electrons
- describe how alkali metals (group 1) form positive ions
- describe how the halogens (group 7) form negative ions
- write formulae for ionic compounds
- represent the electronic structure of the ions in certain ionic compounds.

Happiness is a stable electron structure

Helium, neon and argon belong to a family of elements found in **group 0** of the periodic table called the noble gases. They have this name because they keep to themselves and do not join in chemical reactions at all. What is it that makes them so stable and unreactive?

Science in action

Noble gases are often used to fill light bulbs with heated filaments (such as xenon headlights) because they do not react with the hot wire.

In the first chemistry module (lesson C1 1.3), you saw how, for the first 20 elements, **electrons** stack up around the **nucleus** in **energy levels** (**electron shells**): up to two in the first, eight in the second, and eight in the third. As you learned in lesson C1 1.4, noble gas atoms have eight electrons in the outer shell. After the first 20 elements, things get a bit more complicated.

Table 1 Energy levels.

Element number	Name	Number of electrons	Electron configuration
2	helium	2	2
10	neon	10	2,8
18	argon	18	2,8,8

Krypton, xenon and radon have eight electrons in the outer shell. This appears to be a very stable arrangement that is not easy to disrupt. That is why the noble gases are so unreactive. In fact, this is such a stable arrangement that atoms of other elements take part in chemical reactions to achieve a similar result. In the process, they often have to team up.

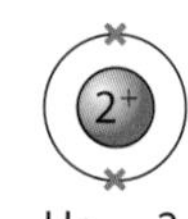

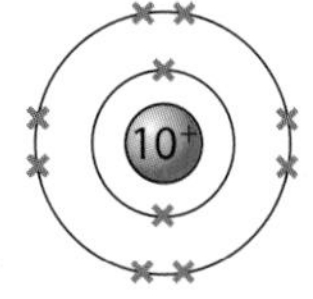

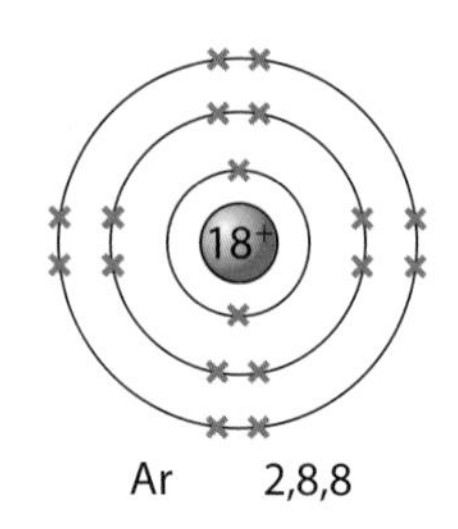

Figure 1 The electronic structure of the noble gases makes them very unreactive.

Examiner feedback

It may help you to remember facts by 'humanising' atoms – but *do not* write things like 'atoms *want* eight electrons in the outer shell' in an examination.

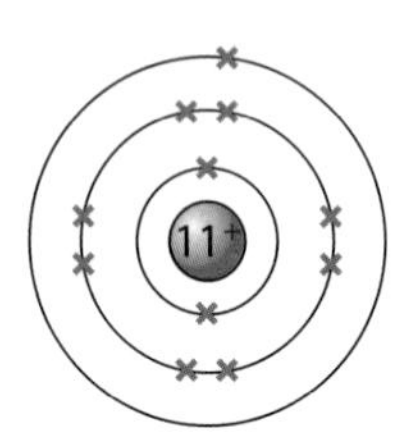

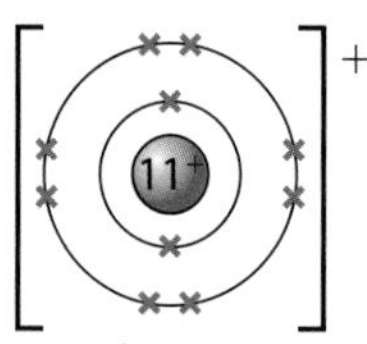

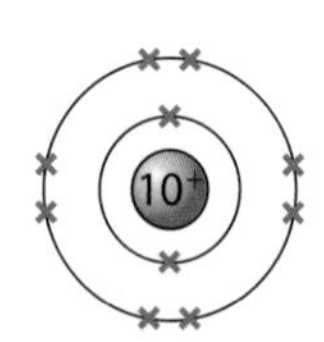

Figure 2 Alkali metal ions have the stable electronic structure of a noble gas, but have a positive electrical charge.

Fast and loose: the single outer electron

Group 1 of the periodic table contains the alkali metals, such as sodium (see lesson C1 1.4). These elements have just one electron in their outer energy level. It is quite easy to lose this electron as not much energy is needed to pull it away from its atom. If this happens, the atom is no longer neutral. It has one more positive **proton** in the nucleus than it has negative electrons. This means that overall the particle now has a single positive charge. Charged particles like this are called **ions**.

A sodium ion is smaller than a sodium atom as it has one electron shell fewer. Ions like this have the electronic structure of a noble gas, so they are very stable.

On the scrounge for electrons

Group 7 of the periodic table contains non-metallic elements called the halogens. Group 7 elements, such as chlorine, all have seven electrons in their outer energy level. Group 7 atoms do not easily *lose* electrons, but they can *capture* an extra electron to form a stable ion that has the electronic structure of a noble gas. The ion that is formed has one extra electron, so it has a single negative charge. Such ions are called halide ions.

Metals and halogens: made for each other

All metal atoms have easily removable electrons in their outer energy levels. When they lose these electrons they form stable positive ions. Group 2 metals form 2+ ions, group 3 metals form 3+ ions, and so on.

Non-metal atoms with five, six or seven electrons in their outer energy levels gain extra electrons to get a stable structure. Group 5 non-metals have to gain three electrons to form 3− ions; group 6 non-metals have to gain two electrons to form 2− ions. Forming such ions enables the atoms to attain the stable electronic structure we see in the atoms of the noble gases (group 0).

Getting hitched: the ionic bond

You may have spotted the obvious connection. Metallic atoms become stable positive ions by losing electrons. Non-metallic atoms become stable negative ions by gaining electrons. Together, these oppositely charged ions form a compound, an entirely new substance, that does not necessarily resemble either of the elements that have come together to form it. Often, the compound is a crystal, a solid that consists of an enormous **lattice** of ions held together by the forces of attraction between them (see lesson C1 1.2). This method of joining is called an **ionic bond**. The compound is called an **ionic compound**.

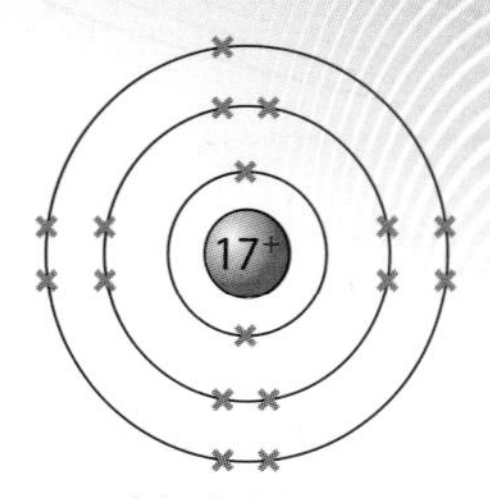

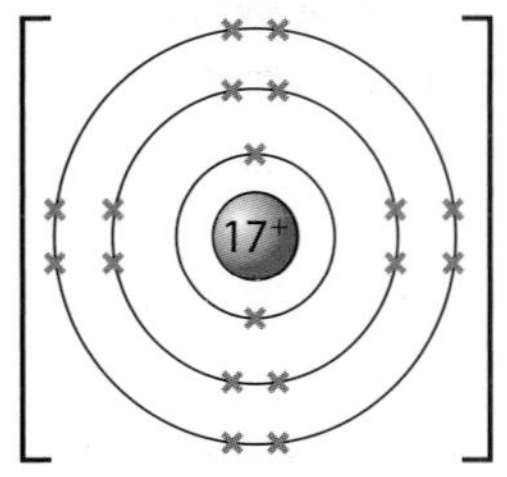

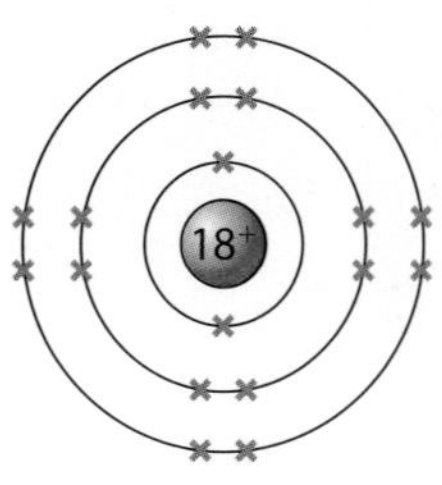

Figure 3 Halide ions have the stable electronic structure of a noble gas, but with a negative electrical charge.

Route to A*

With a clear understanding of the logic behind ionic bonding, you will be able to work out any combination of metal/non-metal – rather than simply remembering the ones you have seen before.

Science skills

These electronic structure diagrams are a good example of how scientists use simple models to help understand complex problems. Atoms are, of course, three-dimensional and far more complex than shown, but these simple diagrams are very powerful, and help us to understand and predict the properties of the elements.

Questions

1. Atoms of an unknown element X have eight electrons in their outer shell. What predictions can you make about the properties of this element X? Explain your answer.
2. What ions will be formed by the alkali metals lithium (Li), potassium (K) and rubidium (Rb)?
3. Draw electron configuration diagrams to show how potassium (K – element 19) becomes an ion.
4. Aluminium is in group 3. What is the charge of an aluminium ion and why?
5. Atoms of iodine have seven electrons in their outer energy shells. Is iodine a metal or non-metal and what ion does it form?
6. What do sodium ions (Na^+), chloride ions (Cl^-) and neon atoms all have in common?
7. Suggest a reason why metals are able to conduct electricity.
8. Sodium is in group 1, calcium is in group 2 and chlorine is in group 7. Explain why sodium chloride is NaCl, but calcium chloride is $CaCl_2$.

Rules for ionic bonding

Learning objectives

- describe how ionic bonds form between metals and non-metals
- understand why the charges need to balance out in ionic compounds
- understand and be able to apply the rules for working out the formulae of ionic compounds.

The salt on your table

As we saw in lesson C2 1.1, sodium is a soft and dangerously reactive metal in group 1 of the periodic table. Chlorine is a green poisonous gas in group 7. If sodium is burnt in chlorine the violent reaction leaves white crystals of a new compound: sodium chloride. This is the common salt you put on your chips.

The sodium and chlorine have combined chemically to form a new compound with new properties. Each sodium atom has lost an electron to become a positive sodium ion (Na^+). Each chlorine atom has gained an electron to form a negative chloride ion (Cl^-). The oppositely charged particles are attracted to form an ionic bond. The charges balance out so the compound is neutral. We can write equations for this.

$$\text{sodium} + \text{chlorine} \longrightarrow \text{sodium chloride}$$
$$2Na + Cl_2 \longrightarrow 2NaCl$$

The balanced chemical equation represents the proportion of the different ions present in the crystal, in this case 1:1. The ions do not actually pair up in this way. They stack up in a giant structure called an ionic lattice (see lesson C2 2.2). This regular stacking pattern gives rise to the typical cubic shape of salt crystals.

Reactive sodium burns in toxic chlorine.

Could you tell that common salt is a compound of the dangerously reactive elements sodium and chlorine?

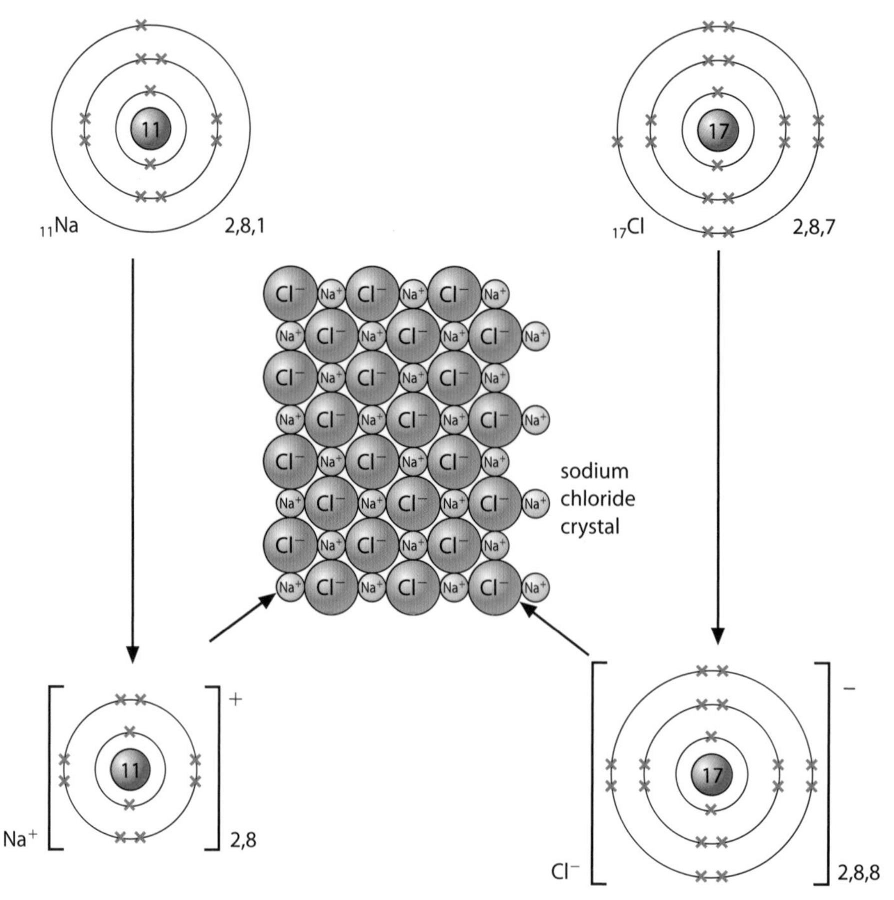

Figure 1 Common salt contains ions stacked in an ionic lattice.

Any group 1 metal will react with any group 7 halogen in the same way. For example, potassium and bromine will react to form the stable salt potassium bromide (KBr), which is made from K^+ and Br^- ions.

Examiner feedback

Remember that there is no such thing as an NaCl particle. This is just shorthand for the simplest ratio of the ions.

Keeping it in balance

Group 2 metals such as calcium have two electrons in their outer shell. If they lose these to form the stable noble gas structure, they become ions with a 2+ charge. If calcium reacts with chlorine, for example, you need two chloride (Cl^-) ions to balance the charge of one calcium (Ca^{2+}) ion and make a neutral compound. The simple chemical formula for calcium chloride is therefore $CaCl_2$, and the balanced equation is:

$$Ca + Cl_2 \longrightarrow CaCl_2$$

Group 6 non-metals such as oxygen have six electrons in their outer shell. They need to gain two electrons to achieve the stable noble gas structure, becoming ions with a 2− charge. So if oxygen reacts with sodium, for example, you need two sodium (Na^+) ions to balance the charge of one oxide (O^{2-}) ion. The simple chemical formula for sodium oxide is therefore Na_2O, and the balanced equation is:

$$4Na + O_2 \longrightarrow 2Na_2O$$

Of course, if calcium reacts with oxygen, the charges on the Ca^{2+} and O^{2-} ions cancel out at a simple 1:1 ratio again:

$$2Ca + O_2 \longrightarrow 2CaO$$

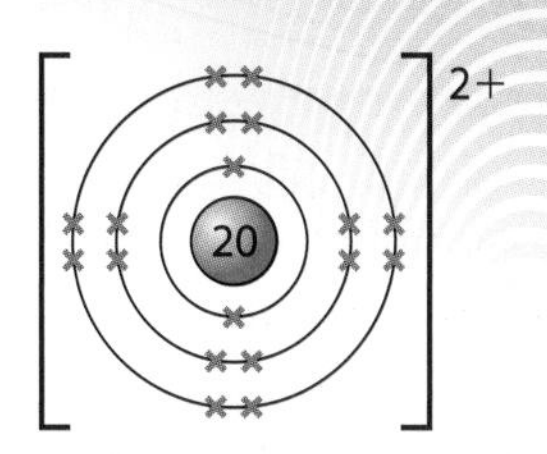

Figure 2 The calcium Ca^{2+} ion has the electron structure of argon: 2, 8, 8.

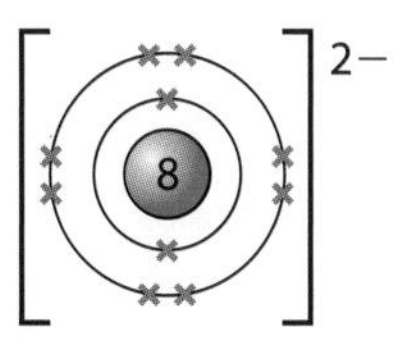

Figure 3 The oxide O^{2-} ion has the electron structure of neon: 2, 8.

Finding the formulae of ionic compounds

You can work out the formulae of ionic compounds using Table 1. This includes some common compound ions that are made up of several atoms joined together. For example, the nitrate ion (NO_3^-) has one nitrogen atom joined to three oxygen atoms and has a charge of −1.

To work out the formula (Table 2), you must make the total charge of the positive and negative ions balance. If the formula contains more than one of a compound ion, the compound ion is put in brackets for clarity.

Table 1 Ion charges.

Positive ions		Negative ions	
aluminium	Al^{3+}	bromide	Br^-
calcium	Ca^{2+}	chloride	Cl^-
copper(II)	Cu^{2+}	fluoride	F^-
hydrogen	H^+	iodide	I^-
iron(II)	Fe^{2+}	oxide	O^{2-}
iron(III)	Fe^{3+}	sulfide	S^{2-}
lithium	Li^+	carbonate	CO_3^{2-}
magnesium	Mg^{2+}	hydrogen carbonate	HCO_3^-
potassium	K^+		
sodium	Na^+	hydroxide	OH^-
zinc	Zn^{2+}	nitrate	NO_3^-
ammonium	NH_4^+	sulfate	SO_4^{2-}

Table 2 Working out the formulae of some ionic compounds.

	Potassium oxide	Calcium sulfide	Aluminium oxide	Copper carbonate	Calcium hydroxide	Ammonium sulfate
Positive ions	K^+ K^+	Ca^{2+}	Al^{3+} Al^{3+}	Cu^{2+}	Ca^{2+}	NH_4^+ NH_4^+
Negative ions	O^{2-}	S^{2-}	O^{2-} O^{2-} O^{2-}	CO_3^{2-}	OH^- OH^-	SO_4^{2-}
Formula	K_2O	CaS	Al_2O_3	$CuCO_3$	$Ca(OH)_2$	$(NH_4)_2SO_4$

Questions

1. Aluminium is in group 3. Explain why aluminium forms an Al^{3+} ion.
2. Calcium loses two electrons to form a 2+ ion. Why is calcium chloride $CaCl_2$?
3. Why is the formula for sodium oxide Na_2O?
4. Calcium and chloride ions both have the electron configuration 2,8,8. What is different about them?
5. Magnesium oxide has the formula MgO. Explain why this is.
6. What is unusual about the ammonium ion compared with all of the other compound ions shown?
7. Use the table of ion charges to write the formula of: **(a)** potassium bromide, **(b)** iron(III) oxide, **(c)** aluminium hydroxide, **(d)** calcium nitrate, **(e)** zinc sulfate and **(f)** calcium hydrogen carbonate.
8. Some transition elements such as iron can vary the number of electrons in their outer energy level. Iron forms two different sulfates, for example, $FeSO_4$ and $Fe_2(SO_4)_3$. For each case, how many electrons have been lost from the outer energy level of the iron atom? Explain your answer.

Covalent bonding

Learning objectives

- explain how non-metals form molecules
- identify the bonds formed between atoms
- understand that some elements form molecules on their own, but most molecules are compounds
- use different models to show covalent bonds for different purposes.

No electrons to spare?

Non-metals cannot form ionic compounds on their own. Their atoms need extra electrons to get a stable 'noble gas' electron configuration, but without metal atoms where can they get the electrons?

The answer is to share. Non-metal atoms can join their outer energy levels together and share one or more pairs of electrons. The shared electrons form very strong **covalent bonds**. When atoms share electrons, they stay together. We can imagine them looking like soap bubbles stuck together. These arrangements are called **molecules**. Because no electrons have been gained or lost, molecules carry no electrical charge. This arrangement can be very stable.

'Give and take'

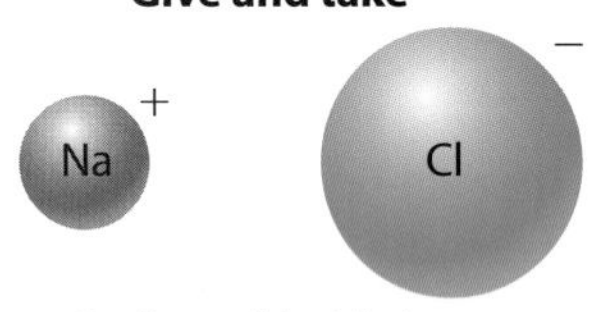

Sodium chloride is an ionic compound.

'Sharing'

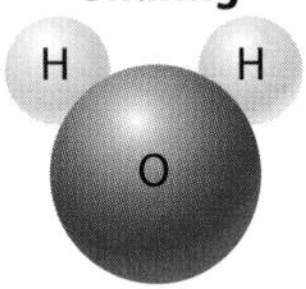

Water is a covalent compound.

Figure 1 Ions stay separate, but covalent bonds make atoms stick together like bubbles.

Choosing the right model

Atoms and molecules are far too small to see, so we can't draw accurate pictures of them. Instead we use models and symbols to help us visualise what is going on in chemical reactions. As we may want to focus on different aspects of the reaction we use different models for different purposes. Don't let this confuse you. Just use the best model for what you want to think about (see Figure 2).

- Molecular models help us to visualise the shape of molecules.
- Electron energy level diagrams show how covalent bonds are formed.
- Structural formulae show covalent bonds clearly.
- Simple formulae show the atoms involved to help us balance reaction equations.

The elements that won't go out alone

Chlorine atoms are one electron short of that stable 'noble gas' arrangement, so they need to share just one pair of electrons to form a Cl_2 molecule. You can show it in an electron shell diagram. Draw the outer shells of the two atoms slightly overlapping, and arrange the electrons in pairs around the circles. Use dots for electrons on one, and crosses for electrons on the other. You will end up with one dot and one cross in the overlap where the two atoms have joined together, and you will see that each chlorine atom now has eight electrons in its outer shell, like argon. The shared pair of electrons makes the covalent bond. The two electrons are exactly the same; using a dot and a cross just helps you to see how the shared pair is formed. A simpler version of this diagram removes the circles altogether and just shows the dots and crosses.

Similarly, oxygen atoms are two electrons short of a stable shell. In the O_2 molecule they share two electron pairs to make a double covalent bond.

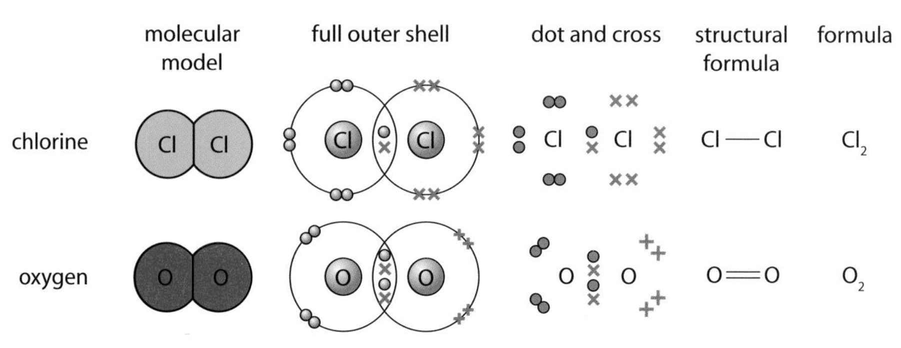

Figure 2 Some different ways to show covalent bonds.

Examiner feedback

Remember that the sharing of electrons can result in very strong covalent bonds.

Route to A*

As an A* student, you will need to be able to switch easily between models, recognising which is most useful to you at the time.

Examiner feedback

Most gaseous elements form covalent molecules like chlorine and oxygen, for example hydrogen (H_2) and nitrogen (N_2). The exceptions are the noble gases – they do not need to.

Compounds made through sharing

Atoms of different non-metallic elements can also join together to make covalent compounds. Chlorine can form a single covalent bond with hydrogen to make hydrogen chloride (HCl). Oxygen needs to share two electrons and so can join with two hydrogen atoms to make hydrogen oxide, or water (H_2O). Nitrogen is three electrons short of a stable shell. It can share a pair of electrons with each of three hydrogen atoms to make an ammonia molecule (NH_3). Carbon is four electrons short of a stable shell. It can share a pair of electrons with each of four hydrogen atoms to make a methane molecule (CH_4).

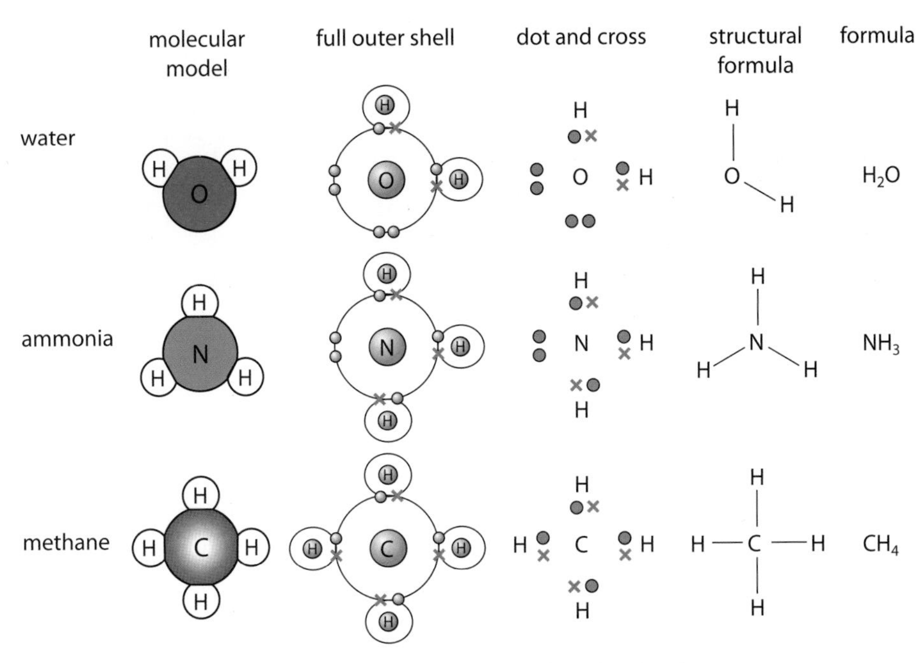

Figure 3 Common compounds with different numbers of covalent bonds.

Giant molecules

Some elements and compounds, such as carbon and silicon dioxide, exist as giant structures (lattices) in which every atom is linked to its neighbours by strong covalent bonds (see lesson C2 2.1). These are called **giant covalent structures**.

Taking it further

The nitrogen atom in ammonia has a 'lone pair' of electrons not involved in covalent bonding. A passing hydrogen ion can be captured here, the nitrogen providing both electrons for a strong covalent bond. This forms the ammonium ion (NH_4^+), which carries the positive charge from the **hydrogen ion** and can form salts just like an alkali metal. An example is the **fertiliser** salt ammonium nitrate, NH_4NO_3.

Science in action

As every single atom in a diamond is attached to its neighbours by strong covalent bonds, diamonds are very hard indeed. In the oil industry, diamond-studded drills are used to cut through rock.

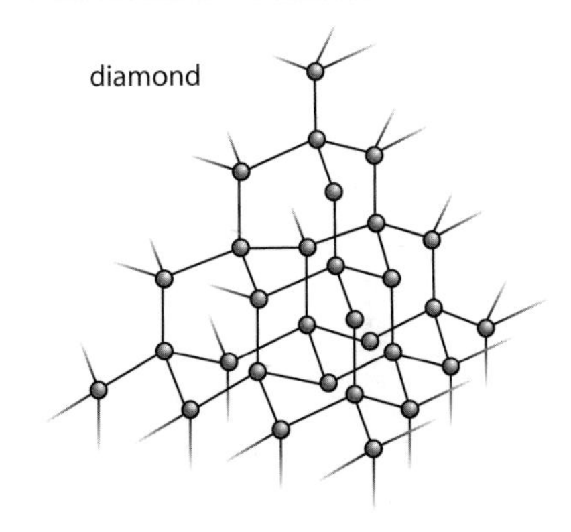

Figure 4 Diamond is a giant covalent structure built from carbon atoms.

Questions

1. Many gases, such as hydrogen and oxygen, exist as molecules (H_2, O_2). What kind of bonding is there within these molecules?
2. Why do the noble gases not need to form covalent bonds?
3. Draw an electronic structure diagram for an H_2 molecule. (Remember, the first energy level can only take two electrons.)
4. Explain in simple terms the difference between ionic and covalent bonding.
5. Draw an electronic structure diagram for an HCl molecule.
6. Suggest a reason why there are not any strong forces between simple molecules.
7. Sand is made from silicon dioxide (SiO_2). What is it about the structure of this compound that makes sand hard?
8. Explain carefully how a carbon atom combines with two oxygen atoms to form carbon dioxide. Draw dot-and-cross and structural formulae diagrams for a CO_2 molecule.

Metals

Learning objectives

- describe and explain some of the properties of metallic substances
- describe the structure of metallic substances
- represent the bonding in metals using a diagram
- explain why metals have their particular properties.

Useful metals

Just over three-quarters of the elements are metals. Humans have depended on metals since the Bronze Age. Today, we make much use of metals such as iron, aluminium and copper.

Metals all have electrons in their outer shells that are easily lost, giving stable positive ions. These 'loose' outer electrons give metals their special properties (see lesson C1 3.5).

1	2											3	4	5	6	7	0
									H								He
Li	Be											B	C	N	O	F	Ne
Na	Mg											Al	Si	P	S	Cl	Ar
K	Ca	Sc	Ti	V	Cr	Mn	Fe	Co	Ni	Cu	Zn	Ga	Ge	As	Se	Br	Kr
Rb	Sr	Y	Zr	Nb	Mo	Tc	Ru	Rh	Pd	Ag	Cd	In	Sn	Sb	Te	I	Xe
Cs	Ba	La	Hf	Ta	W	Re	Os	Ir	Pt	Au	Hg	Tl	Pb	Bi	Po	At	Rn
Fr	Ra	Ac															

Group number; metals; non-metals

Figure 1 All the elements shaded blue are metals.

The structure of metals

In the solid, the metal atoms are packed close together in a regular **metallic structure.** The outer shell electrons are lost from each atom and become free to move throughout the metal. This leaves a giant structure (lattice) of positive metal ions surrounded by **delocalised** electrons.

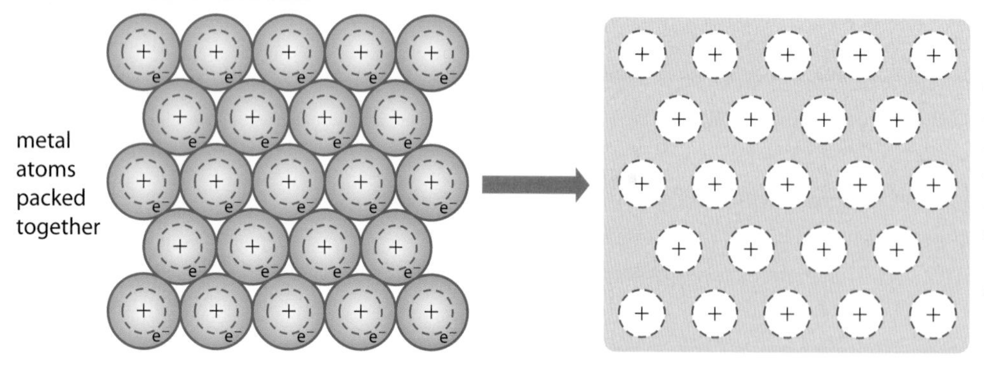

Outer energy level (shell) electrons are lost from atoms leaving a lattice of positive metal ions surrounded by delocalised electrons. The shaded area represents the delocalised electrons.

Figure 2 The structure of metals.

The structure is held together by the electrostatic attraction between the positive metal ions and the negative delocalised electrons. This attraction is strong, and so metals have high melting and boiling points.

Metals conduct heat

Metals conduct energy, in the form of heat, because the delocalised electrons are free to move and can carry energy with them.

- Metals feel cold to the touch because they conduct the energy given out away from your hand, which cools down.
- Computer microprocessors use metal fins to carry energy given out away from the processor and stop it from overheating.
- Pans are often made from metals because they conduct heat from the flame or hotplate to your food.

Science skills

Table 1 The atomic radii of the group 4 elements.

Element	Atomic radius/10^{-12}m
C	70
Si	110
Ge	125
Sn	145
Pb	154

Use this to explain why group 4 elements change from metals to non-metals down the group.

Delocalised electrons transfer energy and cool the computer chip.

Metals conduct electricity

Delocalised electrons also allow metals to conduct electricity very well. Electric cables are usually made of copper, which is an excellent conductor. Any impurities in the copper can cause irregularities in the metallic structure and impede the flow of electrons, so copper for cables has to be at least 99.9% pure.

Even the vibrations of the metal ions can affect the flow of electricity. As a metal gets hot, the ions vibrate more and the current is reduced. If you cool the metal down, it conducts electricity much better. At very low temperatures, certain materials become superconductors. They have no electrical resistance and can carry enormous currents. Supercooled superconducting metal coils are used for the powerful electromagnets in **magnetic resonance imaging** (**MRI**) **scanners** and in particle accelerators such as the **large hadron collider**.

Supercooled metals are used for powerful electromagnets like those in the large hadron collider.

Metals are strong and easy to shape

Metals are strong enough to build machines and bridges with, but they are also easy to shape as they can be bent and hammered into shape. This is because the layers of metal ions can slide over each other while keeping the strong attraction between the positive metal ions and the delocalised electrons.

Hot steel can be squeezed out to make girders.

Science in action

The powerful magnetic fields needed to contain the superheated core of a nuclear fusion reactor are produced by the enormous electric currents that can flow through supercooled, superconducting electromagnets.

Questions

1. **(a)** Describe the structure of a metal. **(b)** Describe the bonding in a metal.
2. Why do metals have high melting points?
3. Explain why metals can be used to make: **(a)** pans; **(b)** electrical wires.
4. Explain why metals can be bent and shaped.
5. Suggest a reason why metals are usually heated before being beaten or pressed into shape.
6. Copper is used to make electrical wires. **(a)** Explain why copper is used to make electrical wires. **(b)** Draw a diagram to show the structure of copper. **(c)** Use your diagram to explain why copper conducts electricity. **(d)** Use your diagram to explain why copper can be stretched into wires. **(e)** Why does copper have to be 99.9% pure to conduct electricity efficiently?
7. Gold is a better conductor than copper. Suggest a reason why it isn't used for household wiring.
8. Explain carefully how temperature affects the way electricity can flow though a metal. Give an example of how this effect has been put to good use.

C2 2.1 Simple molecular substances

Learning objectives

- explain that substances consisting of simple molecules only have weak forces between the molecules and so have low melting and boiling points
- explain that substances consisting of simple molecules do not conduct electricity because the molecules are neutral.

Molecules

Many substances are made up of molecules. This includes some elements such as oxygen (O_2) and hydrogen (H_2), and many compounds such as water (H_2O) and glucose ($C_6H_{12}O_6$). Many elements that are non-metals are made of molecules. Compounds made from non-metals are also made of molecules.

A molecule is two or more atoms joined by covalent bonds. In a covalent bond, two electrons are shared between two atoms. This shared electron pair joins the atoms together.

The formula of a simple molecular substance tells us how many atoms of each type are in one molecule. For example, the formula CH_4 (methane) tells us that a methane molecule is made up of one carbon atom and four hydrogen atoms.

This photo shows water as ice, water and steam. All three states are made up of H_2O molecules.

Intermolecular forces

Covalent bonds within molecules are very difficult to break. If they are broken, this constitutes a chemical change as different substances are formed. For example, if the covalent bonds in water are broken, hydrogen and oxygen are formed.

Between the molecules there are no bonds. However, there are weak forces called **intermolecular forces**. These forces are far weaker than the three types of bonding (ionic, covalent and metallic).

Melting and boiling points

When simple molecular substances melt or boil, it is the weak forces between molecules that are overcome. The covalent bonds do not break. For example, the molecules in water as a solid (ice), liquid (water) and gas (steam) are all H_2O molecules. The molecules in methane as a solid, liquid and gas are all CH_4 molecules.

Table 1 The structure of the simple molecular substance methane as a solid, a liquid and a gas.

State	Solid	Liquid	Gas
Space-filling diagrams (better represent what molecules look like)			
Stick diagrams			

Simple molecular substances have low melting and boiling points because the forces between the molecules are weak. This means that many simple molecular substances are gases, liquids, or solids with low melting points.

Table 2 Melting and boiling points of some simple molecular substances.

Substance	Ethanol	Iodine	Methane	Naphthalene	Oxygen	Water
Formula	C_2H_5OH	I_2	CH_4	$C_{10}H_8$	O_2	H_2O
Melting point/°C	−114	114	−183	80	−219	0
Boiling point/°C	78	184	−162	218	−183	100
State at room temperature	liquid	solid	gas	solid	gas	liquid

Electrical conductivity

An electric current is the flow of electrically charged particles. Molecules have no electric charge – they are neutral. This means that simple molecular substances do not conduct electricity in any state.

Questions

1 **(a)** What is a molecule? **(b)** What is a covalent bond?

2 **(a)** Name three elements made from molecules. **(b)** Name three compounds made from molecules.

3 Which of the following compounds are made from molecules? Calcium oxide (CaO), hydrogen sulfide (H_2S), methanol (CH_3OH), copper sulfate ($CuSO_4$), sodium chloride (NaCl), silane (SiH_4), ammonia (NH_3).

4 **(a)** The formula of glucose is $C_6H_{12}O_6$. What does this formula tell us about glucose? **(b)** How are the atoms held together in a molecule of glucose?

5 Bromine molecules have the formula Br_2. They contain two bromine atoms joined by a single covalent bond. Bromine is a liquid at room temperature that easily turns into a gas due to a low boiling point. Explain why bromine has a low boiling point.

6 Simple molecular substances, including pure water, do not conduct electricity. Explain why.

7 Look at the data in Table 3 and decide which of the substances are made of simple molecules.

Table 3 Properties of simple molecules.

Substance	A	B	C	D	E
Melting point/°C	−85	808	39	41	3550
Boiling point/°C	−20	1465	701	182	4827
Conductivity as solid	insulator	insulator	conductor	insulator	insulator

8 Describe and explain the properties of simple molecular substances by a discussion of their structure and bonding.

A*

Examiner feedback

It is the weak forces *between* molecules that are overcome when covalent substances melt or boil. This is why they have low melting and boiling points. The strong covalent bonds *within* the molecules do not break.

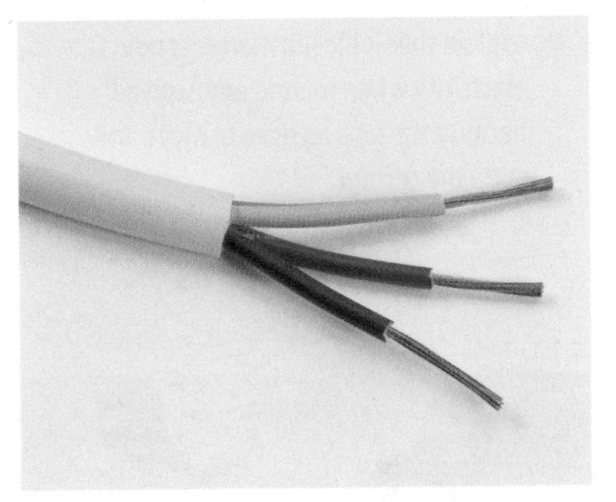

The molecules in the plastic coating on this electrical cable do not conduct electricity. This protects us from electric shocks.

Science skills

Table 4 Data on some simple molecular substances.

Substance	Relative mass of molecule (M_r)	Boiling point/°C
methane (CH_4)	16	−162
ethane (C_2H_6)	30	−89
propane (C_3H_8)	44	−42
butane (C_4H_{10})	58	0
pentane (C_5H_{12})	72	36
hexane (C_6H_{14})	86	69

a Plot a graph of boiling point against relative mass of molecules and draw a line of best fit.

b Describe the relationship between the mass of the molecule and the boiling point.

Ionic substances

Learning objectives

- describe the structure of ionic substances as a giant lattice of positive and negative ions
- explain that ionic substances have high melting and boiling points because there is a strong attraction between the positive and negative ions
- explain that ionic substances conduct electricity when melted or dissolved because the ions are free to move and carry the current.

Ions

Many compounds are made up of ions. Ions are particles that are electrically charged because they contain different numbers of protons and electrons. Ionic substances are compounds made from both metals and non-metals. S imple examples include sodium chloride (common salt), copper sulfate and calcium carbonate.

Ionic lattice

All ionic compounds are solids at room temperature. Inside the compound there are billions of ions from one edge of the solid right across to the other in all directions. The ions are packed together in an ordered, regular structure. This is a giant lattice.

Each ion is surrounded by several ions of opposite charge. Opposite electrical charges attract each other. So each ion is attracted by strong **electrostatic attraction** to all the ions surrounding it. This electrostatic attraction between positive and negative ions is known as ionic bonding.

The formula of an ionic compound tells us the ratio of the ions in the compound. For example, the formula NaCl tells us that there is one sodium ion (Na^+) for every chloride ion (Cl^-) in the structure. The formula Al_2O_3 means that there are two aluminium ions (Al^{3+}) for every three oxide ions (O^{2-}) in the structure.

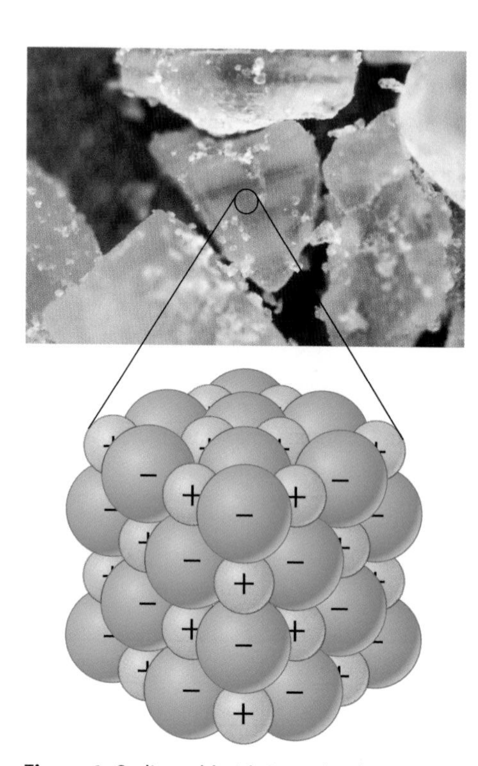

Figure 1 Sodium chloride is made of a giant lattice of positive and negative ions.

Melting and boiling points

The attraction between positive and negative ions is strong. In order to melt and boil ionic compounds lots of these strong attractive forces between ions of opposite charges have to be overcome. This takes a lot of energy and so ionic compounds have high melting and boiling points. This means that all ionic compounds are solids at room temperature.

Table 1 The melting and boiling points of some ionic compounds.

Substance	Sodium chloride	Magnesium bromide	Aluminium oxide	Potassium carbonate
Formula	NaCl	$MgBr_2$	Al_2O_3	K_2CO_3
Melting point/°C	808	711	2040	896
State at room temperature	solid	solid	solid	solid

Table 2 The structure of an ionic compound as a solid, a liquid and a gas.

Solid	Liquid	Gas

Examiner feedback

You need to be able to explain why molecular compounds are gases and liquids at room temperature, while ionic compounds are always solids at room temperature.

Electrical conductivity

Ions are electrically charged particles. An electric current is the flow of electrically charged particles. As a solid, the ions are vibrating in fixed positions and cannot move around. When the ionic compound is melted, the ions can move around and so can conduct electricity.

Many ionic compounds dissolve in water. When they dissolve the ions are free to move around in the solution. This means that ionic compounds that are soluble in water also conduct electricity when dissolved.

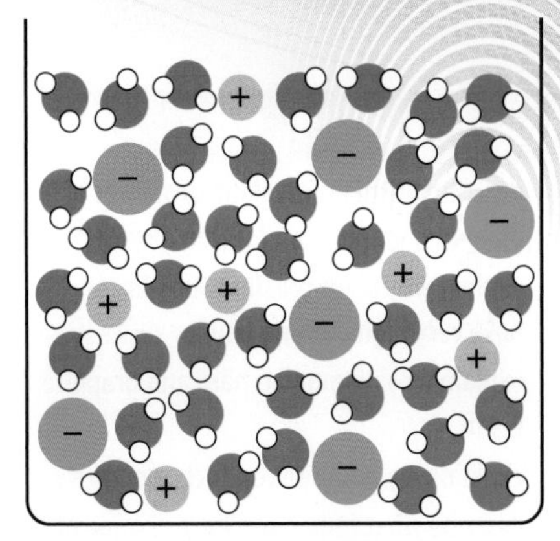

Figure 2 When sodium chloride dissolves in water the positive and negative ions separate and mix in with the water molecules.

Science in action

When ionic compounds are in solution they break down into simpler substances and can conduct electricity. This is called **electrolysis**. Electrolysis has many uses. The electrolysis of sodium chloride solution is a major industrial process producing hydrogen, chlorine and sodium hydroxide from the sodium chloride solution. Each of these products has many uses.

Examiner feedback

When ionic compounds conduct electricity, it is the ions that are moving to carry the current, not electrons.

Questions

1. What are ions?
2. What is ionic bonding?
3. Which of the following compounds have ionic structures? Zinc oxide (ZnO), butane (C_4H_{10}), nickel sulfate ($NiSO_4$), calcium chloride ($CaCl_2$), fluorine (F_2) and hydrazine (N_2H_4).
4. Calcium oxide is an ionic compound with the formula CaO and a melting point of 2600 °C. **(a)** Explain what this formula means. **(b)** Explain why calcium oxide has a high melting point.
5. Give the formula of the following ionic compounds (see lesson C2 1.3 for ion charges). **(a)** Potassium oxide. **(b)** Magnesium carbonate. **(c)** Aluminium sulfate.
6. Potassium iodide, KI, is an ionic compound. **(a)** Explain why it does not conduct electricity as a solid. **(b)** Explain why it does conduct electricity when it is melted. **(c)** Explain why it does conduct electricity when it is dissolved in water.
7. Look at the data in Table 3 and decide which of the substances are ionic compounds.

Table 3 Compounds and their properties.

Substance	A	B	C	D	E
Melting point/°C	763	3550	1453	1520	358
Boiling point/°C	1452	4827	2785	1680	684
Conductivity as solid	insulator	insulator	conductor	insulator	insulator
Conductivity as liquid	conductor	insulator	conductor	conductor	conductor

8. Describe and explain the properties of ionic substances with a discussion of their structure and bonding.

A*

Taking it further

The melting point of an ionic compound is a good indication of the strength of ionic bonding. The stronger the attraction between the ions, the higher the melting point. Two factors that affect the strength of ionic bonding are the size and charge of ions.

From the data below, how do you think the size and charge of an ion affect bond strength?

1 picometre (pm) = 1×10^{-12} m.

Substance	Formula	Melting point (°C)
Calcium oxide	CaO	2572
Magnesium oxide	MgO	2852
Sodium oxide	Na_2O	1132

Ion	Formula	Radius of ion (pm)
Calcium	Ca^{2+}	100
Magnesium	Mg^{2+}	72
Sodium	Na^{+}	102

Covalent structures

Learning objectives

- describe the giant covalent structures of diamond, graphite and silicon dioxide
- explain why giant covalent substances have very high melting points
- explain why diamond is hard and graphite is soft
- explain why graphite conducts electricity
- describe what fullerenes are and some of their uses
- explain that most fullerenes are simple molecular rather than giant covalent substances.

Different forms of carbon

Diamond and graphite are forms of carbon. Both have very high melting points, but diamond is hard and an insulator while graphite is soft and conducts electricity. An understanding of their structure is needed to explain the similarities and differences.

Giant covalent structures

Atoms can join together by covalent bonding to make molecules. In a simple molecular substance, there are millions of separate but identical molecules. For example, water is made up of water molecules (H_2O) in which two hydrogen atoms are joined to one oxygen atom. The molecules themselves are not joined together.

Atoms can also be joined by covalent bonding to form a giant covalent structure. In such a structure all the atoms are covalently bonded in a massive network. An example is diamond. In diamond, the carbon atoms are linked together in a giant lattice, which extends all through the material. Giant covalent substances are sometimes called **macromolecular** substances, but they are not molecules.

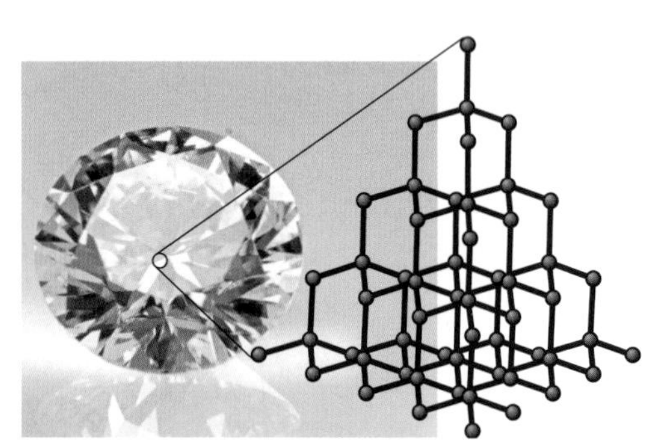

Figure 1 Diamond (a type of carbon) forms a giant covalent structure.

Other examples of giant covalent substances include graphite (another form of carbon), and silicon dioxide (silica, SiO_2). The formula of silicon dioxide shows that the ratio of silicon to oxygen atoms in the structure is 1 : 2.

Melting and boiling points

In order to melt or boil a giant covalent substance, lots of covalent bonds have to be broken. These bonds are very strong and need a lot of energy in the form of heat to break them. For this reason, giant covalent structures have very high melting points. Diamond, for example, melts at over 3500 °C.

Diamond and graphite

Although diamond and graphite are both forms of carbon and have giant covalent structures, they do not have the same properties.

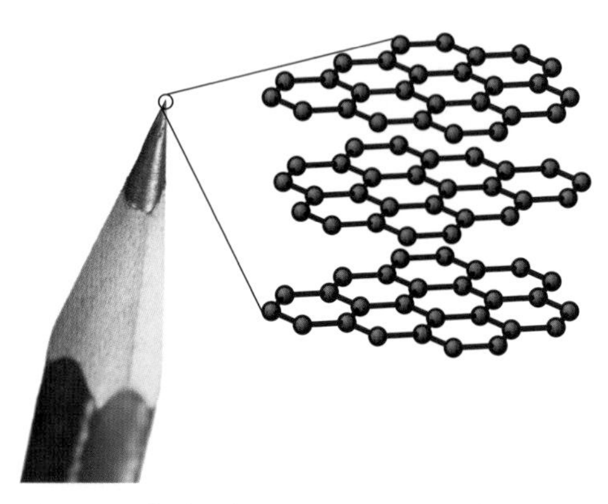

Figure 2 The lead in a pencil is made of graphite, another form of carbon. In graphite, the carbon atoms form layers.

Table 1 Diamond and graphite.

	Diamond	Graphite
Structure	Each C atom is joined to four others by covalent bonds.	Each C atom is joined to three others by covalent bonds. This forms layers, which are free to slide over each other. The layers are not bonded to each other.
Melting point	Very high, because lots of strong covalent bonds have to be broken.	Very high, because lots of strong covalent bonds have to be broken.
Hardness	Very hard, because the atoms are bonded in a rigid network.	Soft and slippery, because the layers can slide over each other (there are weak forces between the layers).
Electrical conductivity	Insulator, because there are no electrons free to move around.	Conductor, because there is one electron from each carbon atom free to move along the layers (it has delocalised electrons).

Fullerenes

In the 1980s a third form of carbon was discovered that forms large molecules, but not giant covalent structures. **Fullerenes** are molecules made up of linked carbon rings. The first to be discovered was a molecule containing 60 carbon atoms in a spherical shape. It was called buckminsterfullerene, after the American architect Buckminster Fuller who built domes with a similar structure.

The domes at the Eden Project have a hexagonal structure similar to that of fullerenes.

Since the original discovery, more fullerenes have been made in different sizes and shapes. They are proving very useful, for example for drug delivery into the body, in lubricants and as catalysts. **Nanotubes** (see lesson C2 2.6) are used for reinforcing materials such as tennis rackets.

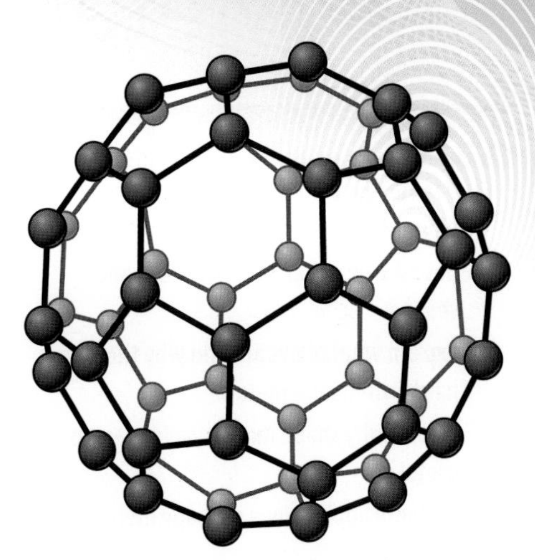

Figure 3 Fullerenes are made up of carbon rings joined together. This is buckminsterfullerene (C_{60}).

Examiner feedback

It is very important to realise that giant covalent structures are not made of molecules. Some students think they are big molecules, but they are not. In a substance made of molecules, each molecule has the same number of atoms in them, whereas in a giant covalent substance there is one continuous network of atoms linked by covalent bonds.

Route to A*

For diamond and graphite, it is important to be able to explain the similarities in (e.g. melting point) and differences between (e.g. hardness, conductivity) their properties by an understanding of their structures.

Science in action

Some drill bits are tipped with diamond. This is because diamond is the hardest known substance and so the drill bits can be used to drill through very hard substances.

Questions

1. **(a)** What is a covalent bond? **(b)** Which two types of structures contain atoms joined by covalent bonds?
2. Give three examples of substances that have a giant covalent structure.
3. Explain why all giant covalent substances have very high melting points.
4. Explain clearly why diamond is hard but graphite is soft and slippery.
5. Explain clearly why diamond is an insulator but graphite is a conductor.
6. Look at the data in Table 2 and decide which of the substances are giant covalent substances.

Table 2 Properties of substances.

Substance	A	B	C	D	E
Melting point/°C	3550	−36	564	3727	850
Conductivity as solid	insulator	insulator	conductor	conductor	insulator

7. **(a)** What are fullerenes? **(b)** Give some uses for fullerenes.
8. Carbon exists in different forms. Some are giant covalent structures, such as diamond and graphite, while fullerenes are simple molecular structures. Explain the difference between simple molecular and giant covalent structures.

Metals and alloys

Learning objectives

- describe how metals conduct heat and electricity
- describe how metals can be bent and shaped
- explain what alloys are and why they are harder than pure metals
- discuss why shape memory alloys are so useful
- evaluate applications of shape memory alloys.

Science in action

Duralumin is an alloy of aluminium used to make aircraft. Pure aluminium has a low density and resists corrosion, making it good for aircraft, but it is not strong enough on its own. By making it into a duralumin alloy, it is made strong enough for use in aircraft.

Thermal and electrical conductivity

There are over 100 elements, most of which are metals. As you learned in lesson C2 1.5, one very important property is that metals conduct heat and electricity. Saucepans are often made from metal because this allows heat to pass easily into the food. Electrical wires are made of metals, often copper, because metals are good electrical conductors.

Metals all have the same type of structure in which the outer shell electrons are delocalised, which means that the electrons can move through the metal. This is why metals conduct heat and electricity so well.

Alloys

Metals can be bent and hammered into shape. This is because the layers of metal atoms can slide over each other while maintaining the metallic bonding that holds the atoms together. However, pure metals are often too soft to be useful, because the layers slide over each other too easily.

Metals can be made harder by turning them into alloys. An **alloy** is a mixture of a metal with small amounts of other elements, usually other metals or carbon. The atoms of the added elements are a different size from the metal atoms. They 'jam up' the metal structure, stopping the layers of atoms from sliding past each other so easily. This makes alloys harder.

Examiner feedback

You will need to be able to explain why alloys are harder than the pure metal.

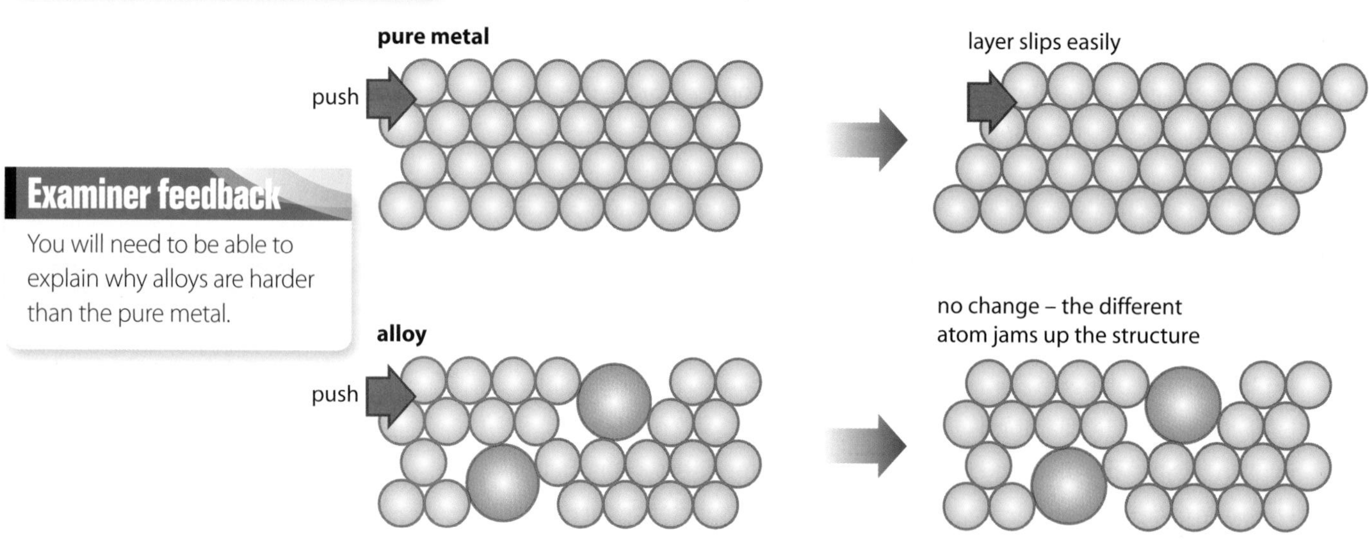

Figure 1 How an alloy can be harder than a pure metal.

The purity of gold is measured in carats. Pure gold is 24 carats. The purest gold used for most jewellery is 22 carats.

Steel is the most common alloy. It is a mixture of iron with small amounts of carbon or other elements. There are many different types of steel, containing different amounts of various alloying elements. Stainless steels, for example, contain chromium for rust resistance, while tool steels have added tungsten for hardness.

Pure gold is too soft to make practical jewellery with. Other metals, such as copper and silver, are added to make a harder and stronger alloy.

Shape memory alloys

A **shape memory alloy** is an alloy that can return to its original shape after being deformed. One example is **nitinol**, which is an alloy of nickel and titanium. Objects made from nitinol are cold-forged in a particular shape. If the object is bent or deformed in some way, then warming or heating it will return it to its original shape.

One common use for nitinol is in the wires in dental braces. In the warmth of the mouth, the wires try to return to their original size and shape. This pulls or pushes the teeth into position. Unlike stainless steel, nitinol braces do not have to be frequently replaced or tightened. The teeth are corrected faster and the braces are more comfortable.

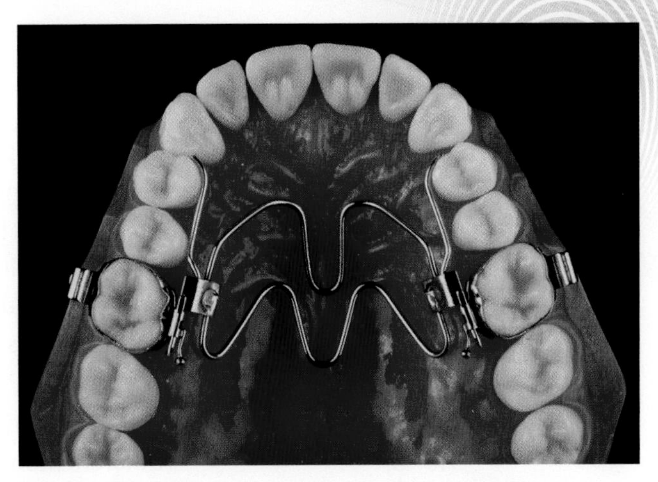

One use of the shape memory alloy nitinol is for dental braces.

Science in action

Another use of nitinol is to repair collapsed arteries. A squashed tube made of nitinol mesh, called a stent, is slid into the artery. As it warms up in the body it expands to its original size and opens the artery up.

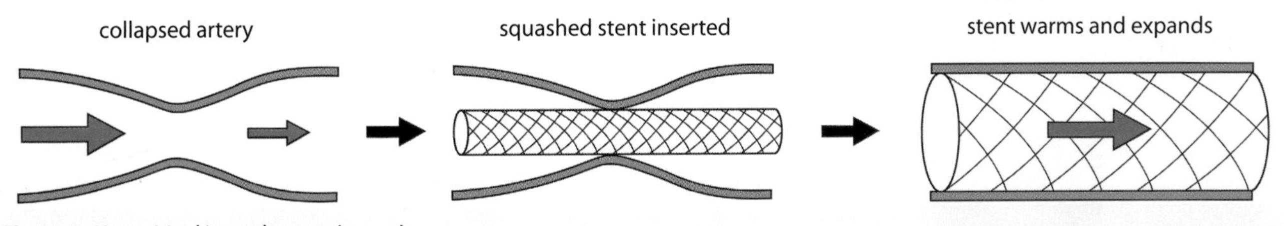

Figure 2 How nitinol is used to repair arteries.

Questions

1. Explain why metals conduct heat.
2. Explain why metals conduct electricity.
3. Explain why metals can be bent and hammered into shape.
4. Pure iron is too soft to be useful. Explain why pure iron is soft.
5. **(a)** The properties of iron are improved by making it into alloys called steel. What are alloys? **(b)** Why are alloys harder than pure metals?
6. Stainless steel is a very useful alloy of iron. Why is this alloy so useful compared with iron and other steels?
7. Tungsten is used to make steels which are used to make tools such as hammers. What property is needed for tool steels?
8. Nitinol is a shape memory alloy. What is a shape memory alloy? Give one example of a use of nitinol. Explain why nitinol is better in this use than a traditional metal or alloy.

A*

Science skills

Here are some data about the hardness of some gold alloys. The hardness is measured by the Vickers Hardness scale (HV), where the higher the value the greater the hardness.

Gold alloy	% gold	Hardness (HV)
24 carat	100	55
22 carat	91.7	138
21 carat	87.5	190
18 carat	75.0	212

a Plot a graph of hardness against percentage gold in the alloy and draw a line of best fit.

b Describe the relationship shown.

C2 2.5

Polymers with different properties

Learning objectives

- explain what affects the properties of a polymer
- describe and explain the differences in properties between thermosoftening and thermosetting polymers in terms of their structure.

$$\left[-\overset{\displaystyle H}{\underset{\displaystyle H}{\overset{|}{\underset{|}{C}}}} - \overset{\displaystyle OH}{\underset{\displaystyle H}{\overset{|}{\underset{|}{C}}}} - \right]_n$$

Figure 1 The structure of poly(ethenol), also known as polyvinyl alcohol or PVA.

One major use of poly(ethenol) is as an adhesive.

Science in action

A common use of thermosetting polymers is in places where the polymer may get hot. A thermosetting polymer will keep its shape, and not soften or melt on heating. For example, they are used in electrical fittings, such as plugs, sockets and light bulb holders, in some kitchen utensils and for kitchen worktops.

Types of polymer

There are many different **polymers**, with a wide range of uses. These range from simple plastic bags and bottles to non-stick coatings, breathable clothing fibres, water-absorbing hydrogels and shape memory polymers. The properties of polymers are determined by their structure. This can depend on the monomers the polymer is made from, or on the way the polymer is made.

Different starting materials

Different monomers produce polymers with different properties. For example, plastic shopping bags are often made from poly(ethene), common name polythene, made from ethene $CH_2{=}CH_2$. A different polymer, poly(ethenol) or PVA, is made from ethenol $CH_2{=}CHOH$. Poly(ethenol), unlike poly(ethene), dissolves in water. Some hospital laundry bags are made from poly(ethenol) so that the bag dissolves in the wash.

Changing the polymerisation conditions

Different polymers with different properties and uses can also be made from the same monomer. This is done by changing the reaction conditions when the polymer is made. For example, there are two forms of poly(ethene): high-density poly(ethene) (HDPE) and low-density poly(ethene) (LDPE). Different conditions of temperature and pressure and a different catalyst are used to make HDPE and LDPE.

Table 1 LDPE and HDPE.

Polymer	Reaction conditions	Difference in structure	Properties	Uses
low-density poly(ethene) LDPE	temperature: 200 °C pressure: 2000 atm catalyst: trace of oxygen	molecules are highly branched and therefore loosely packed	flexible, soft	bags, cling film
high-density poly(ethene) HDPE	temperature: 60 °C pressure: 2 atm catalyst: Ziegler-Natta (titanium based catalysts)	molecules are less branched and therefore tightly packed (this makes the polymer more rigid)	stiffer, harder	buckets, bottles

Thermosoftening and thermosetting polymers

In most polymers, the long polymer chains are not joined to each other. These polymers are called **thermosoftening** polymers (thermoplastics). They soften and melt on heating because there are weak forces between the polymer chains. This means that such polymers can be recycled, as they can be melted and remoulded.

In **thermosetting** polymers (thermosets) the polymer chains are joined to each other by strong covalent bonds called **cross-links**. These polymers do not soften or melt on heating and so cannot be recycled. Thermosetting polymers are usually hard and rigid because of the cross-links, whereas thermosoftening polymers are more flexible.

Table 2 Thermosoftening and thermosetting polymers.

Type of polymer	Thermosoftening	Thermosetting
Examples	poly(ethene), poly(propene), poly(chloroethene)	melanine, bakelite
Effect of heating	softens and melts	does not soften/melt (chars or decomposes if hot enough)
Structure	no cross-links between polymer chains (the chains are often tangled up)	cross-links between polymer chains

Practical

A type of slime can be made by reacting a solution of poly(ethenol) with borax. The borax forms cross-links between the poly(ethenol) chains, making it more viscous. The amount of borax used can be changed and the viscosity measured.

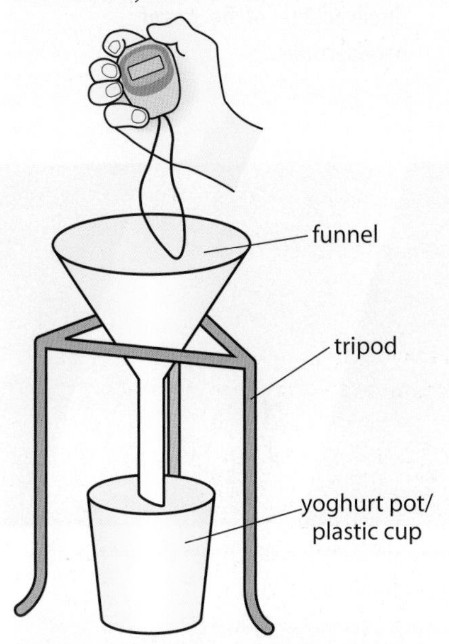

Figure 2 This apparatus can be used to measure the viscosity of slime by timing how long the slime takes to pass through the funnel.

Questions

1 What two things can be changed to make polymers with different properties?

2 **(a)** Why are plastic buckets not made out of low-density poly(ethene)?
(b) Why is cling film not made out of high-density poly(ethene)?

3 High- and low-density poly(ethene) are both made from ethene. What is done differently to make these two polymers with different properties?

4 Decide whether each of the following polymers is a thermosoftening or a thermosetting polymer. **(a)** Cyanoacrylate glue (superglue) sets to form a hard adhesive that does not soften or melt on heating. **(b)** Vulcanised rubber is used to make car tyres. It does not soften as the tyres become hot, so the tyres are long-wearing. **(c)** Polycarbonates are used instead of glass in many greenhouses. Polycarbonate melts at 267 °C.

5 Poly(propene) is a thermosoftening polymer. Melamine is a thermosetting polymer. Explain why poly(propene) can be recycled but melamine cannot.

6 How could you test a polymer to see if it is thermosoftening or thermosetting?

7 When poly(ethenol) reacts with borax, slime is formed. Explain why slime becomes more viscous if more borax is used.

8 Thermosoftening polymers and thermosetting polymers have different properties. Describe and explain these differences.

Route to A*

It is important to be able to explain the difference in structure and properties of thermosoftening and thermosetting polymers.

Examiner feedback

To explain the difference in properties between thermosoftening and thermosetting polymers, it is essential to discuss the interactions between polymer chains. In thermosoftening polymers there are weak forces between the chains, whereas in thermosetting polymers there are strong covalent bonds between the chains.

Nanoscience

Learning objectives

- describe what nanoscience is and what nanoparticles are
- explain why nanoparticles have different properties to the bulk material
- evaluate the advantages and disadvantages of the uses of nanoparticles.

What is nanoscience?

Nanoscience is a relatively new area of science. There are many potential benefits of nanoscience but also some concerns. You may already be using products containing nanoparticles.

Nanoscience is the study of **nanoparticles**. A **nanometre** (nm) is one millionth of a millimetre. A nanoparticle is a particle between 1 and 100 nm in size. These nanoparticles contain a few hundred atoms. They are too small to be seen, even with the most powerful light microscope. For comparison, a human hair is about 80 000 nm thick, many times thicker than a nanoparticle.

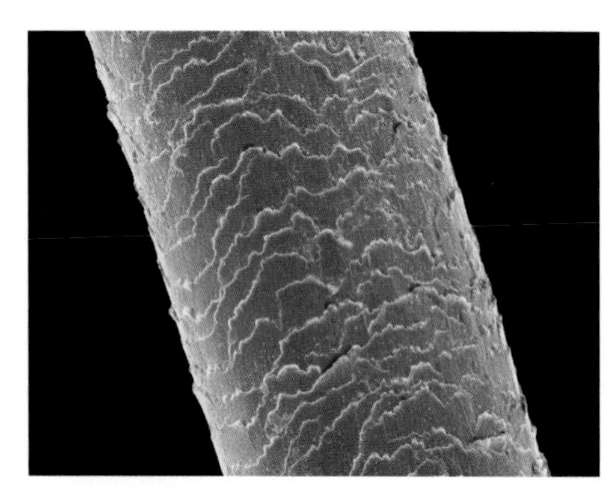

This human hair is 800 times thicker than the biggest nanoparticles.

Nanoparticles can have very different properties from the same materials in bulk. For example, nanoparticles:

- may be a different colour
- may have a different strength
- may react differently
- may have different electrical/thermal conductivity.

The main reason why nanoparticles have different properties from large pieces of the same material (in bulk) is that the nanoparticles have a much larger surface area to volume ratio. This means that a much higher fraction of the atoms are on the surface.

Examiner feedback

You may be asked to explain why nanoparticles have different properties to the same materials in bulk. The key is to discuss the idea that a greater fraction of the atoms is exposed at the surface in a nanoparticle than in the bulk material. An alternative way to describe this is to say that the nanoparticle has a greater surface area to volume ratio.

Uses of nanoparticles

Sunblock lotions contain titanium dioxide to block out harmful ultraviolet (UV) rays from the sun. In traditional sunblocks, the large particles of titanium dioxide used also reflected light, giving the sunblock a white appearance. Now many sunblocks use nanoparticles of titanium dioxide. They still block out UV rays, but the sunblock is colourless. Titanium dioxide and other nanoparticles are also starting to be used in cosmetics.

Bulk silver metal is very unreactive. However, nanoparticles of silver can kill bacteria. Clothes manufacturers incorporate silver nanoparticles into some clothes, to kill bacteria and prevent smells. Some deodorant sprays may soon contain silver nanoparticles.

Science skills

a Calculate the surface area to volume ratio of cubes with 1 cm, 2 cm and 3 cm side lengths.

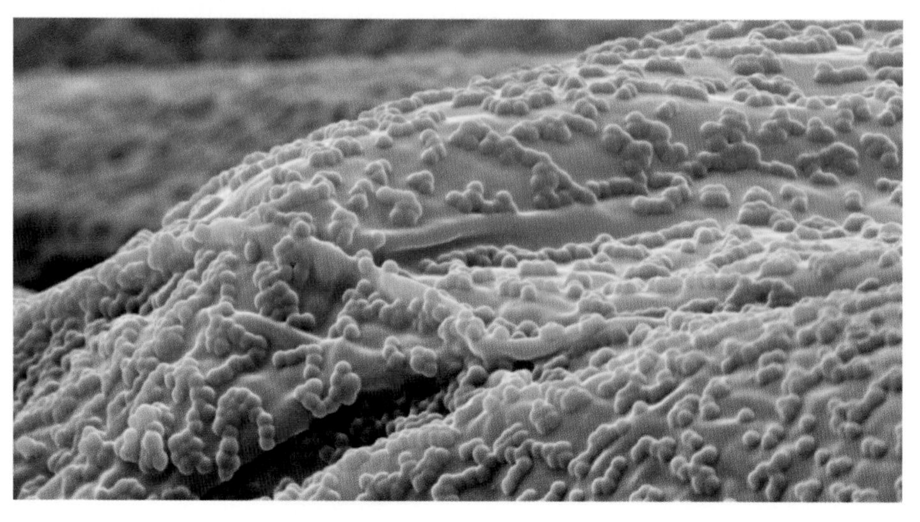

The beads on this cotton cloth are nanoparticles of a water-repellent polymer, creating waterproof cotton.

Nanoparticles are also being used as coatings. Self-cleaning windows have a coating of titanium dioxide nanoparticles, which causes rain to wash off dirt. Some new refrigerators have a coating of silver nanoparticles on the inside to kill bacteria.

Nanoparticles of gold appear red rather than the yellow colour of the bulk material. Gold is unreactive as a bulk material, but gold nanoparticles can act as a catalyst in some chemical reactions. A catalyst is a substance that speeds up a reaction without getting used up. Many other new catalysts are being made from nanoparticles.

Nanoparticles also have potential use as sensors. For example, gold nanoparticles have been used to detect toxic lead ions. These sensors can be very selective, only detecting a specific substance.

Carbon nanotubes are examples of fullerenes (see lesson C2 2.3) and have many potential uses. One property of these nanotubes is that they can be used as semiconductors. They have the potential to replace silicon in microchips, making computers faster and more powerful. They may also be useful in the treatment of cancer. Carbon nanotubes are very strong and are being used to make stronger but lighter construction materials. They have been used in tennis rackets and golf clubs, and could also be used to strengthen steel and concrete.

Concerns about nanoparticles

There are some concerns about the effects of nanoparticles on people. Given that nanoparticles have different properties to the bulk material, it is possible that some nanoparticles may be toxic even if the bulk material is not. For example, nanoparticles are usually more reactive than the bulk material and can pass through the skin and cell membranes. Further research is needed to find out what the effects might be.

Questions

1. Explain what the term nanoparticle means.
2. Explain what the term nanoscience means.
3. List some ways in which the properties of nanoparticles can differ from the bulk material.
4. Why do nanoparticles of a substance have different properties to the bulk material?
5. Carbon nanotubes are very strong and are used in making some golf clubs and tennis racquets. Carbon nanotubes are fullerenes. What are fullerenes?
6. Why are some clothes being made that contain nanoparticles of silver?
7. Give one example of the use of nanoparticles, and explain the benefit of using them, in each of the following areas: **(a)** construction materials **(b)** coatings **(c)** cosmetics such as sunblocks **(d)** catalysts **(e)** sensors **(f)** computers.
8. Describe some of the potential advantages and disadvantages of using nanoparticles.

A*

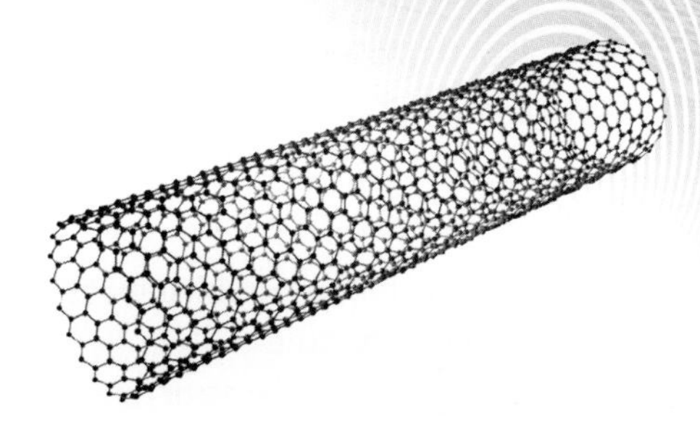

Figure 1 The structure of a single-walled carbon nanotube. Some types have a double wall.

Science in action

You can now buy pyjamas containing silver nanoparticles from high street shops. Some of the marketing for these pyjamas has been for hospital patients, with some evidence that they may help to protect patients from infection by the MRSA bacterium.

Science in action

Self-cleaning windows have a surface coating of nanoparticles. When there is daylight on the windows, energy from the Sun allows the nanoparticles to break down the dirt. The nanoparticles also affect rain so that rain droplets are spread out in a layer across the surface. This layer of water washes away the broken-down dirt.

Assess yourself questions

1 Which of the following substances are:

(a) elements? **(b)** compounds? *(2 marks)*

Mg, CO, Co, S_8, Br_2, H_2S, C_4H_{10}, Ar, C_2H_6O, silver nitrate, chromium, sulfur dioxide

2 Which of the following substances have:

- an ionic structure
- a simple molecular structure
- a metallic structure
- a giant covalent structure?

(a) magnesium oxide (MgO)
(b) lead (Pb)
(c) diamond (C)
(d) ammonia (NH_3)
(e) carbon disulfide (CS_2)
(f) nickel (Ni)
(g) silicon dioxide (SiO_2)
(h) buckminsterfullerene (C_{60})
(i) calcium bromide ($CaBr_2$)
(j) iron sulfide (FeS) *(10 marks)*

3 Calcium chloride ($CaCl_2$) is an ionic compound. It has a high melting point of 772 °C. It does not conduct electricity as a solid, but does conduct electricity when melted or dissolved in water.

(a) Explain why calcium chloride has a high melting point. *(2 marks)*

(b) Explain why calcium chloride does not conduct electricity as a solid but does when melted or dissolved in water. *(3 marks)*

(c) Copy and complete Figure 1 to give the electronic structure of the calcium ions in calcium chloride. *(1 mark)*

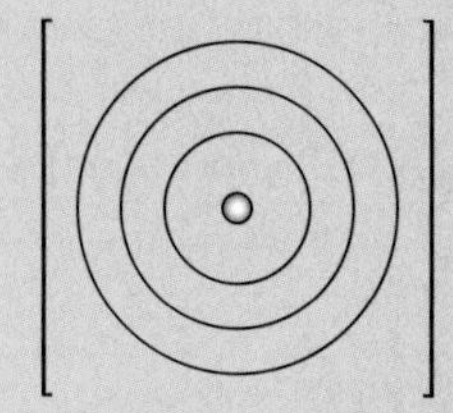

Figure 1 Electron shells.

(d) Copy and complete Figure 1 again to give the electronic structure of the chloride ions in calcium chloride. *(1 mark)*

4 Write the formula of the following ionic compounds.

(a) potassium oxide
(b) aluminium sulfide
(c) magnesium bromide
(d) sodium carbonate
(e) magnesium hydroxide
(f) silver nitrate
(g) iron(III) sulfate
(h) ammonium carbonate *(8 marks)*

5 (a) The electronic structure of the Na^+ ion is 2,8. Give the electronic structure of the following ions:

Mg^{2+}, F^-, S^{2-}, Al^{3+}, K^+, Cl^- *(6 marks)*

(b) What do the electronic structures of all common ions have in common? *(1 mark)*

6 Potassium reacts with iodine to form an ionic compound. Potassium is a group 1 element. Iodine is a group 7 element.

(a) What name are the elements in group 1 known by? Give the charge on the potassium ion and explain why it has this charge. *(2 marks)*

(b) What name are the elements in group 7 known by? Give the charge on the iodide ion and explain why it has this charge. *(2 marks)*

(c) Give the formula of potassium iodide. *(1 mark)*

7 (a) Draw 'stick' diagrams of the following molecules.

(i) H_2O **(ii)** CO_2 **(iii)** NH_3 **(iv)** N_2

(b) A 'stick' diagram is shown of each of the following molecules in Figures 2, 3 and 4. Draw a 'dot-cross' diagram for each to show the outer shell electrons.

(i)

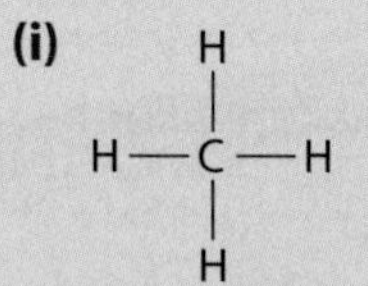

Figure 2 Methane.

(ii) Cl—Cl

Figure 3 Chlorine.

(iii) O═O

Figure 4 Oxygen.

8 Use the data in Table 1 to decide whether the following substances have simple molecular, ionic, metallic or giant covalent structures.

Table 1 Properties of substances.

	Melting point/°C	Boiling point/°C	Electrical conductivity as solid	Electrical conductivity as liquid	Electrical conductivity as solution
A	735	1435	✗	✓	✓
B	1610	2230	✗	✗	insoluble
C	7	81	✗	✗	insoluble
D	370	914	✗	✓	insoluble
E	114	183	✗	✗	✗
F	1455	2837	✓	✓	insoluble

(7 marks)

9 Graphite, diamond and buckminsterfullerene are all forms of the element carbon. Diamond and graphite have a giant covalent structure whereas buckminsterfullerene is a simple molecular substance with the formula C_{60}.

(a) Diamond and graphite both have very high melting points. Explain why. *(2 marks)*

(b) Graphite conducts electricity but diamond does not. Explain this difference. *(2 marks)*

(c) Diamond is very hard while graphite is soft. Explain this difference. *(3 marks)*

(d) Buckminsterfullerene has a lower melting point than diamond. Explain why. *(2 marks)*

10 **(a)** Iron is a metal. Metals have high melting points, conduct electricity as solids and can be bent and shaped.

(i) Explain why iron conducts electricity. *(2 marks)*

(ii) Explain why iron has a high melting point. *(2 marks)*

(iii) Explain why iron can be bent and shaped. *(2 marks)*

(b) Steels are alloys of iron. Alloys are harder than pure metals.

(i) What is an alloy? *(2 marks)*

(ii) Why are alloys harder than pure metals? *(2 marks)*

(c) The data in Table 2 shows how the strength of steel varies with the percentage of carbon in the steel. Strength is measured here as tensile strength which is where the steel is stretched until it breaks. The units are meganewtons (MN), where 1 MN = 1 000 000 N.

Table 2 Carbon in steel.

Percentage C in steel	0.15	0.3	0.6	0.8
Tensile strength/MN	15	39	69	92

(i) Plot a graph of strength against percentage of carbon. Draw a line of best fit. *(4 marks)*

(ii) Describe the relationship between the percentage of carbon and the strength of the steel. *(1 mark)*

(iii) The tensile strength of steel containing 3% carbon is 20 MN. Comment on this value in the light of the data on your graph. *(2 marks)*

(d) Shape memory alloys (SMAs) are examples of smart alloys. These alloys will return to their original shape when heated. A good example is nitinol, an alloy of nickel and titanium. Give a use of SMAs and describe how the special properties of SMAs allow this use. *(2 marks)*

11 Some clothes are now made containing nanoparticles of silver. Silver as a bulk material is an unreactive metal, but nanoparticles of silver have different properties and are added to clothes to kill bacteria. Some people are concerned that silver nanoparticles might be harmful even though bulk silver is safe.

(a) What are nanoparticles? *(1 mark)*

(b) Why do nanoparticles have different properties to bulk materials? *(1 mark)*

(c) Why might nanoparticles of silver be harmful even if bulk silver is safe? *(1 mark)*

12 **Table 3** Some information about four polymers.

Polymer	low-density poly(ethene)	high-density (poly)ethene	isotactic poly(propene)
Monomer	ethene	ethene	propene
Conditions under which polymer is made	temperature: 200 °C pressure: 2000 atm catalyst: trace of oxygen	temperature: 60 °C pressure: 2 atm catalyst: Ziegler-Natta	temperature: 60 °C pressure: 2 atm catalyst: Ziegler-Natta
Properties	flexible, soft	hard, strong	strong
Uses	bags	buckets, bottles	rope, carpet

(a) High-density poly(ethene) and low-density poly(ethene) have different properties and uses. High-density poly(ethene) is harder and stronger. In what way is their production different that leads to this difference in properties? *(1 mark)*

(b) High-density poly(ethene) and isotactic poly(propene) have different properties and uses. Isotactic poly(propene) is more flexible. In what way is their production different that leads to this difference in properties? *(1 mark)*

(c) The three polymers in Table 3 are thermosoftening. These polymers soften and melt on heating, and can be remoulded. Some other polymers are thermosetting, and do not soften or melt on heating and cannot be remoulded. Explain this difference in terms of structure and bonding.
In this question you will be assessed on using good English, organising information clearly and using specialist terms where appropriate. *(6 marks)*

13 **(a)** Describe some simple tests to show that a substance is ionic.

(b) Describe some simple tests to show that a substance is simple molecular.

C2 3.1

The structure and mass of atoms

Learning objectives

- describe the structure of atoms in terms of protons, neutrons and electrons
- explain what isotopes are
- explain how the masses of atoms are measured relative to ^{12}C atoms.

Examiner feedback

Mass numbers are whole numbers (integers) as they are the number of protons and neutrons in an atom. Relative atomic masses are not whole numbers as they are the average of the mass numbers, taking into account all the isotopes of an element.

The structure of atoms

Atoms are made of three smaller subatomic particles called **protons**, **neutrons** and **electrons**. Table 1 shows the relative mass and electric charge of these particles. Both mass and charge are measured relative to a proton, which is assigned a mass of 1 and a charge of +1.

Table 1 The mass and charge of the main subatomic particles.

Subatomic particle	Relative charge	Relative mass
proton	+1	1
neutron	0 (neutral)	1
electron	−1	0.0005 (negligible)

At the centre of the atom is a tiny **nucleus**. The nucleus contains all the atom's protons and neutrons, so most of the mass of the atom is concentrated there. The electrons move around the nucleus occupying energy levels (shells).

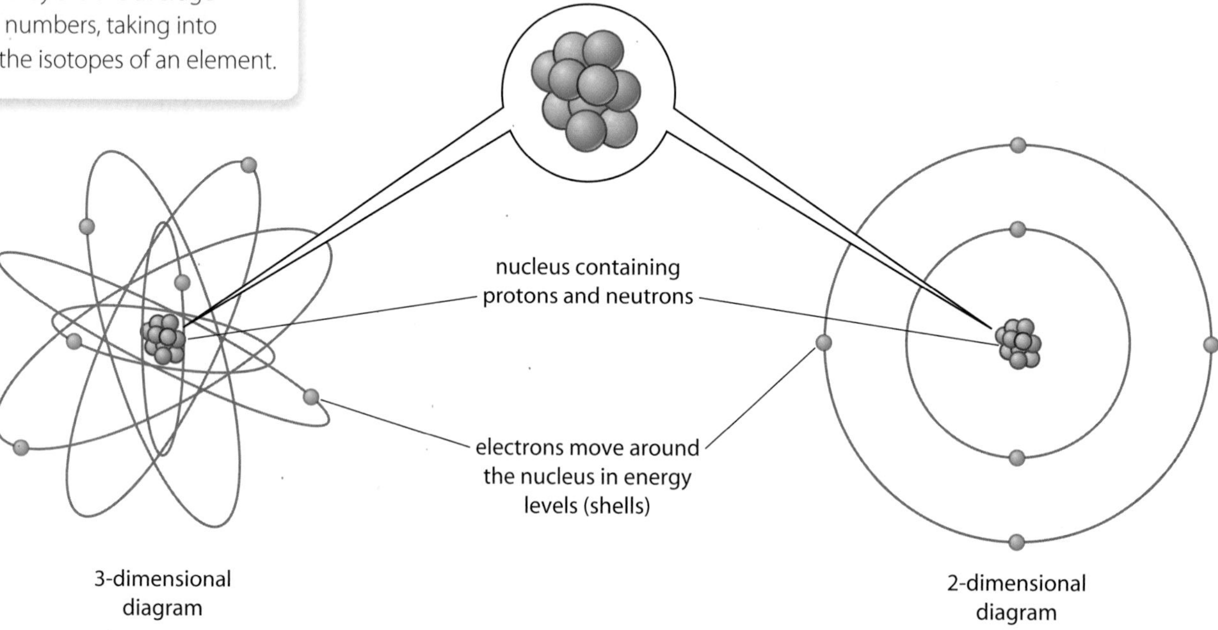

Figure 1 Close-up of an atom. Note that the sizes of the protons, neutrons and electrons are not to scale.

Examiner feedback

Electrons do have mass, but they are about 2000 times lighter than protons and neutrons. This means that over 99.9% of the mass of the atom is from the protons and neutrons, so when looking at the mass of atoms, the mass of the electrons is negligible.

Atomic mass

Different atoms have different masses because they contain different numbers of subatomic particles. The mass of atoms is important because it allows scientists to work out what mass of chemicals to use in reactions.

Atoms are described by their **mass number** and **atomic** (**proton**) **number**, which can be shown as follows.

Mass number = number of protons + number of neutrons ⟶ 23

$^{23}_{11}Na$

Atomic number (proton number) = number of protons ⟶ 11

Figure 2 Definitions of mass number and atomic number.

The mass number and atomic number can be used to work out how many protons, neutrons and electrons there are in an atom. Remember that atoms are neutral, so the number of electrons always equals the number of protons.

For example, in $^{23}_{11}Na$, the number of:

protons = atomic number = 11

neutrons = mass number − atomic number = 23 − 11 = 12

electrons = atomic number = 11

Isotopes

The number of protons in an atom determines which element it is. Atoms of different elements have different numbers of protons. For example, all atoms with 17 protons are chlorine atoms.

Isotopes are atoms of the same element with different mass numbers. In other words, they are atoms with the same number of protons but different numbers of neutrons. For example, there are two isotopes of chlorine, ^{35}Cl and ^{37}Cl.

Relative atomic mass (A_r)

The mass number of atoms tells us the relative mass of an individual atom. The mass of an atom is measured relative to one atom of the main carbon isotope, ^{12}C, which has a mass of exactly 12.

Most elements consist of a mixture of isotopes, so the relative mass of an element is an average figure based on the abundance of each isotope. It is called the **relative atomic mass** (**A_r**). For example, about three-quarters of chlorine atoms are ^{35}Cl with a relative mass of 35, while one-quarter of chlorine atoms are ^{37}Cl atoms with a relative mass of 37. The average (mean) mass of all chlorine atoms, the relative atomic mass (A_r), is 35.5.

Questions

1. Explain what the mass number and the atomic number of an atom represent.
2. Explain why all atoms are neutral.
3. Give the number of protons, neutrons and electrons in the following atoms.
   $^{19}_{9}F$ $^{40}_{18}Ar$ $^{39}_{19}K$
4. $^{79}_{35}Br$ and $^{81}_{35}Br$ are isotopes of bromine. **(a)** By considering the number of subatomic particles in each atom, explain why these two atoms are isotopes. **(b)** Explain why both atoms are bromine atoms. **(c)** About half of all bromine atoms are ^{79}Br and half are ^{81}Br. What is the relative atomic mass of bromine?
5. Identify the element: **(a)** whose atoms contain 16 protons; **(b)** whose atoms contain 19 electrons; and **(c)** with some atoms having atomic number 6 and mass number 14.
6. Explain why the mass number of an atom is a whole number but relative atomic mass is not.
7. Explain why the mass of electrons is regarded as being negligible when considering the mass of atoms.
8. Describe the structure of atoms in detail.

A*

Taking it further

Scientists now know that protons and neutrons are made up of even smaller particles. These smaller particles are called quarks.

Table 2 The two isotopes of chlorine.

Isotope	$^{35}_{17}Cl$	$^{37}_{17}Cl$
Atomic number	17	17
Mass number	35	37
Protons	17	17
Neutrons	18	20
Electrons	17	17

Science in action

Ideas about the structure of the atom have changed a great deal over time. The model changed as first electrons, then protons and the nucleus, energy levels and neutrons were discovered in turn. The current model is still developing as scientists learn more.

Taking it further

Isotopes of the same element have the same chemical properties. This is because they have the same electronic structure. For example, ^{35}Cl and ^{37}Cl atoms have the same chemical properties as they both have the electronic structure 2,8,7.

The mole

Learning objectives

- calculate the relative formula mass (M_r) of a substance
- calculate the percentage by mass of an element in a compound
- recall that the mass of 1 mole of a substance equals the M_r in grams
- use the equation: mass = $M_r \times$ moles.

Relative formula mass (M_r)

The **relative formula mass** (M_r) of a substance is the sum of the relative atomic masses (A_r) of all the atoms shown in its formula. For example, H_2O contains two hydrogen atoms ($A_r = 1$) and one oxygen atom ($A_r = 16$) and so has a relative formula mass of 18.

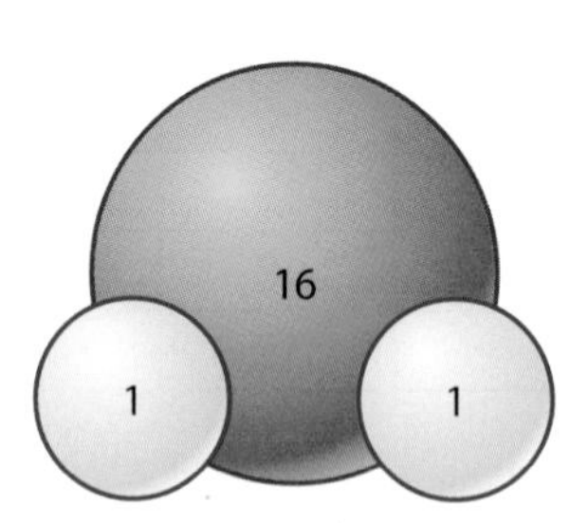

Calculate the M_r of water, H_2O.
(Relative atomic masses: H = 1, O = 16.)
$M_r = (2 \times 1) + 16 = 18$

Figure 1 Calculating the relative formula mass of water.

Percentage by mass of elements in compounds

Water has a relative formula mass of 18. Most of this mass is from the oxygen, which has a relative mass of 16. This means that 16 18ths, which is 88.9%, of the mass of the water molecule, is oxygen.

We can work out the percentage by mass of any element in a compound using the following equation:

$$\text{Percentage by mass of an element in a compound} = 100 \times \frac{\text{relative mass of all the atoms of that element}}{M_r}$$

e.g. percentage of O in $H_2O = 100 \times \frac{16}{18} = 88.9\%$

e.g. percentage O in $Ca(NO_3)_2 = 100 \times \frac{(6 \times 16)}{164} = 58.5\%$

This man is drinking about 15 moles of water molecules.

One mole of a substance

A pair is two of something. A dozen is twelve of something. And in science, a **mole** is 602 204 500 000 000 000 000 000 of something! Scientists count particles using moles. A mole of particles of many substances fits well into a boiling tube or beaker. For example, a mole of water molecules has a mass of 18 g. There are about 15 moles of water in a glass of water.

The number 602 204 500 000 000 000 000 000 was carefully chosen so that the mass of that number of particles equals the relative formula mass (M_r) in grams. For example, the M_r of water is 18 and so the mass of 1 mole of water molecules is 18 g. The M_r of carbon dioxide is 44, so the mass of 1 mole of carbon dioxide molecules is 44 g.

If 1 mole of water molecules has the mass 18 g, then 2 moles has the mass 36 g (2×18). The general equation is shown below. This can be remembered by thinking of 'Mr Moles'.

$$\text{Mass (g)} = M_r \times \text{moles}$$

When calculating the number of moles, the mass must be measured in grams. Some common conversion factors are shown in Table 1.

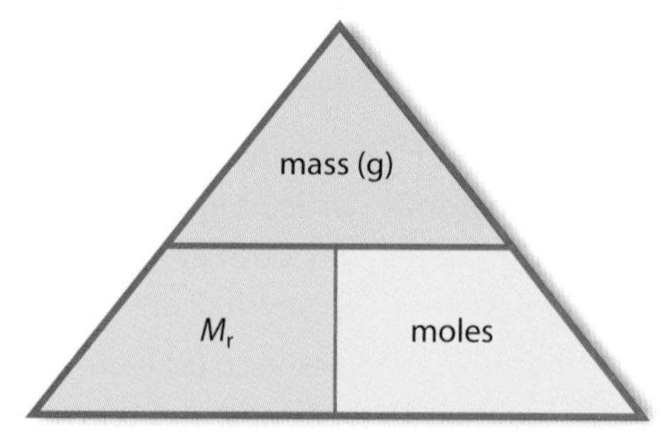

Figure 2 This triangle relates mass, M_r and moles. By covering the quantity you need to calculate you can find the equation to use.

Table 1 Conversion factors.

1 mg	= 0.001 g
1 kg	= 1000 g
1 tonne	= 1 000 000 g

Table 2 shows some example calculations using this equation.

Table 2 Calculations using mass (g) = $M_r \times$ moles.

Substance	Formula	M_r	Mass	Moles
Water	H_2O	18	36 g	$= \frac{\text{mass (g)}}{M_r} = \frac{36}{18} = 2$
Calcium carbonate	$CaCO_3$	100	1 tonne (1 million g)	$= \frac{\text{mass (g)}}{M_r} = \frac{1\,000\,000}{100} = 10\,000$
Carbon dioxide	CO_2	44	$= M_r \times$ moles $= 44 \times 0.25$ $= 11$ g	0.25
Ammonium nitrate	NH_4NO_3	80	$= M_r \times$ moles $= 80 \times 150$ $= 12\,000$ g	150
Unknown gas	unknown	$= \frac{\text{mass (g)}}{\text{moles}} = \frac{3.2}{0.10} = 32$	3.2 g	0.10
Unknown solid	unknown	$= \frac{\text{mass (g)}}{\text{moles}} = \frac{0.61}{0.005} = 122$	0.61 g	0.005

Questions

1 Write the following masses in grams: **(a)** 20 kg **(b)** 5 tonnes **(c)** 50 mg.

2 Calculate the relative formula mass (M_r) of the following substances. (A_r values: H = 1, N = 14, O = 16, Na = 23, Mg = 24, Al = 27, S = 32, Cl = 35.5, K = 39, Ca = 40, Fe = 56, Cu = 63.5.)
(a) O_2 **(b)** Na **(c)** S_8 **(d)** NH_3 **(e)** $FeCl_3$ **(f)** $Ca(OH)_2$
(g) $Mg(NO_3)_2$ **(h)** NH_4NO_3 **(i)** $Al_2(SO_4)_3$ **(j)** $CuSO_4.5H_2O$

3 Calculate the percentage by mass of the element shown in the following: (A_r values: H = 1, N = 14, O = 16, Mg = 24, S = 32, K = 39, Ca = 40, Fe = 56, Cu = 63.5, Br = 80.)
(a) Mg in $MgBr_2$ **(b)** Fe in Fe_2O_3 **(c)** N in NH_4NO_3
(d) O in $Ca(NO_3)_2$ **(e)** O in $CuSO_4.5H_2O$

4 Calculate the mass of 1 mole of the following substances (A_r values: H = 1, N = 14, O = 16, Na = 23, Mg = 24, Al = 27, S = 32, Cl = 35.5.)
(a) Mg **(b)** Cl_2 **(c)** Na_2O **(d)** $Al(OH)_3$ **(e)** $(NH_4)_2SO_4$

5 Rewrite the equation linking mass, moles and M_r to show: **(a)** moles as a subject **(b)** M_r as the subject.

6 How many moles are there in each of the following substances? (A_r values: C = 12, O = 16, Na = 23, S = 32, Fe = 56.)
(a) 8 g of SO_3 **(b)** 371 g of Na_2CO_3 **(c)** 1 kg of Fe_2O_3

7 What mass would the following quantities have? (Relative atomic masses: H = 1, N = 14, O = 16, Mg = 24, Cl = 35.5.)
(a) 2.5 moles of Cl_2 **(b)** 20 moles of NH_4OH **(c)** 0.02 moles of $Mg(NO_3)_2$

8 What is the M_r of the following substances? **(a)** Cyclohexane, for which 0.05 moles has a mass of 4.2 g. **(b)** Aspirin, for which 0.001 moles has a mass of 0.18 g.

A*

Examiner feedback

In calculations, a good general rule in chemistry is to work to 3 significant figures.

Science in action

The relative formula mass (M_r) of a substance can be measured by mass spectroscopy. This is an instrumental method of analysis (see lesson C2 3.6).

Examiner feedback

In all chemical calculations, most of the marks are for the method, rather than the final answer. Always show full working and write correct mathematical statements. If you make a slip in a calculation and get the wrong answer, you will get no marks if you have not shown your working, whereas you may only lose 1 mark if you have shown full working.

C2 3.3

Reacting-mass calculations

Learning objectives

- calculate the masses of substances involved in chemical reactions.

What a chemical equation tells us

It is important that scientists know what masses of chemicals to use in a reaction to make a certain amount of product. These masses can be calculated.

The equation below shows the reaction between hydrogen and nitrogen to make ammonia. It tells you how many particles of each substance are involved in the reaction: one mole of nitrogen (N_2) molecules reacts with three moles of hydrogen (H_2) molecules to make two moles of ammonia (NH_3) molecules.

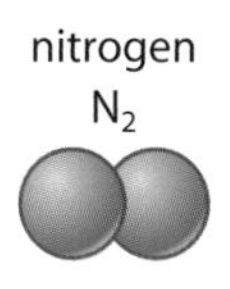

Figure 1 Nitrogen and hydrogen react together to make ammonia.

Calculating reacting masses

There are two general ways to calculate the mass of chemicals that react together or are produced in a reaction. One method uses ratios based on relative formula masses and the other method calculates the number of moles taking part in the reaction.

Table 1 Calculating masses.

Using ratios	Using moles
• Calculate the M_r of the substance whose mass you are given and the substance whose mass you are calculating. • Multiply these M_r values by the balancing numbers in the chemical equation to find the mass ratio. • Work out the mass that would react with/be made from 1 g of the substance whose mass you are given using the mass ratio. • Scale this up from 1 g to the mass you are given.	• Calculate the number of moles of the substance whose mass is given (moles = mass (g) ÷ M_r). • Use the chemical equation to work out how many moles of the substance asked about are used or made. • Calculate the mass of the substance asked for (mass (g) = M_r × moles).

Example 1

Iron is made when aluminium reacts with iron oxide. This reaction is used to weld railway lines together. What mass of aluminium is needed to react with 640 g of iron oxide? (Relative atomic masses: O = 16, Al = 27, Fe = 56.)

$Fe_2O_3 + 2\,Al \longrightarrow 2\,Fe + Al_2O_3$

M_r Fe_2O_3 = 160, Al = 27	Moles of $Fe_2O_3 = \frac{\text{mass (g)}}{M_r} = \frac{640}{160} = 4$
Fe_2O_3 reacts with 2 Al	Moles of Al = moles of $Fe_2O_3 \times 2 = 4 \times 2 = 8$
160 g of Fe_2O_3 reacts with 54 g (2 × 27) of Al	Mass of Al = M_r × moles = 27 × 8 = 216 g
1 g of Fe_2O_3 reacts with $\frac{54}{160}$ g of Al	
640 g of Fe_2O_3 reacts with $640 \times \frac{54}{160}$ g of Al = 216 g	

Example 2

Calcium hydroxide (slaked lime) is used by farmers to **neutralise** acidic soil. Calcium hydroxide is made by adding water to calcium oxide (quicklime). What mass of calcium hydroxide is made from 14 kg of calcium oxide? (Relative atomic masses: H = 1, O = 16, Ca = 40.)

$CaO + H_2O \longrightarrow Ca(OH)_2$

M_r CaO = 56, $Ca(OH)_2$ = 74 CaO makes $Ca(OH)_2$ 56 g of CaO makes 74 g of $Ca(OH)_2$ 1 g of CaO makes $\frac{74}{56}$ g of $Ca(OH)_2$ 14 kg of CaO makes $14 \times \frac{74}{56}$ kg of $Ca(OH)_2$ = 18.5 kg	Moles of CaO $= \frac{\text{mass (g)}}{M_r} = \frac{14\,000}{56} = 250$ Moles of $Ca(OH)_2$ = moles of CaO = 250 Mass of $Ca(OH)_2 = M_r \times$ moles $= 74 \times 250$ = 18 500 g = 18.5 kg

Example 3

Titanium is a metal. One of its uses is to make replacement hip joints. Titanium can be made by reacting titanium chloride with sodium. What mass of titanium chloride reacts with 460 g of sodium? (Relative atomic masses: Na = 23, Cl = 35.5, Ti = 48.)

$TiCl_4 + 4\,Na \longrightarrow Ti + 4\,NaCl$

M_r Na = 23, $TiCl_4$ = 190 4Na reacts with $TiCl_4$ 92 g (4 × 23) of Na reacts with 190 g of $TiCl_4$ 1 g of Na reacts with $\frac{190}{92}$ g of $TiCl_4$ 460 g of Na reacts with $460 \times \frac{190}{92}$ g of $TiCl_4$ = 950 g	Moles of Na $= \frac{\text{mass (g)}}{M_r} = \frac{460}{23} = 20$ Moles of $TiCl_4$ = moles of Na ÷ 4 = 20 ÷ 4 = 5 Mass of $TiCl_4 = M_r \times$ moles $= 190 \times 5 = 950$ g

Questions

1. Describe in words what this balanced equation means: $2H_2 + O_2 \longrightarrow 2\,H_2O$
2. Calcium oxide is formed when calcium carbonate is heated. What mass of calcium oxide is formed from 50 g of calcium carbonate? (Relative atomic masses: C = 12, O = 16, Ca = 40.)
   $CaCO_3 \longrightarrow CaO + CO_2$
3. What mass of hydrogen is produced when 96 g of magnesium reacts with hydrochloric acid? (Relative atomic masses: H = 1, Mg = 24.)
   $Mg + 2HCl \longrightarrow MgCl_2 + H_2$
4. What mass of oxygen reacts with 46 g of sodium? (Relative atomic masses: O = 16, Na = 23.)
   $4Na + O_2 \longrightarrow 2Na_2O$
5. What mass of water is formed when 1 kg of methane burns? (Relative atomic masses: H = 1, C = 12, O = 16.)
   $CH_4 + 2O_2 \longrightarrow CO_2 + 2H_2O$
6. Propane (C_3H_8) is often used as the fuel in gas fires and barbecues. What mass of oxygen is needed to burn 110 g of propane? (Relative atomic masses: H = 1, C = 12, O = 16.)
   $C_3H_8 + 5O_2 \longrightarrow 3CO_2 + 4H_2O$
7. What mass of aluminium reacts with 10.65 g of chlorine to make aluminium chloride? (Relative atomic masses: Al = 27, Cl = 35.5.)
   $2Al + 3Cl_2 \longrightarrow 2AlCl_3$
8. What mass of aspirin, $C_6H_4(OCOCH_3)COOH$, can be made from 1 g of salicylic acid, $C_6H_4(OH)COOH$? (Relative atomic masses: H = 1, C = 12, O = 16.)
   $C_6H_4(OH)COOH + (CH_3CO)_2O \longrightarrow C_6H_4(OCOCH_3)COOH + CH_3COOH$

Reaction yields

Learning objectives

- explain why reactions do not give 100% yield
- calculate the percentage yield for a reaction
- explain that some reactions are reversible.

When you bake a cake, some of the ingredients get left in the bowl or on the cake tin. The same sort of thing happens in a chemical reaction.

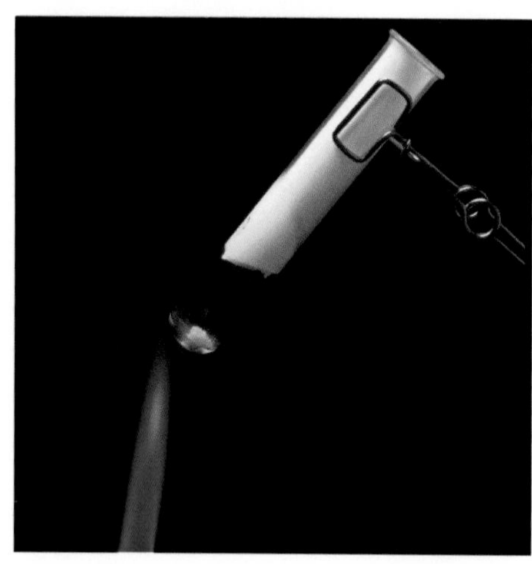

As the white ammonium chloride is heated at the bottom of the tube it breaks apart into ammonia and hydrogen, which react together to re-form ammonium chloride in the cooler parts of the tube.

Why don't reactions produce as much as expected?

If you react 4 g (2 moles) of hydrogen with 32 g (1 mole) of oxygen, you should in theory make 36 g (2 moles) of water. In a reaction, the atoms in the **reactants** are rearranged to make the **products**, so there should be the same mass of products as of reactants. However, if you do this reaction you will almost certainly end up with less than 36 g of water.

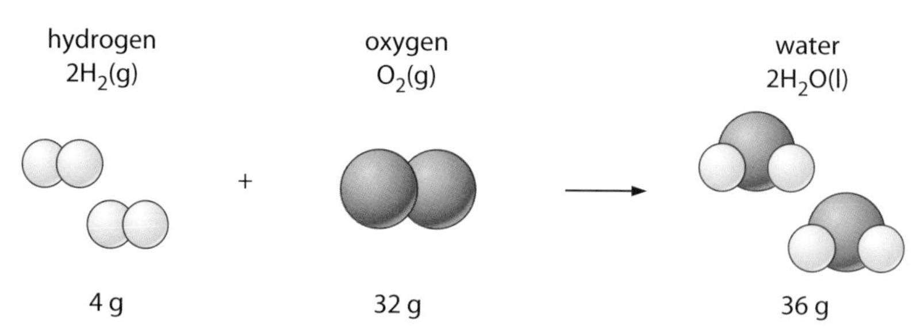

Figure 1 Chemical reaction to form water.

There are a number of reasons why you do not get as much product as you might expect.

1 When you carry out a reaction in the laboratory, some chemicals are lost along the way. For example, some material may get stuck to the sides of a test tube or flask.

2 Sometimes other reactions can take place as well the reaction you want. For example, when methane is burned not all of it reacts to form carbon dioxide because other reactions make carbon monoxide or carbon (soot) instead.

3 Some reactions are **reversible** and the products can turn back into the reactants.

Percentage yield

The amount of product obtained from a reaction is known as the yield. The **percentage yield** compares the amount that is actually produced with the amount expected in theory.

$$\%\ \text{yield} = \frac{\text{mass of product obtained}}{\text{maximum theoretical mass of product}} \times 100$$

For example, if you only obtain 27 g of water from the reaction in Figure 1, where the maximum theoretical yield is 36 g, then the percentage yield is 75%. This means that only 75% of the water that could have been produced has been obtained.

$$\text{Percentage yield} = \frac{27}{36} \times 100 = 75\%$$

The higher the percentage yield of a reaction, the better. A high yield means that fewer raw materials are used to make the same amount of product. This means that the process is more sustainable. Scientists work hard to improve their processes and techniques to obtain the highest possible yield.

Reversible reactions

Some chemical reactions are reversible. This means that both the forward and reverse reactions can take place. For example, the gases ammonia and hydrogen chloride react together to form the white solid ammonium chloride. On heating, ammonium chloride breaks up into ammonia and hydrogen chloride.

ammonia + hydrogen chloride $\rightleftharpoons$ ammonium chloride

$NH_3 + HCl \rightleftharpoons NH_4Cl$

Route to A*

Some questions may ask you to work out the theoretical yield before the actual yield. The calculation of a theoretical yield is a reacting mass calculation like in lesson C2 3.3. It is important to work carefully to perform this reacting mass calculation before calculating the percentage yield.

Questions

1 Why is it desirable to have a high yield in a chemical reaction?

2 Calculate the percentage yield in each of the following.

	Theoretical maximum mass of product	Mass of product obtained
(a)	10g	4g
(b)	2g	1.6g
(c)	50g	39g

3 When 6kg of hydrogen reacts with 28kg of ammonium, you might expect to produce 34kg of ammonia. In practice, only about 10kg is formed. **(a)** Explain why you would expect to produce 34kg of ammonium in this reaction. **(b)** Give three possible reasons why less than 34kg of product is formed. **(c)** Calculate the percentage yield for this reaction.

4 Quicklime (CaO) is made by the thermal decomposition of calcium carbonate:

$CaCO_3 \longrightarrow CaO + CO_2$

(a) Calculate the theoretical mass of quicklime that can be formed from 200g of calcium carbonate. (Relative atomic masses: C = 12, O = 16, Ca = 40.) **(b)** In a reaction, 108g of quicklime was obtained from 200g of calcium carbonate. Calculate the percentage yield.

5 Aluminium is extracted from aluminium oxide, which comes from the ore bauxite, by electrolysis:

$2Al_2O_3 \longrightarrow 4Al + 3O_2$

(a) Calculate the theoretical mass of aluminium that can be formed from 1kg of aluminium oxide. (Relative atomic masses: O = 16, Al = 27.) **(b)** In a reaction, 480g of aluminium was obtained from 1kg of aluminium oxide. Calculate the percentage yield.

6 Hydrogen is made by the reaction of methane with steam:

$CH_4 + H_2O \longrightarrow 3H_2 + CO_2$

(a) Calculate the theoretical mass of hydrogen that can be formed from 1kg of methane. (Relative atomic masses: H = 1, C = 12.) **(b)** In a reaction, 250g of hydrogen was obtained from 1kg of methane. Calculate the percentage yield.

7 Sulfur trioxide is formed by reaction of sulfur dioxide with oxygen:

$2SO_2 + O_2 \rightleftharpoons 2SO_3$

(a) Calculate the theoretical mass of sulfur trioxide that can be formed from 100g of sulfur dioxide. (Relative atomic masses: O = 16, S = 32.) **(b)** In a reaction, 105g of sulfur trioxide was formed from 100g of sulfur dioxide. Calculate the percentage yield. **(c)** Look at the equation for this reaction and give the main reason why the yield is less than 100%.

8 The production of aspirin has a 75% yield. Explain clearly what this means and suggest reasons why it is less than 100%.

Empirical formulae

Learning objectives

- explain what the empirical formula of a compound represents
- calculate the empirical formula of a compound from the mass or percentage of the elements in the compound.

What is an empirical formula?

All substances have an **empirical formula**. This represents the simplest whole number ratio of atoms (or ions) of each element in a substance. For example, the empirical formula of silicon dioxide (silica) is SiO_2. This means that the ratio of silicon (Si) atoms to oxygen (O) atoms is 1:2; there are twice as many oxygen atoms as silicon atoms.

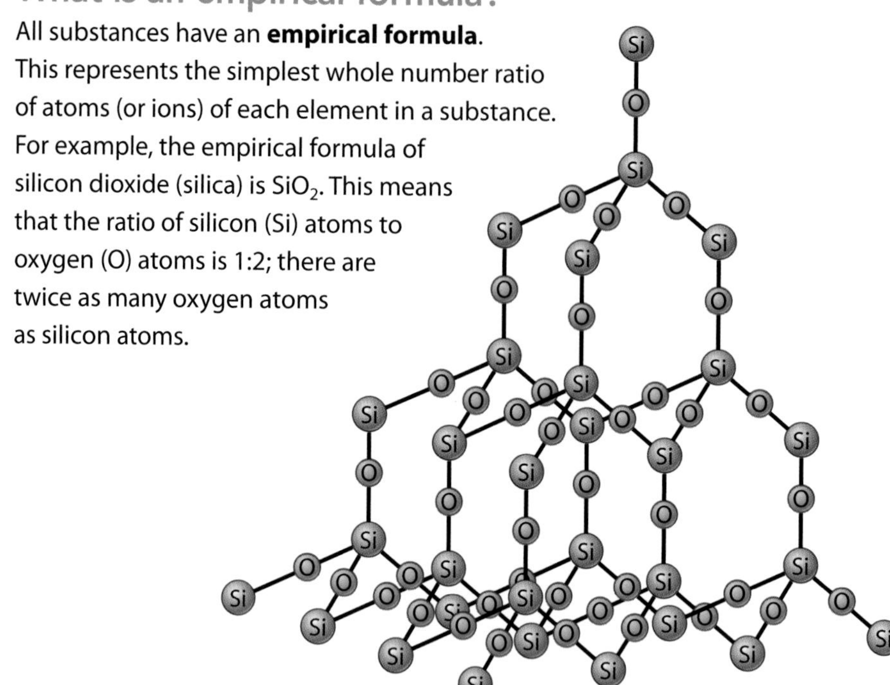

Figure 1 Silicon dioxide (empirical formula = SiO_2).

Examiner feedback

Only substances made of molecules can have a molecular formula. Substances like sodium chloride (NaCl), which is ionic, and copper (Cu), which is metallic, do not have a molecular formula as they are not made of molecules.

Substances that are made of molecules also have a **molecular formula**. This represents the number of atoms of each element in one molecule. For example, ethene has the molecular formula C_2H_4 – this means that there are two carbon (C) and four hydrogen (H) atoms in one molecule. It has the empirical formula CH_2, meaning that the simplest ratio of carbon to hydrogen atoms is 1:2.

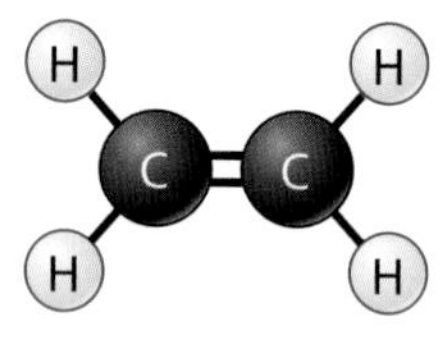

Figure 2 Ethene (molecular formula = C_2H_4, empirical formula = CH_2).

For some molecules, the molecular formula is the same as the empirical formula. For example, the molecular formula of water is H_2O; there are two hydrogen (H) atoms and one oxygen (O) atom in each molecule. The empirical formula is also H_2O, meaning that the simplest ratio of hydrogen to oxygen atoms is 2:1.

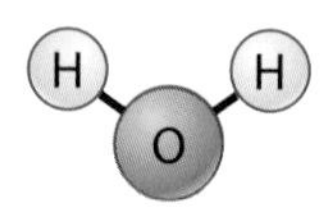

Figure 3 Water (molecular formula = H_2O, empirical formula = H_2O).

Calculating an empirical formula

Substances can be analysed to find which elements they are made from. This analysis gives the mass (or percentage by mass) of each element in the substance. The empirical formula can be calculated from this information as follows.

(a) Make a column for each element.

(b) Divide the mass (or percentage) of each element by its relative atomic mass (A_r).

(c) Simplify this ratio by dividing all the answers by the smallest answer.

(d) Find the simplest whole number ratio. The numbers are from real experiments so may not be exact whole numbers. However, you may need to multiply the answers by 2, 3 or 4 to get close to whole numbers.

(e) Write the empirical formula.

Science in action

Compounds can be analysed by a method called instrumental microanalysis. This provides information about which elements a compound contains and the percentage by mass of each element. This information is used to calculate the empirical formula.

Example 1

Analysis of a compound found in an iron **ore** determined that it contained 70% iron and 30% oxygen by mass. Find the empirical formula of the iron oxide in this ore. (A_r: Fe = 56, O = 16.)

	Fe	O
(a) Make a column for each element.	Fe	O
(b) Divide each percentage by the A_r.	$\frac{70}{56} = 1.25$	$\frac{30}{16} = 1.875$
(c) Simplify this ratio.	$\frac{1.25}{1.25} = 1$	$\frac{1.875}{1.25} = 1.5$
(d) Find the simplest whole number ratio.	$1 \times 2 = 2$	$1.5 \times 2 = 3$
(e) Write the empirical formula.	Fe_2O_3	

Example 2

Analysis of a compound found that it contained 2.4 g of carbon, 0.4 g of hydrogen and 3.2 g of oxygen. Find the empirical formula of this compound. (A_r: C = 12, H = 1, O = 16.)

	C	H	O
(a) Make a column for each element.	C	H	O
(b) Divide each mass by the A_r.	$\frac{2.4}{12} = 0.2$	$\frac{0.4}{1} = 0.4$	$\frac{3.2}{16} = 0.2$
(c) Simplify this ratio.	$\frac{0.2}{0.2} = 1$	$\frac{0.4}{0.2} = 2$	$\frac{0.2}{0.2} = 1$
(d) Find the simplest whole number ratio.	1	2	1
(e) Write the empirical formula.	CH_2O		

Questions

1. **(a)** Propene has the molecular formula C_3H_6. Explain what this means.
   (b) Propene has the empirical formula CH_2. Explain what this means.
2. A compound is found to contain 40% sulfur and 60% oxygen by mass. Find the empirical formula of the compound. (Relative atomic masses: O = 16, S = 32.)
3. A compound is found to contain 8.9 g of lead and 6.1 g of chlorine. Find the empirical formula of the compound. (Relative atomic masses: Cl = 35.5, Pb = 207.)
4. A compound is found to contain 18.2% potassium, 59.4% iodine and 22.4% oxygen by mass. Find the empirical formula of the compound. (Relative atomic masses: O = 16, K = 39, I = 127.)
5. A compound is found to contain 2.61 g of carbon, 0.65 g of hydrogen and 1.74 g of oxygen. Find the empirical formula of the compound. (Relative atomic masses: H = 1, C = 12, O = 16.)
6. A hydrocarbon is found to contain 81.8% carbon by mass. Find the empirical formula of the hydrocarbon. (Relative atomic masses: H = 1, C = 12.)
7. 1.0 g of aluminium reacts with chlorine to form 4.94 g of aluminium chloride. Calculate the empirical formula of aluminium chloride. (Relative atomic masses: Al = 27, Cl = 35.5.)
8. Some iron wool is placed in a crucible and heated until the mass stops increasing. Data is shown below. (Relative atomic masses: O = 16, Fe = 56.)
   Mass of empty crucible = 25.27 g
   Mass of crucible plus iron wool = 25.67 g
   Mass of crucible and contents at the end = 25.84 g
   (a) Why does the mass of the crucible and its contents increase?
   (b) Why was the crucible heated until the mass stopped increasing?
   (c) Calculate the empirical formula of the iron oxide formed.

What is the empirical formula of the iron oxide in this ore?

Practical

Magnesium oxide can be made by heating magnesium ribbon in a limited oxygen supply, using the apparatus in Figure 4. If the reactant and product are weighed, the empirical formula for magnesium oxide can be calculated.

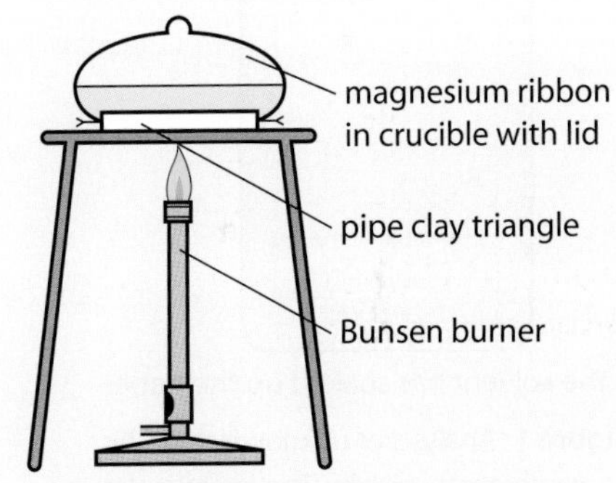

Figure 4 Finding the empirical formula of magnesium oxide.

Taking it further

For substances made from molecules, the molecular formula can be worked out using the relative formula mass and the empirical formula.

Analysis

Learning objectives

- explain why instrumental methods are used to analyse substances
- explain how artificial colourings in foods can be analysed
- describe how gas chromatography can be used to separate mixtures
- describe how substances can be analysed by mass spectroscopy.

Why analyse?

It is very important that we can analyse substances to identify what they contain and to measure quantities. Substances that are commonly tested include samples of air, water, food and medicines. Air and water may be tested to check levels of pollutants, while food and medicines are analysed to make sure they do not contain anything that could be harmful.

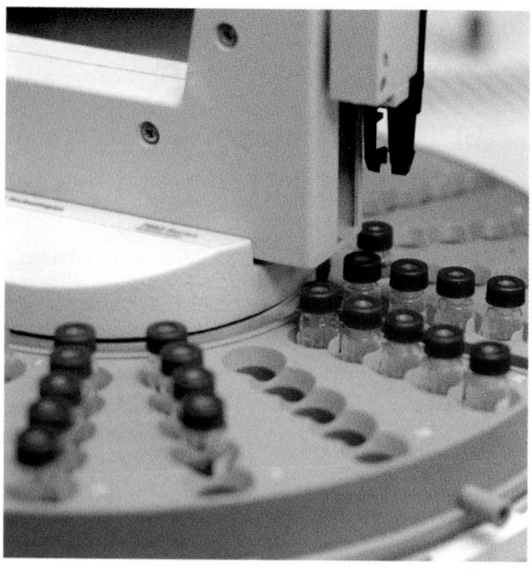
Instrumental analysis.

There are many different methods of analysis. They include different types of chromatography.

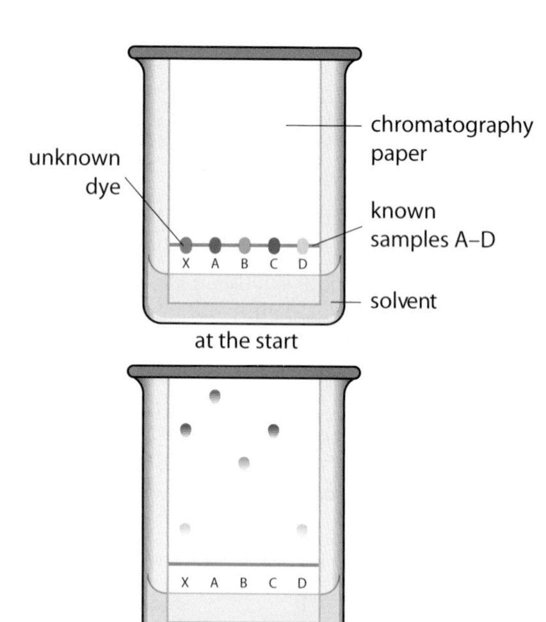

Figure 1 Analysis of unknown dye X by paper chromatography. The analysis shows that X contains two separate dyes, the known substances C and D.

Paper chromatography

Paper chromatography has been used for many years to analyse coloured substances. Chromatography separates the different dyes or **pigments** in a coloured substance.

Artificial colourings added to foods can be analysed in this way. A sample of the colour from the food is placed on a piece of chromatography paper along with samples of known dyes, then a solvent is added. The solvent soaks up the paper, taking the dyes with it. The more soluble a colouring is, the further up the paper it travels. Different colourings move different distances up the paper.

Instrumental methods

Many **instrumental techniques** have been developed by scientists to analyse substances. In an instrumental method, a sample is placed inside a device that performs some sort of analysis. These methods are fast, accurate and very sensitive. They can be used to test very small samples.

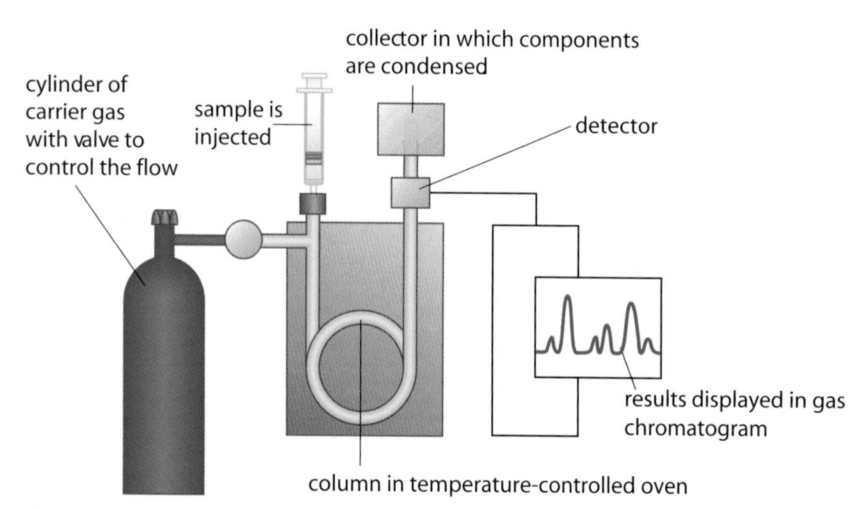

Figure 2 Gas chromatography.

Gas chromatography

A very common instrumental method is **gas chromatography**. The sample is injected into the machine and vaporised. The sample passes through a long column packed with a solid that is wound into a coil. An inert (unreactive) gas, such as nitrogen, is passed through the column to move the sample through. Different substances in the mixture travel through the column at different speeds. This means that they reach the end of the column at different times and so are separated.

The time taken for a substance to reach the detector at the end of the column is called its **retention time**. A gas chromatogram is produced that shows how many compounds are in the sample and the retention time of each one. The retention time can help to identify the substance. In the example in Figure 3, three compounds have been separated and detected.

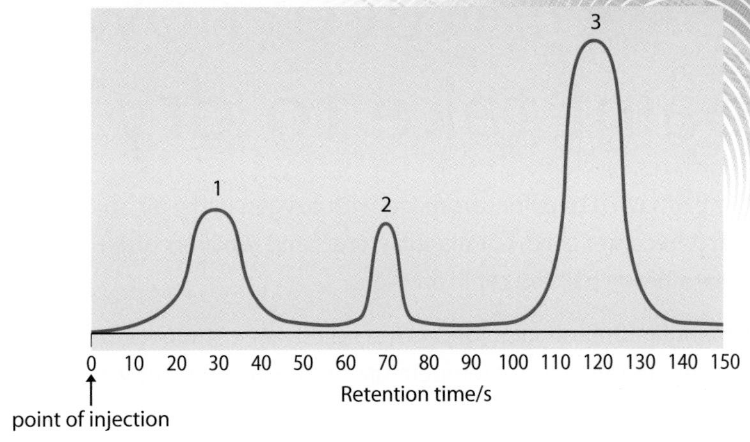

Figure 3 This sample contains three substances.

Mass spectroscopy

Another instrumental technique is **mass spectroscopy**. A mass spectrometer can identify tiny amounts of a substance very quickly and accurately. It does this by measuring the mass of the particles in the substance.

In mass spectroscopy the molecules lose an electron to form a **molecular ion**. The relative formula mass of the substance equals the mass of the molecular ion. Often, the molecular ion breaks apart and other lighter fragments are detected as well.

A mass spectrometer is often attached to a gas chromatography machine. After a sample has been separated by chromatography, the mass spectrum of each compound can be recorded, allowing it to be identified.

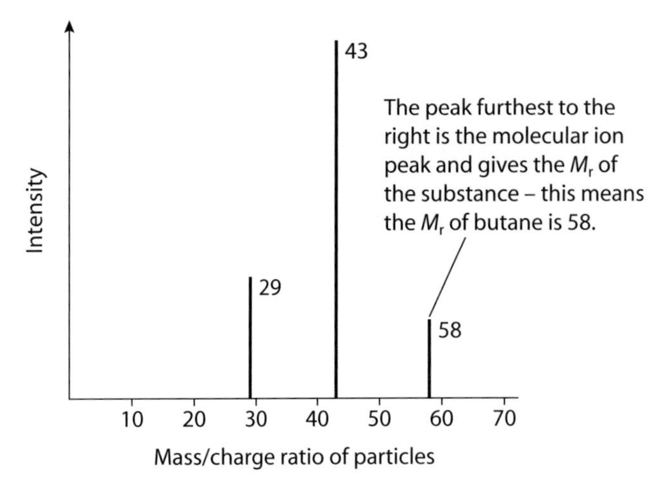

Figure 4 An example of a mass spectrum. This is butane.

Questions

1 Look at Figure 5. On the left it shows the chromatogram of the artificial colourings used to colour some sweets. On the right it shows the chromatogram of the sweets themselves. **(a)** Which sweets contain one colouring? **(b)** Which sweet contains the most colourings and which ones does it contain? **(c)** Which colourings are in the yellow sweets?

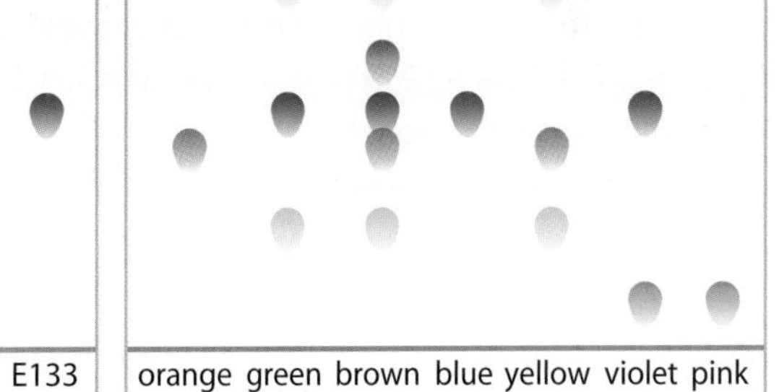

Figure 5 Chromatograms of sweets.

2 How does paper chromatography separate the substances in a mixture?

3 How does gas chromatography separate the substances in a mixture?

4 Look at the gas chromatogram in Figure 3. **(a)** Record the retention time of each substance. **(b)** Use the data table opposite to identify which substances are in the sample. **(c)** Which substance is there most of in the sample?

Substance	P	Q	R	S	T
Retention time / s	70	135	56	118	30

5 In mass spectroscopy, a molecular ion is formed. What is a molecular ion?

6 What key piece of information do we find from mass spectroscopy in order to help identify a compound?

7 How can mass spectroscopy be used with gas chromatography to help identify substances?

8 What is instrumental analysis, what is it used for and why is it so useful?

ISA practice: how long does paint take to dry?

The paint used to paint cars reacts with oxygen in the air, so that it becomes solid. Car manufacturers and repairers often place a newly painted car in an oven.

A manufacturer of paint for cars has asked some students to investigate the effect of temperature on the time it takes for paint to become solid.

Section 1

1 Write a hypothesis about how you think temperature will affect the rate of reaction. Use information from your knowledge of rates of reaction to explain why you made this hypothesis. *(3 marks)*

2 Describe how you could carry out an investigation into this factor.

You should include:

- the equipment that you would use
- how you would use the equipment
- the measurements that you would make
- how you would make it a fair test.

You may include a labelled diagram to help you explain the method.

In this question you will be assessed on using good English, organising information clearly and using specialist terms where appropriate. *(6 marks)*

3 Think about the possible hazards in the investigation.

(a) Describe *one* hazard that you think may be present in the investigation. *(1 mark)*

(b) Identify the risk associated with the hazard that you have described, and say what control measures you could use to reduce the risk. *(2 marks)*

4 Design a table that you could use to record all the data you would obtain during the planned investigation. *(2 marks)*

Total for Section 1: 14 marks

Section 2

Study Group 1 was two students, who carried out an investigation into the hypothesis. They used car spray paint and decided the paint was solid when a match would no longer scratch the surface. Figure 1 shows their results.

Paint samples dried at different temperatures

20°C: paint solidified after 120 min

50°C: paint solidified after 50 min

70°C: paint solidified after 34 min

90°C: paint solidified after 26 min

110°C: paint solidified after 20 min

Figure 1 Study Group 1's results.

5 **(a)** Plot a graph of these results. *(4 marks)*

(b) What conclusion can you make from the investigation about a link between the temperature and the rate of reaction?
You should use any pattern that you can see in the results to support your conclusion. *(3 marks)*

(c) Look at your hypothesis, the answer to question 1. Do the results support your hypothesis?

Explain your answer. You should quote some figures from the data in your explanation. *(3 marks)*

Below are the results from three more study groups.

Study Group 2 was another group of two students. Figure 2 shows their results.

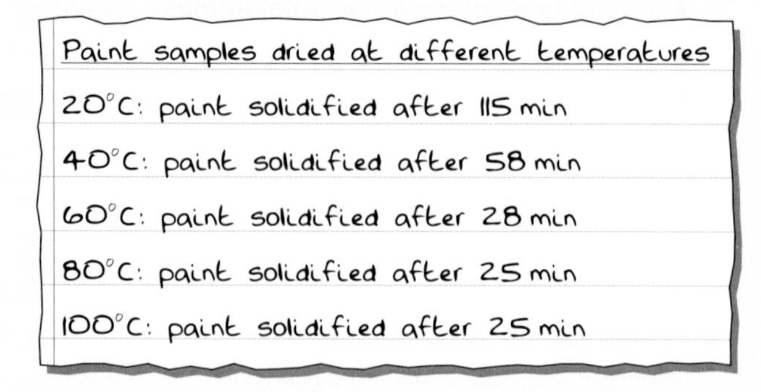
Paint samples dried at different temperatures

20°C: paint solidified after 115 min

40°C: paint solidified after 58 min

60°C: paint solidified after 28 min

80°C: paint solidified after 25 min

100°C: paint solidified after 25 min

Figure 2 Study Group 2's results.

Study Group 3 was a third group of students. Their results are given in Table 1.

Table 1 Results from Study Group 3.

Temperature at which paint is solidified/°C	Time for paint to solidify/min			
	Test 1	Test 2	Test 3	Mean of tests
15	145	142	138	142
30	95	98	92	95
45	75	102	73	83
60	38	36	34	36
75	29	26	27	27

Study Group 4 is a group of researchers who looked on the internet and found a graph showing how temperature affects the rate of reaction between a metal and sulfuric acid (Figure 3).

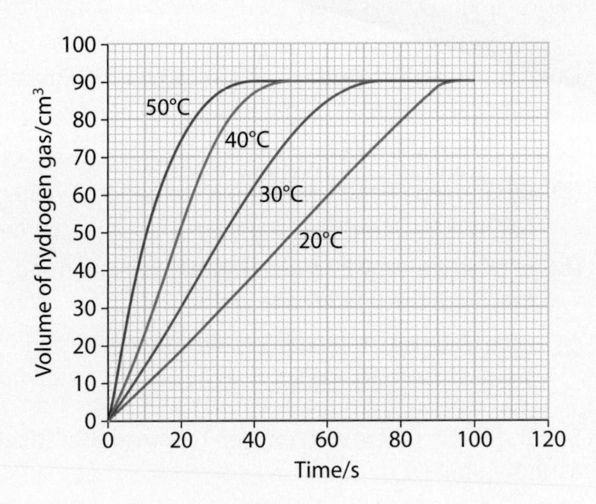

Figure 3 Graph found by Study Group 4.

6 **(a)** Draw a sketch graph of the results from Study Group 1. *(3 marks)*

(b) Look at the results from Study Groups 2 and 3. Does the data support the conclusion you reached about the investigation in question 5(a)? Give reasons for your answer. *(3 marks)*

(c) The data contain only a limited amount of information. What other information or data would you need in order to be more certain whether the hypothesis is correct or not?
Explain the reason for your answer. *(3 marks)*

(d) Look at Study Group 4's results. Compare them with the data from Study Group 1. Explain how far the data supports or does not support your answer to question 5(b). You should use examples from Study Group 4's results and from Study Group 1. *(3 marks)*

7 **(a)** Compare the results of Study Group 1 with Study Group 2. Do you think that the results for Study Group 1 are *reproducible*?
Explain the reason for your answer. *(3 marks)*

(b) Explain how Study Group 1 could use results from other groups in the class to obtain a more *accurate* answer. *(3 marks)*

8 Applying the results of the investigation to a context.
Suggest how ideas from the original investigation and the other studies could be used by the manufacturers to decide on the best temperature at which to operate the car paint oven. *(3 marks)*

Total for Section 2: 31 marks

Total for the ISA: 45 marks

Assess yourself questions

1 Atoms of fluorine have an atomic number of 9 and a mass number of 19. Explain what the terms *atomic number* and *mass number* mean. *(2 marks)*

2 An atom of phosphorus can be represented as

$^{31}_{15}P$

(a) What is the atomic number of this atom? What is the mass number? *(2 marks)*

(b) How many protons, neutrons and electrons are there in this atom? *(3 marks)*

3 Magnesium atoms consist of a mixture of isotopes. The main isotopes are ^{24}Mg, ^{25}Mg and ^{26}Mg.

(a) How many protons do these three isotopes contain? How many neutrons do these contain? *(2 marks)*

(b) What are isotopes? *(2 marks)*

(c) **(i)** The relative atomic mass of magnesium is 24.3. Explain why this is not a whole number. *(1 mark)*

(ii) The masses of all atoms are measured relative to another atom. Which atom is this? *(1 mark)*

4 Identify the following atoms by giving the symbol with mass number (e.g. ^{24}Mg). You will need to use the periodic table to help.

(a) An atom with 13 protons and 14 neutrons.

(b) An atom with 53 protons and 74 neutrons.

(c) An atom with 35 protons and 44 neutrons.

(d) An atom with atomic number 18 and mass number 40. *(4 marks)*

5 Using the periodic table calculate the relative formula mass, M_r, of the following substances.

(a) CO_2

(b) H_2SO_4

(c) Br_2

(d) $Al(OH)_3$

(e) $Fe_2(SO_4)_3$ *(5 marks)*

6 Calculate the percentage by mass of:

(a) Mg in $MgBr_2$

(b) Fe in Fe_2O_3

(c) N in NH_4NO_3

(d) O in $Fe(OH)_3$

(e) O in $Ca(NO_3)_2$ *(5 marks)*

7 Calculate the mass of 1 mole of:

(a) Na

(b) Na_2CO_3

(c) $(NH_4)_2SO_4$ *(3 marks)*

8 Propane (C_3H_8) is often used as the fuel in gas fires and barbecues. What mass of oxygen is needed to burn 110 g of propane? *(3 marks)*

$C_3H_8 + 5O_2 \longrightarrow 3CO_2 + 4H_2O$

9 Ammonium sulfate is a very good fertiliser. It is made by the reaction of ammonia with sulfuric acid. How much ammonium sulfate fertiliser can be made from 1 kg of ammonia? *(3 marks)*

$2NH_3 + H_2SO_4 \longrightarrow (NH_4)_2SO_4$

10 Aluminium is made by the electrolysis of aluminium oxide. What mass of aluminium is formed from 100 g of aluminium oxide? *(3 marks)*

$Al_2O_3 \longrightarrow 2Al + 3O_2$

11 A common way to make hydrogen is to react methane with steam. What mass of hydrogen can be formed from 40 g of methane? *(3 marks)*

$CH_4 + H_2O \longrightarrow CO + 3H_2$

12 Quicklime (calcium oxide) is made by heating limestone (calcium carbonate) in a lime kiln.

$CaCO_3 \longrightarrow CaO + CO_2$

(a) What mass of calcium oxide is formed from 150 g of calcium carbonate? *(3 marks)*

(b) In practice, only 80 g of calcium oxide is formed. Calculate the percentage yield. *(1 mark)*

(c) Give *three* possible reasons why the percentage yield is less than 100%. *(3 marks)*

13 Sulfur dioxide reacts with oxygen to form sulfur trioxide as shown below.

$2SO_2 + O_2 \longrightarrow 2SO_3$

(a) What mass of sulfur trioxide is formed from 32 g of sulfur dioxide? *(3 marks)*

(b) In practice, only 35 g of sulfur trioxide is formed. Calculate the percentage yield. *(1 mark)*

14 **(a)** The molecular formula of glucose is $C_6H_{12}O_6$. The empirical formula of glucose is CH_2O. What is an empirical formula? *(2 marks)*

(b) The molecular formulae of some substances are shown. Give the empirical formula of these substances. *(5 marks)*

(i) benzene = C_6H_6

(ii) hydrazine = N_2H_4

(iii) propane = C_3H_8

(iv) ethanoic acid = $C_2H_4O_2$

(v) butanol = $C_4H_{10}O$

15 Use the data to calculate the empirical formula of the following compounds.

(a) N = 30.4%, O = 69.6% *(3 marks)*

(b) Fe = 1.75 g, O = 0.75 g *(3 marks)*

(c) C = 82.5%, H = 17.2% *(3 marks)*

(d) C = 38.7%, H = 16.1%, N = 45.2% *(3 marks)*

(e) P = 4.75 g, H = 0.46 g *(3 marks)*

16 A sample of gallium was burned in oxygen to form gallium oxide. 1.00 g of gallium burned to form 1.34 g of gallium oxide. Calculate the empirical formula of gallium oxide. *(4 marks)*

17 An organic compound containing carbon, hydrogen and oxygen only was found to contain 48.6% carbon and 8.1% hydrogen. Find the empirical formula of this compound. *(2 marks)*

18 Paper chromatography using ethanol as solvent was carried out to find the pigments in a paint. The results are shown in Figure 1. Which pigments does the paint contain? *(1 mark)*

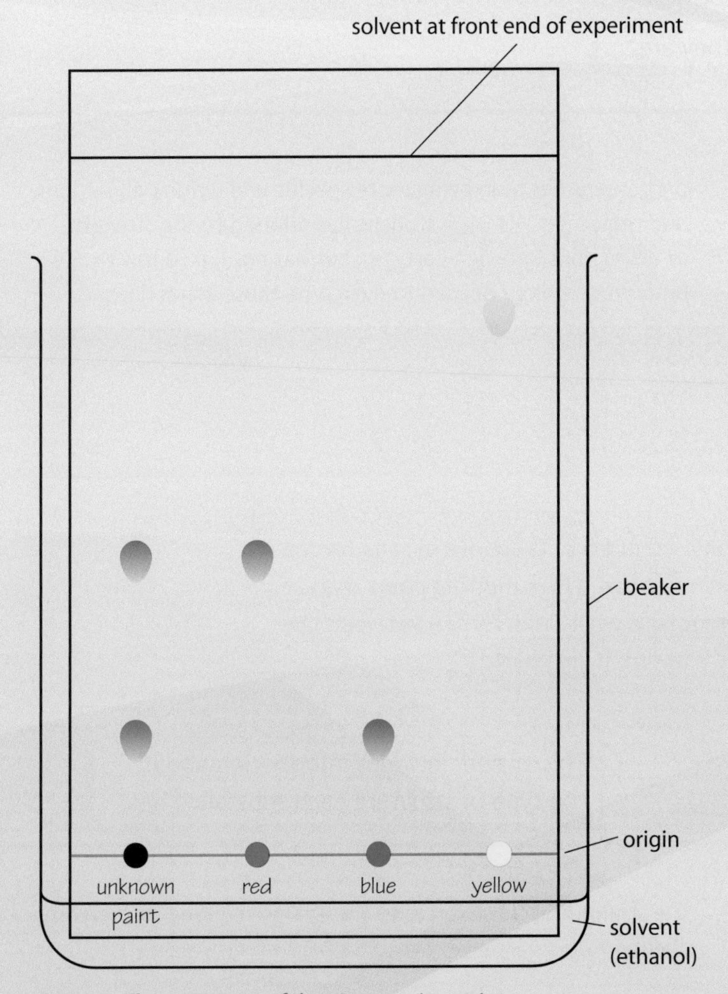

Figure 1 Chromatogram of the pigments in a paint.

19 Figure 2 shows a typical chromatogram for corn oil, a vegetable oil. It shows the fatty acids from the oil. Use the chromatogram to answer the questions that follow.

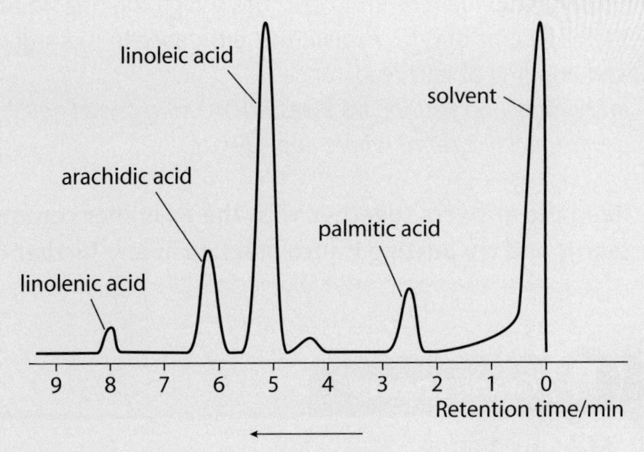

Figure 2 Gas chromatogram of corn oil.

(a) Which fatty acid travelled through the gas–liquid column fastest? *(1 mark)*

(b) How many fatty acids were found in the mixture? Which fatty acid was present in the largest amount? *(2 marks)*

(c) Gas–liquid chromatography is an instrumental method of analysis. Give three advantages of instrumental methods compared with chemical analysis. *(3 marks)*

(d) Name another instrumental method that can be used to analyse the substances separated by gas–liquid chromatography. *(1 mark)*

20 Mass spectroscopy can be used to find the relative formula mass (M_r) of compounds using the peak for the molecular ion. The mass spectrum of methyl ethanoate is shown. Methyl ethanoate is a solvent used in glues.

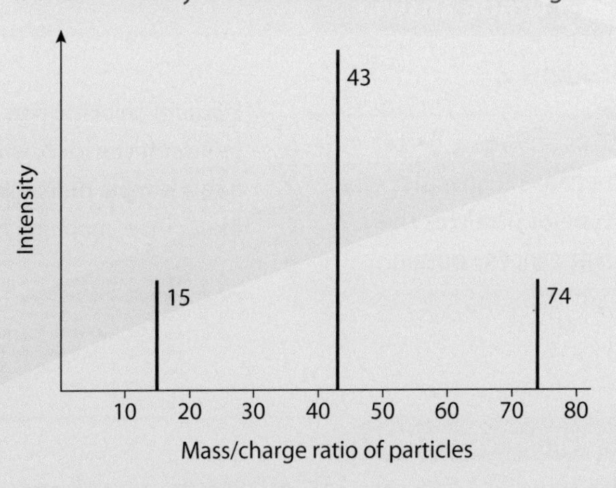

(a) What is a molecular ion? *(1 mark)*

(b) What is the relative formula mass (M_r) of propanone? *(1 mark)*

GradeStudio Route to A*

Here are three students' answers to the following question:

Sodium chloride (common salt) has a high melting point of 801 °C. Water has a low melting point of 0 °C. Explain this difference in melting points by discussing the structure and bonding of each substance.
In this question you will be assessed on using good English, organising information clearly and using specialist terms where appropriate. *(6 marks)*

Read the answers together with the examiner comments. Then check what you have learnt and try putting it into practice in any further questions you answer.

Grade answer

Student 1

Sodium chloride has a high melting point because there are strong forces between the particles. The melting point of water is low because there are weak forces between the particles.

The answer does not say what type of particle they are.

Examiner comment

This answer contains correct basic information but lacks detail. This question is about structure and bonding but the student has not mentioned the types of structure and bonding that sodium chloride and water have. The student only refers to particles rather than being more specific and writing about ions and molecules. Also, the student has referred to the strength of attractions between particles, but has not linked this to the amount of energy needed to overcome those attractions.

Grade answer

Student 2

Sodium chloride has an ionic structure. There are strong forces between the ions, which mean it has a high melting point. Water has a simple molecular structure with weak forces between the molecules, meaning it has a low melting point.

This answer is better because it specifies which type of particles the substances contain.

This answer includes information about the type of structure each substance has.

Examiner comment

This answer is better because it describes the type of structure that each substance has. It also specifies the type of particles that the forces act between. However, the answer does not link the strength of these forces to the energy required to overcome them.

A* Grade answer

Student 3

The structure of sodium chloride is ionic. There are strong attractive forces between the positive and negative ions. It takes a lot of energy to overcome these strong attractions so sodium chloride has a high melting point. The structure of water is simple molecular. There are strong covalent bonds between the atoms in each molecule, but weak forces of attraction between the molecules. When melted, it is the weak forces between the molecules that are overcome, which does not take much energy and so it has a low melting point.

This answer links the strength of the forces between particles to the amount of energy needed to overcome them.

Examiner comment

This answer is excellent. It refers to the type of structure and bonding that each substance has. For sodium chloride, it refers to the 'attractive forces between positive and negative ions' rather than just 'forces between ions'. For water, the answer clearly distinguishes between the strong covalent bonds within the molecule, which are not overcome on melting, and the weak attractive forces between molecules, which are overcome. It also links the strength of these forces to the amount of energy needed to overcome them and so the melting point.

The answer is organised well and makes points in a logical order. It is also well written with good spelling, punctuation and grammar.

- Read the whole question carefully.
- In chemistry, the use of correct terminology can be very important. For example, at GCSE level, rather than referring to particles you should use the terms atoms, molecules or ions as appropriate.
- In questions worth a lot of marks where you are asked to explain something, you must give depth and detail to your answers.
- For longer answer questions, plan your answer before you start writing. Think through and note down the key points you want to make on the exam paper, perhaps by the question, before starting to write your answer.
- Present your answer in a clear, structured way, listing your arguments in a logical order.
- Do your best to use good spelling, punctuation and grammar throughout your answers.
- Avoid using words like 'it' and 'they'. Always be specific by saying what you are referring to.

C2 Rates, salts and electrolysis

In this section, you will find out about how to measure and control the rates of chemical reactions. Temperature, concentration, pressure and surface area are all important factors. Substances called catalysts affect the rates of reactions. They can help to reduce the cost of industrial processes.

You will find out about energy transfers by chemical reactions. Exothermic reactions release energy to the surroundings and are useful for self-heating cans. Endothermic reactions absorb energy from the surroundings and are useful for sports injury packs.

Different salts are useful for different purposes, like ammonium nitrate for fertilisers and copper sulfate for fungicides. You will find out how to make them, and learn how to work out the best way to make a particular salt.

Ionic compounds break down to form simpler substances when electricity is passed through them, if they are molten or dissolved in water. This process of electrolysis is useful for making chemicals from salt and aluminium from aluminium ore. You will find out how to predict what happens when electricity is passed through a particular ionic substance.

Test yourself

1. Describe the structure of an atom.
2. What happens when elements react together?
3. Describe what happens when calcium carbonate reacts with acids.
4. Describe what happens during the complete combustion of methane, a hydrocarbon fuel. Include an equation in your answer.
5. Name the process used to extract aluminium from aluminium oxide, and to purify copper.

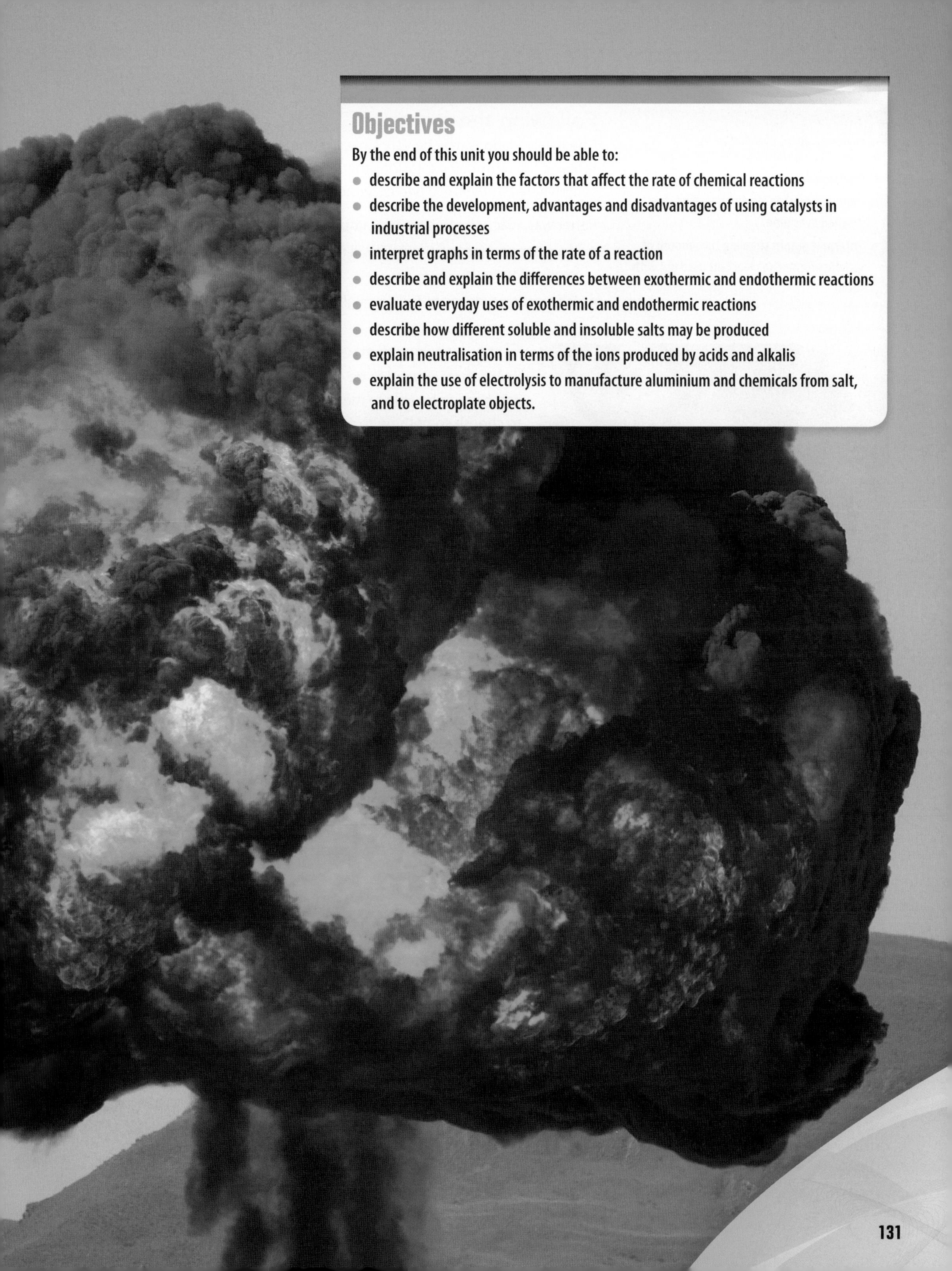

Objectives

By the end of this unit you should be able to:

- describe and explain the factors that affect the rate of chemical reactions
- describe the development, advantages and disadvantages of using catalysts in industrial processes
- interpret graphs in terms of the rate of a reaction
- describe and explain the differences between exothermic and endothermic reactions
- evaluate everyday uses of exothermic and endothermic reactions
- describe how different soluble and insoluble salts may be produced
- explain neutralisation in terms of the ions produced by acids and alkalis
- explain the use of electrolysis to manufacture aluminium and chemicals from salt, and to electroplate objects.

C2 4.1 Reaction rates

Learning objectives

- explain how the rate of a reaction can be found by measuring the amount of reactant used or the amount of product formed over time
- interpret graphs showing the amount of product formed (or reactant used up) with time.

Following the reaction

Chemical reactions may be very slow, like the slow rusting of an old car, or very fast, like the petrol/air explosions that drive the car's engine. Many, of course, lie somewhere between the two extremes.

Either way, reaction rates are very important in industry.

- If a reaction runs too slowly, it will make the process very inefficient and raise production costs.
- If a reaction runs too quickly, it might get out of control and cause an explosion.

Rusting is a relatively slow reaction.

First, we must measure…

Reaction rates need to be controlled, but before you can do that you need to be able to measure them. In practical terms, you need to choose something to measure that indicates the progress of the reaction. Then you keep track of that quantity over a sensible time interval. Examples of useful things to measure are:

- the mass of a reactant used up during a reaction
- the volume of a gas (product) produced during a reaction.

The rate of reaction is 'what happens' divided by 'how long it takes', that is:

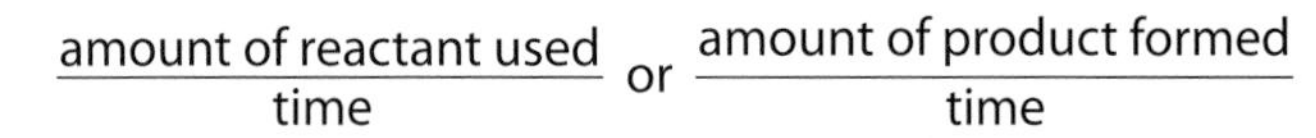

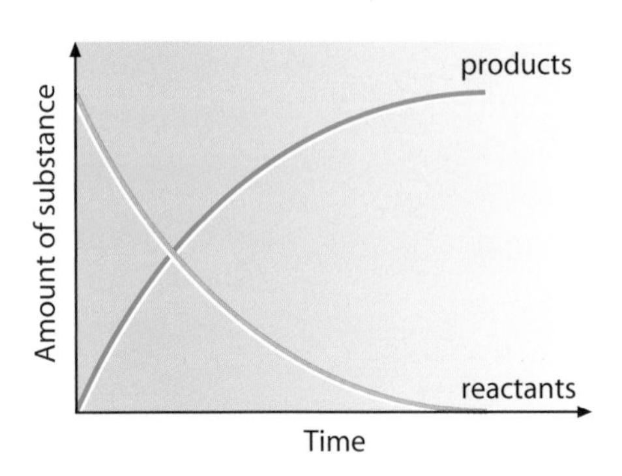

Figure 1 This graph shows how the reactants are used up and the products are formed during a reaction.

Speed it up

Often in industry you are trying to make a reaction go faster. This can make the industrial process more efficient and so help to make more money for your company. The key factors that affect the rate of a reaction are:

- temperature
- **concentration** (in liquids) or pressure (in gases)
- surface area
- the use of a catalyst.

Practical

There are several ways of measuring the rate of a reaction. If the reaction produces a gas there are two simple methods we can use. For example, when calcium carbonate reacts with hydrochloric acid to form carbon dioxide gas:

calcium carbonate + hydrochloric acid ⟶ calcium chloride + water + carbon dioxide

$$CaCO_3 + 2HCl \longrightarrow CaCl_2 + H_2O + CO_2$$

One method is to measure the volume of gas given off over time using a gas syringe. If you plot a graph of gas volume against time, it is steepest at the start. The rate of the reaction is the greatest here, as there are the most acid and marble chips available to react. As the reactants are used up there are less of them. The reaction gets slower and the graph levels off. When the graph is totally flat, the reaction has finished because one or both of the reactants has been used up.

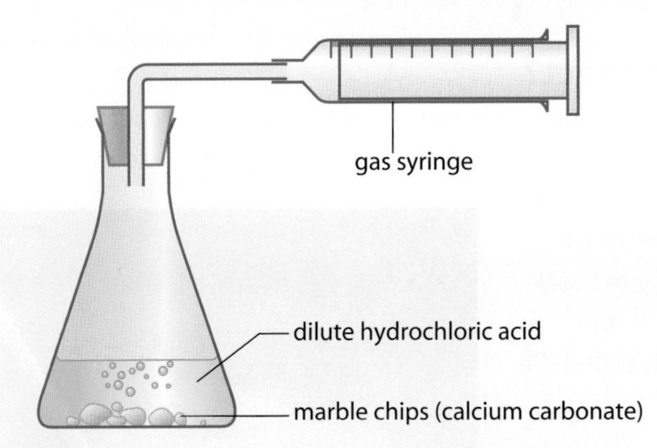

Figure 2 Measuring the volume of gas produced in the reaction between hydrochloric acid and marble chips, a natural form of calcium carbonate.

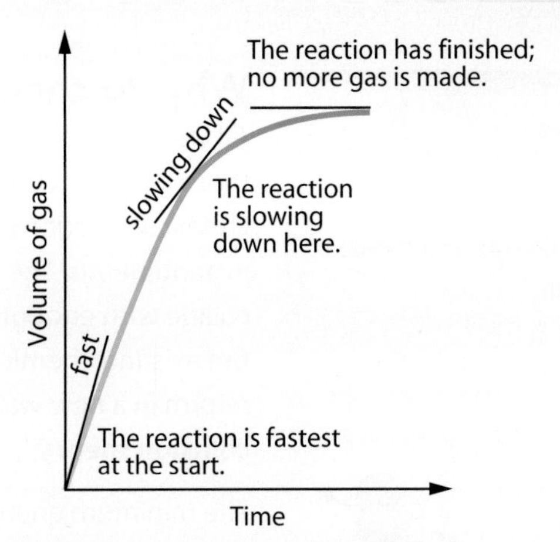

Figure 3 A typical volume/time graph.

A second method is to follow the reaction by measuring how the mass changes as the carbon dioxide is lost into the air. If the experiment is carried out on a balance, the loss of mass can be measured every minute until the reaction finishes. The loss of mass from the flask is the mass of escaping carbon dioxide. A graph of the mass of the carbon dioxide produced over time would look very similar to the volume/time graph, for the same reasons.

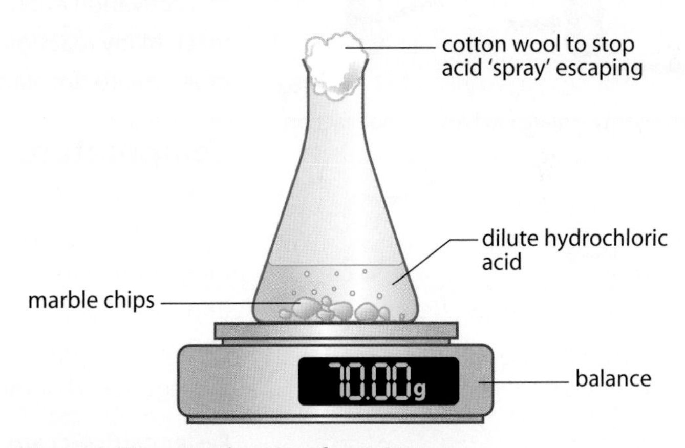

Figure 4 Measuring the mass of escaping gas.

Questions

1. What problems can be caused in industry if chemical reactions are **(a)** too fast or **(b)** too slow?
2. Magnesium ribbon reacts with excess acid. How could you tell if the reaction has finished?
3. Why would it be difficult to measure the rate of reaction for something like the rusting of iron?
4. Gas can be collected in an upside-down measuring cylinder in a bowl of water. Suggest why it is better to use a gas syringe.
5. Copper dissolves in hot nitric acid, giving off nitrogen dioxide gas. How could you measure the rate of this reaction?
6. For the marble chips/acid reaction:
   (a) How do you know that the reaction has finished?
   (b) Marble chips are usually left in the flask at the end of this reaction. Which reactant has been used up? **(c)** Sketch the shape of graph you would get if you plotted the *mass of the apparatus and reactants* against time.
   Explain your graph.
7. Magnesium dissolves in sulfuric acid to give magnesium sulfate, water and hydrogen gas.
   (a) List the reactants and products. **(b)** Describe how you could follow this reaction using a gas syringe.
   (c) Sketch the shape of the graph you would expect to get from this. **(d)** Suggest why it might be more difficult to follow this reaction by mass loss on a balance than the marble/acid reaction.
8. Malachite (copper carbonate) chips dissolve in sulfuric acid to give copper sulfate, water and carbon dioxide. Describe a simple experiment to follow the rate of this reaction using an open beaker and a balance.

The effect of temperature

Learning objectives

- explain the collision theory of chemical reactions
- explain why increasing the temperature speeds up a reaction.

Why do chemicals react?

All chemicals consist of tiny particles. Different chemicals can only react when these particles collide. However, collision alone is not enough. If it was, most chemical reactions would be almost instantaneous. The particles must collide with enough energy to break the existing chemical bonds and reform in a new way. This is called the **collision theory**.

Matches need a kick start of energy before they catch fire: you have to strike them.

The minimum energy required for any given reaction is called its **activation energy**. If the particles do not have this activation energy, they cannot react. Many reactions need a 'kick start' of energy to get started.

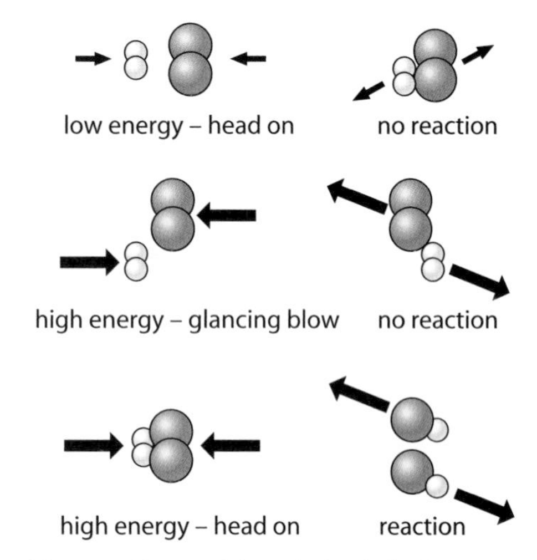

Figure 1 You need the right kind of collision before you get a reaction.

Temperature matters

At any given temperature, collisions of many different types occur. They will range from maximum-energy, 'head on' impacts to gentler glancing blows. The higher the temperature, the higher the average speed of the particles.

Milk turning sour is a chemical reaction.

Faster particles carry more kinetic energy. Collisions that generate enough energy to break the bonds occur more often, so a larger proportion of the particles have a chance to react. This is why raising the temperature speeds up reactions, and cooling slows them down.

Science skills

In the marble chip/acid experiment you could plot a graph of the time it takes to produce the gas against temperature, but it is better to convert the figures to show the rate of production, and plot a graph of that instead, as shown below. The rate of gas production is found by:

$$\text{rate} = \frac{\text{volume of gas/cm}^3}{\text{time taken/s}}$$

Table 1 Results of the experiment.

Temperature/°C	20	32	44	56	68	80
Time to produce 100 cm³/s	400	200	100	50	25	12.5
Rate/cm³/°C	0.25	0.5	1	2	4	8

You may not be surprised to find that the gas is produced more quickly when you raise the temperature.

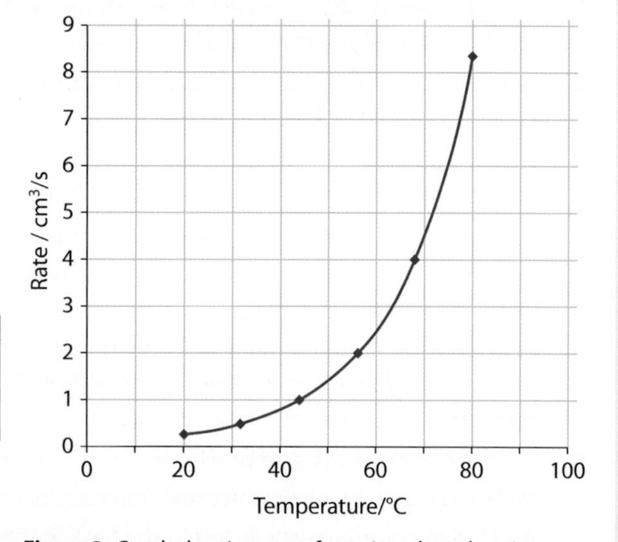

Figure 2 Graph showing rate of reaction plotted against temperature.

Monitoring the reaction

The marble chips (calcium carbonate) in acid experiment (see lesson C2 4.1) can be run at different temperatures by putting the apparatus in a water bath. You could use a gas syringe and time how long it takes to produce 100 cm^3 of gas, for example, at different temperatures.

The pace hots up

Looking at a rate/temperature graph you might expect all industrial reactions to be run at as high a temperature as possible. High temperatures are often used, but there are limits, as the products may themselves break down if heated too much.

Cost is also important. You need a lot of energy to reach very high temperatures, and energy costs money. Industrial chemists are always on the lookout for the most cost-effective ways to run their chemical reactions.

Practical

A common lab experiment uses the reaction between sodium thiosulfate and acid to investigate reaction rates. This reaction produces sulfur, which makes the solution turn cloudy. The end-point for this reaction is when the cross disappears, as in Figure 3.

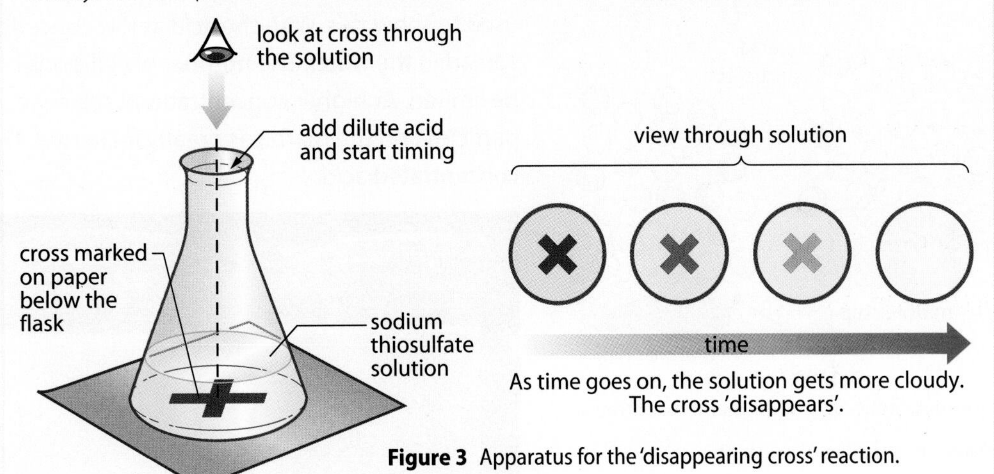

Figure 3 Apparatus for the 'disappearing cross' reaction.

Questions

1 The glasses of milk shown in the photograph on the previous page are the same age. Which do you think was left in a warm room, and which was left in the fridge?

2 **(a)** Why doesn't a Bunsen burner light as soon as you turn on the gas? **(b)** What does a spark provide that sets off this reaction?

3 Why should you never turn on a light switch if you can smell a gas leak?

4 Why does food cook faster in a 'deep fat fryer' than in boiling water?

5 Explain why raising the temperature increases the rate of reaction.

6 In the calcium carbonate/acid reaction shown in lesson C2 4.1: **(a)** What factors, other than temperature, might affect how fast the carbon dioxide is produced? **(b)** How would you ensure that these other factors were not affecting the results?

7 **(a)** From the graph of reaction rate vs temperature shown in Figure 2, approximately what rise in temperature doubles the reaction rate?

(b) Sketch the graph from Figure 2. Now draw two new sketch graphs on the same axes, one for an initially faster reaction and the other for an initially slower reaction.

8 Plot a graph of 'time taken' against temperature using the data in Table 2.

Table 2 Results for the thiosulfate 'disappearing cross' experiment.

Temperature/°C	Time for cross to disappear/s
20	360
30	180
40	91
50	46
60	23

Explain why your graph looks different to the reaction rate versus temperature graph (Figure 2) and from your graph predict how long the reaction would take at 45 °C.

Opportunities for interaction

Learning objectives

- explain how changing the concentration of reactants in solution affects the rate of a reaction
- explain how increasing the pressure of gases increases the rate of a reaction.

Close encounters

The concentration of a solution tells us how much **solute** is dissolved in the water. If there is a large amount of solute in a small amount of water, the solution is **concentrated**. If there is a lot of water and little solute, the solution is **dilute**. Before two particles can react, they must meet. This is more likely to happen in a concentrated solution than in a dilute one, so we should expect reaction rates to increase with the concentration of the solution.

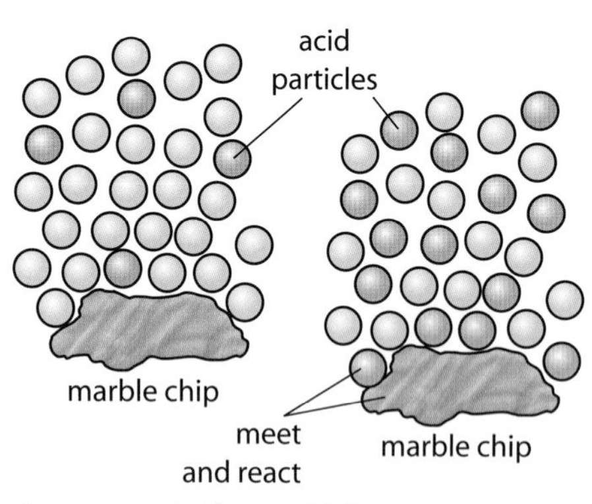

Figure 1 Reaction of acid with marble chips.

The reaction of acid on marble (calcium carbonate) (see lesson C2 4.1) can be used to show this. With the acid at low concentrations, the particles are widely spread in the water. The number of collisions between them and the marble will be limited. At higher concentrations, the chance of a collision between the acid particles and the marble is greatly increased. The marble chips fizz much faster in concentrated acid.

The more concentrated the acid, the more the fizz.

Taking it further

In many simple chemical reactions, doubling the concentration (or pressure) of one reactant will double the rate of reaction. In reactions involving more than two particles, however, the relationship is more complex.

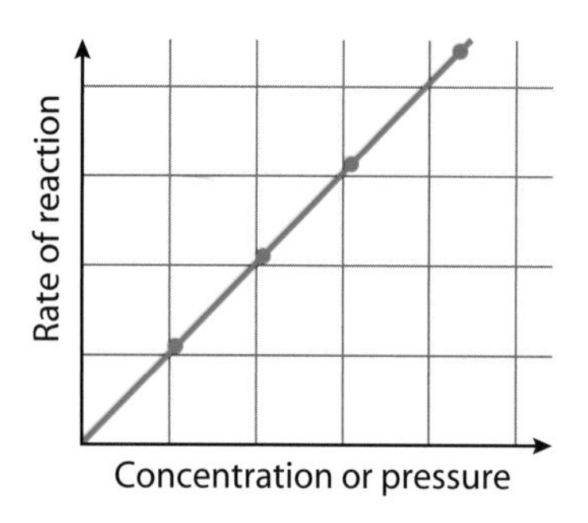

Figure 3 Typical graph of reaction rate against concentration or pressure.

Gas pressure

Similar arguments apply to gases, but here **pressure** takes the place of concentration. All else being equal, the greater the pressure, the greater the number of particles of gas in a given space. If you double the pressure you will squash the particles into half the volume.

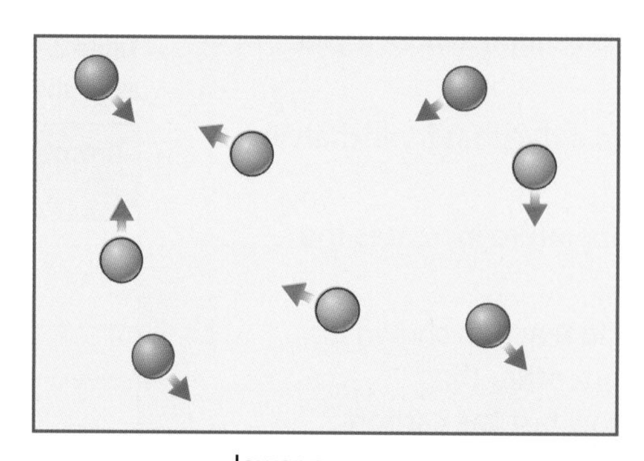

Figure 2 Higher pressure brings gas particles closer together.

The closer the particles are together, the more chance they have of colliding. If you double the pressure, you will double the rate of reaction. Graphs drawn for rate against concentration and rate against pressure look very similar.

Practical

Magnesium ribbon dissolves in sulfuric acid to give hydrogen gas. The gas can be collected in a gas syringe.

magnesium + sulfuric acid ⟶ magnesium sulfate + hydrogen

$$Mg + H_2SO_4 \longrightarrow MgSO_4 + H_2$$

A student dissolved 2-cm strips of magnesium ribbon in different concentrations of acid and timed how long it took to collect 20 cm^3 of hydrogen in each case.

Table 1 The student's results.

Solution	Volume of acid	Volume of water	Time to collect 20 cm^3 of gas
A	10	40	200
B	20	30	100
C	30	20	67
D	40	10	50
E	50	0	40

Look at Table 1.

a Which solution is the most concentrated?

b Which solution gives the fastest reaction?

c Why didn't the student use a solution containing no acid?

d Plot a graph of volume of acid used (cm^3) against time to collect 20 cm^3 of gas. Plot volume of acid on the *x*-axis, time taken on the *y*-axis.

e Use your graph to find the time needed to collect 20 cm^3 of gas if the volume of acid used was: **i** 15 cm^3 **ii** 45 cm^3.

f Which variables should be kept constant during this reaction?

g Add an extra column to your table labelled 'Rate of CO_2 production in cm^3/s'. Work out this rate for A to E.

h Plot a new graph of 'volume of acid' against your newly calculated rate of reaction. How does this compare with the graph from **d**? Explain the difference in appearance.

Questions

1 A student made two different solutions of sodium chloride:
solution A: 2 g dissolved in 200 cm^3 water
solution B: 4 g dissolved in 500 cm^3 water
(a) Calculate the concentrations of these solutions in grams per cm^3. **(b)** Which solution has the greater concentration?

2 Why is dilute acid not as dangerous as concentrated acid?

3 Bleach can be used straight from the bottle to kill germs in the toilet. Why must it be diluted down before being used to bleach clothes?

4 A concentrated lemon drink says 'dilute with five parts water before drinking'. What would you notice if you diluted it with 10 parts of water instead? Explain this effect.

5 In the marble chip/acid reaction, what would happen to the rate of reaction if you doubled the concentration of the acid?

6 Explain how increasing the concentration of reactants increases the rate of a reaction. Sketch a graph to illustrate your answer.

7 Sulfur dioxide gas reacts with oxygen gas to form gaseous sulfur trioxide: $2SO_2 + O_2 \longrightarrow 2SO_3$
(a) How would increasing the pressure change the rate of the reaction? **(b)** Explain your answer in terms of the collision theory.

8 The graph shows the volume of carbon dioxide produced when marble chips were dissolved in excess hydrochloric acid.

Calculate the rate of this reaction over the first 20 seconds, showing your working. Describe what happens to the rate of reaction over the next minute, explaining your answer. The experiment was repeated with all variables the same except that the acid was twice as strong. Make a copy of this graph and sketch the line you would expect to get for this second reaction.

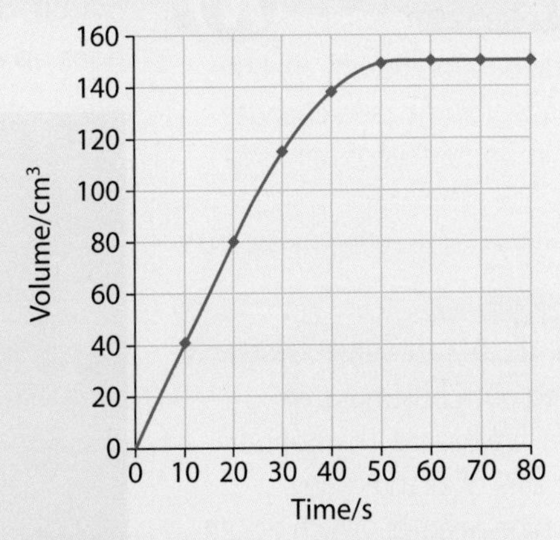

Figure 4 Graph of gas produced against time.

A*

C2 4.4 Where reactions happen

Learning objectives

- explain how the rate of a chemical reaction increases if the solid reactants have a greater surface area
- relate rate of reaction and surface area to collision theory.

Sparklers are made from tiny iron filings glued onto an iron wire. Iron filings burn because they have such a large surface area.

Route to A*

An understanding of the way that the surface area to volume ratio changes with size is important across all three sciences.

Science in action

A dead, dried out Christmas tree can be a real fire hazard. Think about the surface area of all those tiny, flammable pine needles. Which would be more dangerous – the kind that drop their needles and form a layer on the floor, or the kind that keep them on the branches?

Free up some surface

If you cut a potato into small pieces it will cook much faster than a whole potato. The smaller the pieces of potato, the faster they cook. This is because more surface is exposed to the boiling water. It is the same with chemical reactions.

You have probably seen this idea in action when you try to light a campfire. The fire is much easier if you start with small twigs and splinters. You cannot set a large log on fire with just a match. Smaller twigs have a larger surface area than a big log, so they catch fire more easily.

The heart of the matter

Every time you break up a solid, you expose an extra bit of material to react. If you have the same amount of material in pieces of half the width, you will have twice the surface area (for the same amount of material), which will double the reaction rate.

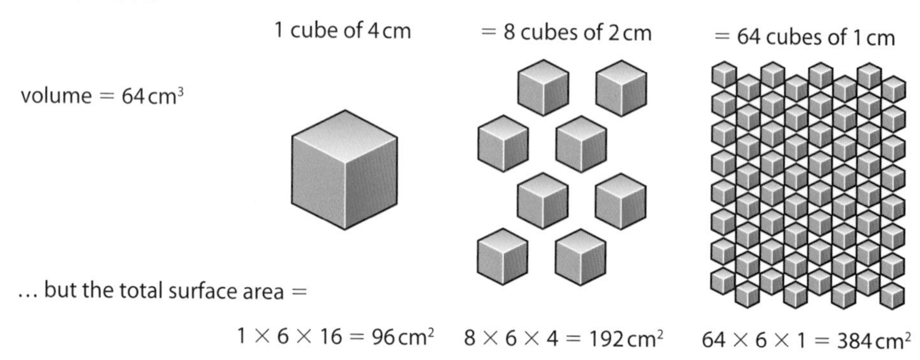

Figure 1 Smaller cubes, larger surface area.

A chip off the block

One large lump of marble (calcium carbonate) dropped into a beaker of acid fizzes steadily. But the same amount of marble in powdered form foams up out of the beaker. Collision theory can be used to explain this. The acid particles can only collide with the carbonate particles exposed at the surface. Crushing a marble chip increases the surface area several thousand times. The reaction rate goes up accordingly.

Which beaker had the powdered marble?

Lost to the air

If this reaction takes place in an open beaker, the carbon dioxide escapes into the air. If this is done on a balance, the total mass will go down as the carbon dioxide is lost. If you measure how much gas is lost in a given time, you can work out the rate of the reaction.

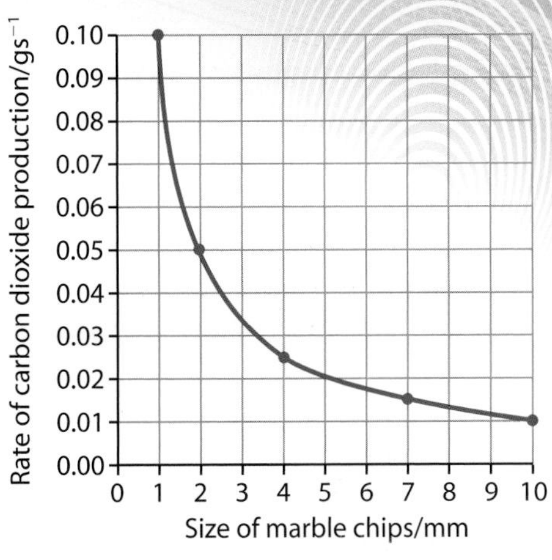

Figure 2 This graph shows how the reaction rate speeds up when you use smaller marble chips.

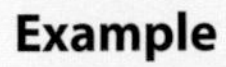

Example

In an experiment adding acid to 1 mm marble chips, the mass fell from 50 g to 45 g in 50 seconds.

Mass lost (= amount of carbon dioxide produced) = 5 g. Time taken = 50 s.

So: rate $= \frac{\text{mass lost}}{\text{time taken}} = \frac{5\text{ g}}{50\text{ s}} = 0.1\text{ g/s}$

Practical

A student carried out an experiment to compare the rate of reaction of 0.1 g magnesium ribbon and 0.1 g magnesium powder when reacting with excess sulfuric acid. She measured the volume of hydrogen produced using a gas syringe.

Table 1 The student's results.

Time/s	Volume of hydrogen from Mg ribbon/cm³	Volume of hydrogen from Mg powder/cm³
0	0	0
10	22	52
20	40	68
30	56	77
40	68	82
50	76	84
60	82	86
70	85	86
80	86	86
90	86	86

a Plot a graph showing these results, with time as the *x*-axis and volume as the *y*-axis.

b Why is the final volume of hydrogen the same in both experiments?

c Why do the graphs flatten off near the end of the reaction?

d Which reaction produced the most hydrogen in the first 30 seconds?

e What is the average rate of production of hydrogen in cm^3 per second for each reaction during this time?

f Explain the difference in the results.

Questions

1. Why does chewing your food well help you to digest it?
2. Why do small twigs catch fire more quickly than a big log?
3. Pythons swallow their food whole, and then take days to digest it. Explain why.
4. Flour mills and biscuit factories can easily produce clouds of flammable dust. **(a)** Why is this dangerous? **(b)** What could they do to overcome it?
5. Using the graph of reaction rate against marble chip size: **(a)** Find the mass loss per second for the 2 mm chips. **(b)** How long would it take the mass to drop by 1 g for these chips? **(c)** How much faster did the 2 mm marble chips react compared with the 4 mm chips?
6. If you broke a 1 cm cube up into 1 mm cubes, by what factor would the surface area increase?
7. Why is it that delicate carvings on the outside of churches seem to suffer more from the effects of acid rain than large blocks made from the same stone?
8. Explain carefully why iron filings on sparklers burn when heated, yet an iron poker can be left red hot in a fire for hours with no obvious reaction.

A*

Catalysts

Learning objectives

- explain what a catalyst is
- evaluate the advantages and disadvantages of using a catalyst in industrial processes.

Science in action

Living things rely on biological catalysts called enzymes. Many industrial processes now use these enzymes to make their products. For example, sugar syrup is made from corn starch using enzymes.

Everlasting activity

We are going to look at one final way of speeding up reactions. Ever helped others get things done just by being there with them and smiling? Some chemicals have the same effect on reactions. Just by being there while the reaction is taking place, they make things go much faster. Here's an example: the breakdown of hydrogen peroxide, an unstable compound of hydrogen and oxygen. Left on its own, hydrogen peroxide will slowly break down into water and oxygen gas:

hydrogen peroxide $\longrightarrow$ water + oxygen

Drop a spatula of powdered manganese dioxide into hydrogen peroxide and the mixture starts to fizz rapidly as oxygen is given off. There is nothing particularly surprising in that, you might think – just another chemical reaction in progress.

However, if you filter the mixture after the reaction has finished, you get back the same amount of manganese dioxide as you put in. The oxygen has escaped and the water is left behind, but the manganese dioxide is unchanged. It can be used all over again.

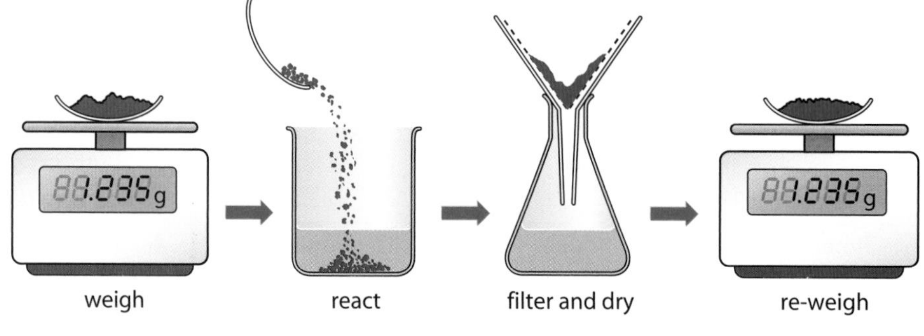

Figure 1 Manganese dioxide speeds up the breakdown of hydrogen peroxide, but remains unchanged itself.

Weakening the bonds

The reaction that occurs is the simple breakdown of hydrogen peroxide to water and oxygen, which would have occurred slowly on its own. The manganese dioxide has simply speeded up this reaction without itself being altered. It has acted as a **catalyst**.

$$2H_2O_2 \xrightarrow{\text{manganese dioxide catalyst}} 2H_2O + O_2$$

Catalysts work by lowering the activation energy needed for the reaction. The process is complex but a simple model is that the existing bonds are weakened on the catalyst surface. This makes it easier for them to be broken during collisions so that new bonds can form. The overall result is that the reaction happens more easily.

Science in action

Many manufacturers are developing fuel cells as replacements for petrol and diesel engines. In a **fuel cell**, hydrogen combines with oxygen to form water and also produce electricity. The reaction is catalysed by a platinum catalyst.

reactants

bonds weaken

catalyst surface

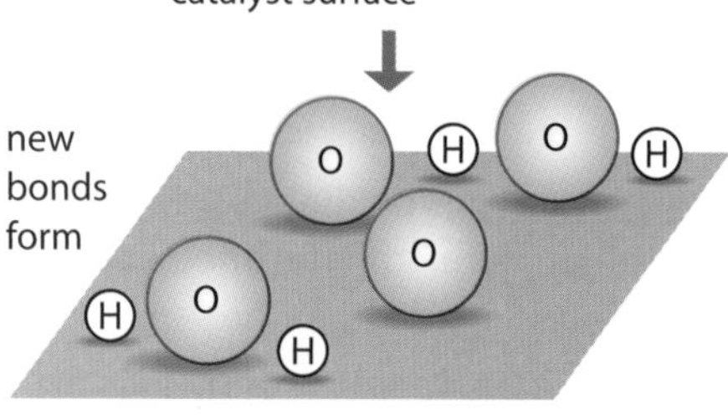

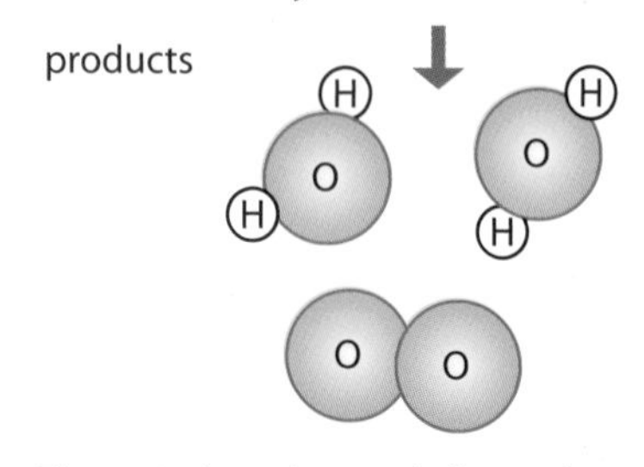

Figure 2 A catalyst works by weakening bonds, which lowers the activation energy.

Surface area matters

This kind of reaction can only happen on the surface of the catalyst. So the catalyst has to be in the smallest pieces possible. Breaking it up into a fine powder may not help, because the particles may just clump together. The answer is to attach small particles to tiny wire grids just millimetres apart. This allows the reactants to get to as large a surface as possible.

Nanotechnology (see lesson C2 2.6) could take this approach to a new level. Create particles of catalyst just a thousandth of a millimetre across and their surface area goes up a thousand times for the same amount of material. This makes the catalyst a thousand times more effective.

The catalyst in a car's catalytic converter is held in a 'millimetre-sized' grid. Nanocatalysts could one day make cars completely pollution-free.

The catalyst for the job

In the simple breakdown of hydrogen peroxide, iron oxide or copper oxide will work too, but manganese dioxide gives the fastest reaction. It is important to find the right catalyst for the particular reaction you want to speed up.

Transition metals or their oxides are often used as catalysts:

- Iron is used in the production of ammonia.
- Nickel is used to turn oils into fats for margarine or chocolate.
- Platinum and rhodium are used in the production of nitric acid.
- Vanadium oxide is used in the production of sulfuric acid.

Disadvantages of some catalysts include:

- They may become poisoned by impure reactants; they then need purifying before being re-used.
- Some, such as the platinum used in fuel cells, are very expensive.

Overall, catalysts are very important in industry. Without them some reactions would be far too slow to be useful so much more fuel would need to be burnt to raise the reaction temperature and speed things up. This would cost the company money, reducing their profitability or making their products more expensive. Burning that extra fuel would also waste valuable resources and increase pollution, giving out much more carbon dioxide, which is linked to global warming.

Making chocolate like this needs a nickel catalyst.

Examiner feedback

Many different catalysts are used in industry. You do not need to know them all, but you should be able to explain what is happening if you are given the details of a particular reaction.

Questions

1. Use the diagrams in Figure 1 to explain how you could show that manganese dioxide is not taking part directly in the hydrogen peroxide reaction.
2. How do catalysts allow reactions to run at lower temperatures?
3. Why is it not good enough to simply grind a catalyst into a fine powder?
4. How could 'nanocatalysts' help revolutionise the way we tackle pollution?
5. Platinum is very expensive. Why might this not matter so much for a catalyst?
6. Exhaust gases flow out from an engine very rapidly and so could escape into the air before they have time to react to form less harmful products. How does a catalytic converter overcome this problem?
7. How could you compare the effectiveness of different transition metal oxides as catalysts for the hydrogen peroxide reaction?
8. In a particular industrial process, using a catalyst means that the reaction can run efficiently at 100 °C instead of 300 °C. What are the potential benefits of using a catalyst in this reaction?

Reactions with energy to spare

Learning objectives

- explain how energy transfers are associated with chemical reactions
- describe how exothermic reactions transfer energy to the surroundings
- explain how this process can be used to our advantage.

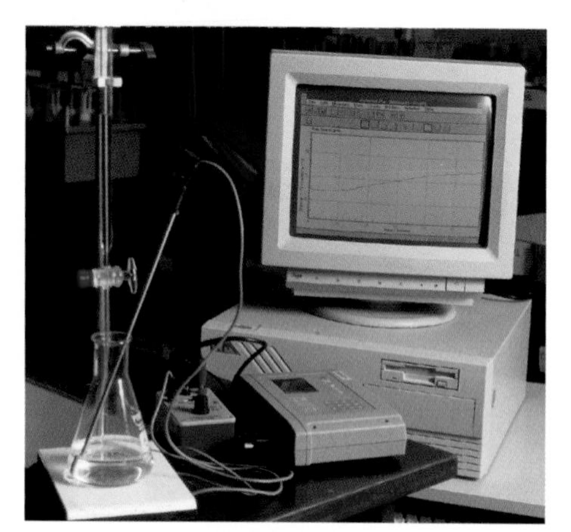

You could follow the rise in temperature during neutralisation with a temperature sensor connected to a computer.

Enough energy is released in the thermit reaction to melt the iron.

Calcium (left) and magnesium (right) both fizz in acid. The test tubes get hot.

Energy on the loose

It is not always easy to tell when a chemical reaction has occurred. For example, if you mix cold, dilute hydrochloric acid and sodium hydroxide solution, you will not see any obvious change. But touch the tube carefully and you will *feel* that something has happened. The tube has been warmed by the energy given out as the reaction takes place.

Chemical changes are often accompanied by changes in temperature, as energy is transferred to or from the surroundings. In the case above, energy is given out during neutralisation, which heats the surroudings. Chemical reactions that give out energy like this are called **exothermic** reactions.

acid + alkali ⟶ salt + water (energy given out)

e.g. $HCl + NaOH \longrightarrow NaCl + H_2O$ (energy given out)

Energy release runs in families

Many chemical reactions are exothermic like this, particularly reactions that start reacting as soon as the reactants are mixed. For example, metals reacting with acids:

metal + acid ⟶ salt + hydrogen (energy given out)

e.g. $Mg + H_2SO_4 \longrightarrow MgSO_4 + H_2$ (energy given out)

Displacement reactions are also exothermic. An example is when aluminium displaces iron in the spectacular thermit reaction:

aluminium + iron oxide ⟶ aluminium oxide + iron (energy given out)

$2Al + Fe_2O_3 \longrightarrow Al_2O_3 + 2Fe$ (energy given out)

Oxidation reactions, such as calcium metal turnings tarnishing as they react with the oxygen in air, also give out energy:

calcium + oxygen ⟶ calcium oxide (energy given out)

$2Ca + O_2 \longrightarrow 2CaO$ (energy given out)

A spark gets things going

Combustion reactions, such as burning fuels, are exothermic. For these reactions it is usually the energy we want, not the products. We burn methane to heat our homes and petrol to run our cars.

fuel (e.g. oil, gas or wood) + oxygen ⟶ carbon dioxide + water (energy given out)

You have to light the fuel with a spark or match to start the reaction going. This provides the activation energy (see lesson C2 4.2) needed for the reaction to begin. After that, the reaction generates enough energy of its own to keep the process going.

Energy out of control

Science in action

Once a fuel starts to burn, the exothermic release of energy can easily get out of control. Fuels such as petrol are highly flammable and have to be handled with special care. Hazard symbols are used to warn of the dangers when transporting or storing fuels.

Highly flammable
These substances easily catch fire.

Explosive
These substances cause an explosion.

Figure 1 Look out for these hazard symbols.

An exothermic reaction that has got out of control.

Combustion reactions often produce light and sound as well as heat.

Science skills

A student had three beakers, each containing 50 cm^3 of copper sulfate solution at room temperature. She added an excess of powdered iron, magnesium and zinc to each in turn, measuring the temperature rise each time. Look at Table 1.

a What has happened?

b Which reaction gave out the most energy?

c Comment on any pattern you see, comparing it to other chemical reactions you know of.

Table 1 The student's results.

Metal added	Temperature rise/°C
Fe	9
Mg	28
Zn	15

Hand warmers and hot coffee

Sports people use heat packs to relax stressed muscles. One type contains a damp mixture of iron filings, salt and charcoal in a sealed bag. When the bag is opened and squeezed, oxygen gets in from the air and reacts with the iron, giving out energy. The other chemicals help to speed up the reaction.

iron + oxygen ⟶ iron oxide (energy given out)

How can you get hot coffee from a cold can? Self-heating coffee cans have a hidden compartment full of calcium oxide. When a button is pressed, water is mixed with this and a strongly exothermic reaction occurs (see lesson C1 2.3). This quickly heats up the coffee.

calcium oxide + water ⟶ calcium hydroxide (energy given out)

Questions

1. If you dip an iron nail into copper sulfate solution it comes out coated in copper. **(a)** What kind of reaction is this? **(b)** How else might you be able to tell that a chemical reaction has occurred?
2. Do you think that iron rusting is an exothermic or endothermic reaction? Explain your answer.
3. Combustion reactions provide energy. What other forms of energy are sometimes given off as well?
4. If you mix acid and alkali solutions in a test tube you are holding, what clue will you get that a reaction has taken place?
5. Suggest some other uses for 'self-heating' cans.
6. The thermit reaction can be started with a simple fuse yet produces enough energy to make iron melt. What does this tell you about the reaction?
7. **(a)** Why do fuels need a 'spark' to get them burning? **(b)** What keeps the reaction going once it has started? **(c)** What problems could be caused if fuels did not need this 'spark'?
8. What are the main 'families' of exothermic reactions? Which can usefully be used as energy sources, and what are the potential dangers that go with their use?

C2 5.2

Reactions that take in energy

Learning objectives

- explain why endothermic reactions need to take in energy from the surroundings
- explain the importance of endothermic reactions.

No ice is needed for this summer drink.

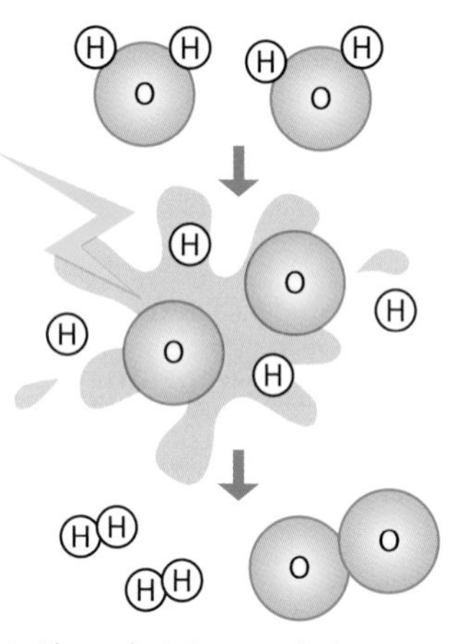

Figure 1 Electrolysis is an endothermic process

Sunlight energy powers this glucose factory.

The drink that cools itself

You can make a pleasant summer drink by adding a spoonful of baking soda to a glass of sweetened lemon juice. When you do this a chemical reaction occurs:

sodium hydrogen carbonate (baking soda) + citric acid (lemon juice) ⟶ sodium carbonate + carbon dioxide + water

If you are holding the glass as you stir in the baking soda, you will feel it get cooler. A temperature change is an indication that a chemical reaction is happening. In this case, the temperature goes down, not up.

Supplying the energy to make things go

Thermal decomposition reactions are endothermic.

Reactions that take energy in from their surroundings are called **endothermic** reactions. It is fairly rare for endothermic reactions to 'go on their own'. You usually need to pump energy into them to make them work. Examples include materials that decompose when heated. We call these reactions thermal decomposition. Green copper carbonate ($CuCO_3$) will not break down on its own. Once heated, though, it quickly decomposes into black copper oxide (CuO) and carbon dioxide (CO_2). The energy transferred by heating you put in breaks the compound apart.

copper carbonate [energy] ⟶ copper oxide + carbon dioxide

The current way to purify metals

Electrolysis is an endothermic process. Electrical energy is used to split compounds into the ions that make them up. You can use electrolysis to extract aluminium from molten aluminium oxide, for example. It can also be used to make hydrogen and oxygen from water:

water [energy] ⟶ hydrogen + oxygen

The chemical energy store

One of the most important endothermic reactions of all is photosynthesis, in which plants use the energy from sunlight to build complex chemicals such as glucose from carbon dioxide and water:

carbon dioxide + wate [energy] ⟶ glucose + oxygen

This process stores up energy from sunlight as chemical energy that we can get back later in exothermic reactions such as combustion. Fossil fuels – coal, oil and gas – got their energy from sunlight in this way millions of years ago. When we burn fossil fuels we are getting that energy back; the problem is that we are using them at an increasing rate and they will soon be gone. However, if we grow plants today, for oils or foodstuffs, we can make new 'energy stores' whenever we need them.

Examiner feedback

Endothermic reactions take in energy from the surroundings whereas exothermic reactions release energy to the surroundings. Both types of reaction have practical and commercial applications.

Future energy storage?

Wind farms provide pollution-free energy. There's one big catch: the wind doesn't blow all the time. Then, at other times, it blows so hard you make more electricity than you need. Similarly, photovoltaic cells make electricity from sunlight by day but are not much use at night. What can be done? How can we store the extra energy?

What's the big problem with wind farms?

Unused electricity can be used to electrolyse water and make hydrogen. This can then be stored until needed, then used as a fuel to generate electricity. On a larger scale, this idea could be used to produce the hydrogen for the next generation of pollution-free cars, lorries and buses.

Science in action

A group of scientists in the United States are developing an energy storage system that can split water directly into hydrogen and oxygen, using sunlight and special catalysts. These gases can then be used to make electricity whenever it is needed, using a fuel cell.

Questions

1 Why is it unusual for endothermic reactions to 'go on their own'?

2 When you make toast, starch molecules break up into complex caramels and sugars. Is this reaction exothermic or endothermic? Explain your answer.

3 In the lemon juice/baking soda drink, what would you *see* happening that would suggest that a chemical reaction was occurring?

4 Write a balanced chemical equation for the decomposition of copper carbonate.

5 The hydrogen and oxygen ions are held together by strong bonds in water. What is the electrical energy doing in this reaction?

6 Balance the chemical equation for this reaction: $H_2O \longrightarrow H_2 + O_2$

7 How could the energy from wind farms or photovoltaic cells be stored?

8 Fossil fuels are used because of their exothermic combustion reactions. Explain how an endothermic reaction is the key to their formation, the 'stored' energy they contain and the way they are used.

A*

Science in action

Sports people use 'cold packs' to relieve injuries. These packs contain ammonium nitrate and a separate bag of water. When you squeeze the pack, the water mixes with the ammonium nitrate which starts to dissolve. The ammonium nitrate breaks up into ions, which is an endothermic process. The pack gets cold, drawing energy transferred by heating from the injured muscle.

warm

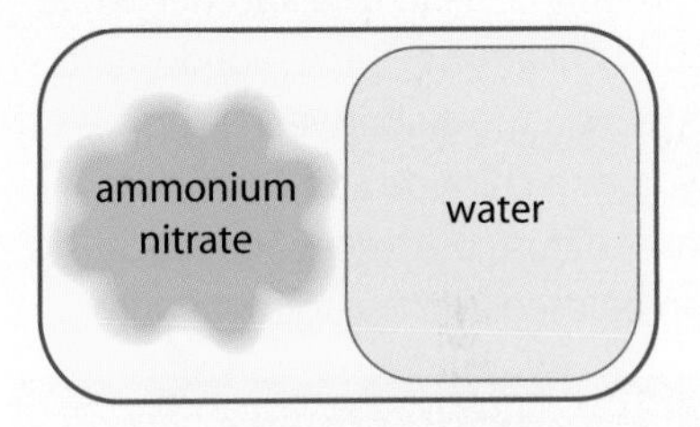

The water is kept separate in a thin plastic bag.

squeeze and burst inner bag

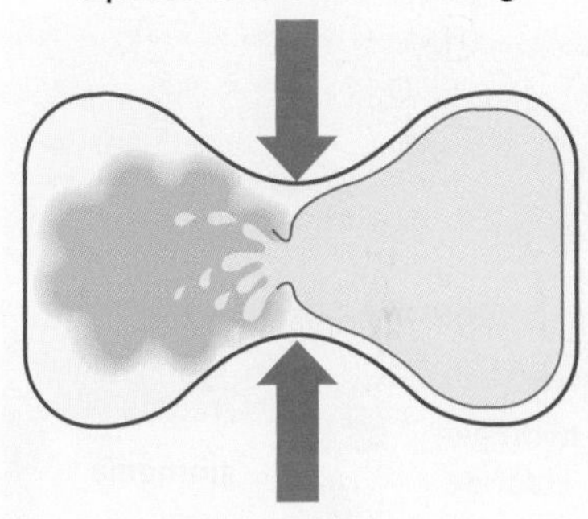

cold

The mixture cools as the ammonium nitrate dissolves.

Figure 2 Squeeze the cold pack to start the endothermic reaction.

Energy in reversible reactions

Learning objectives

- explain the energy changes involved in reversible reactions
- evaluate the importance of energy changes in a range of reversible reactions.

Practical

If you stand open beakers of concentrated hydrochloric acid and ammonia close to one another, white 'clouds' of ammonium chloride start to form in the air between them, as hydrogen chloride and ammonia gases diffuse out, meet and react. This can be used as a test for either gas.

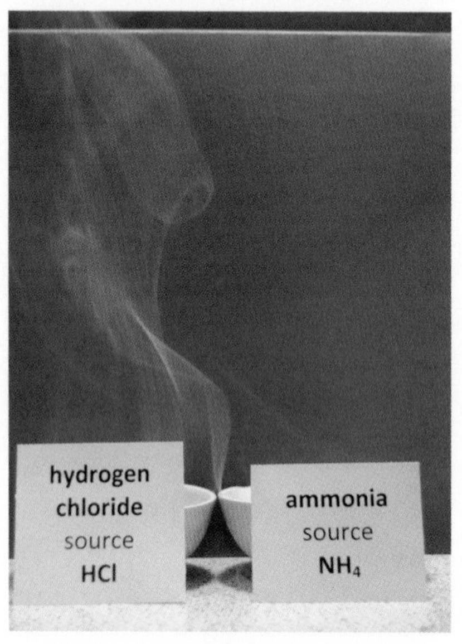

A test for ammonia or hydrogen chloride.

Examiner feedback

Cobalt chloride has similar hydrated (blue) and anhydrous (pink) forms. You could explain the energy changes in the same way as for copper sulfate.

Breaking up and getting back together

If you heat ammonium chloride, the compound breaks up to form ammonia and hydrogen chloride gases. Like all thermal decomposition reactions, this reaction is endothermic. Energy has to be put into the reaction to break the chemical bonds.

$$\text{ammonium chloride} \rightleftharpoons \text{ammonia} + \text{hydrogen chloride}$$

This reaction is easily reversible (see lesson C2 3.4). If the gases are allowed to mix when cool, they recombine to form ammonium chloride. The energy change is also reversed. The energy that had to be put in to break the chemical bonds is now given out as the chemical bonds are formed again. The 'back reaction' is exothermic.

$$NH_4Cl \underset{\text{exothermic}}{\overset{\text{endothermic}}{\rightleftharpoons}} NH_3 + HCl$$

In all reversible reactions, if the reaction in one direction is exothermic, then the reverse reaction will be endothermic. It is also important to note that the same amount of energy is transferred in each case.

The test for water

Blue copper sulfate has water molecules chemically bound up in its crystals. If you heat it, the water is driven out, forming white anhydrous copper sulfate.

If you add water to this white powder, it turns blue again. This is such a clear reaction that it is used as a test for water. The water molecules re-form the chemical bonds with the copper sulfate and new, blue crystals form. If you carry out this reaction in a test tube, you can feel the tube get hot. Energy is released as the bonds re-form.

$$\underset{\text{(blue)}}{\text{hydrated copper sulfate}} \underset{\text{exothermic}}{\overset{\text{endothermic}}{\rightleftharpoons}} \underset{\text{(white)}}{\text{anhydrous copper sulfate}} + \text{water}$$

Adding water to anhydrous copper sulfate causes an exothermic reaction. The colour change is a test for water.

Re-usable hand warmers

The iron-based hand warmers discussed in lesson C2 5.1 can only be used once, as the oxidation reaction is not easily reversible. However, some hand warmers use a reversible reaction and can be used over and over again.

Sodium acetate melts to form a clear liquid if heated in a water bath. This is an endothermic reaction as the compound is breaking up into ions. What is unusual is that this clear liquid can be cooled back down to room temperature without recrystallising, forming a '**supercooled liquid**'. However, it only takes a 'seed' crystal to be dropped in for the crystallisation to start and rapidly spread. When this happens, bonds are re-forming so the reaction is exothermic. Energy is given out and the material heats up.

Re-usable hand warmers.

$$\text{solid sodium acetate} \underset{\text{exothermic}}{\overset{\text{endothermic}}{\rightleftharpoons}} \text{liquid sodium acetate}$$

Sodium acetate hand warmers are filled with the supercooled liquid. If a button on the pack is pressed, it 'seeds' the liquid and crystals start to grow. Energy is given out and the pack heats up. Unlike a one-time hand warmer, a reusable one can be 'reset' by putting it into boiling water for a few minutes, to re-melt the crystals. Once it has cooled down, the hand warmer is ready to be used again when needed.

Getting ready to reverse photosynthesis.

Photosynthesis and respiration

In lesson C2 5.2 we saw how plants build carbon dioxide and water into glucose and oxygen in the endothermic process called photosynthesis. This also has an exothermic 'reverse' equivalent, called **respiration**. You are getting the energy you need for life by running the photosynthesis reaction in reverse.

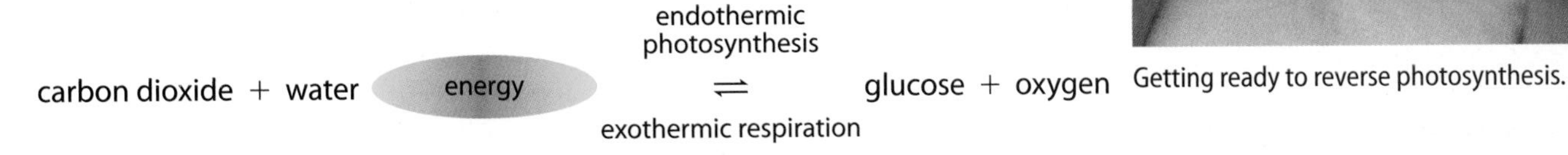

Questions

1. You pour a little water onto some white anhydrous copper sulfate in a cool tube. **(a)** Describe what you would see and feel. **(b)** Explain these effects.
2. It takes 2.5 kJ of energy to turn a quantity of hydrated copper sulfate into anhydrous copper sulfate. How much energy would be released if sufficient water was added to this white powder?
3. Explain why sodium acetate-based hand warmers are re-usable, whereas iron-based hand warmers are not.
4. Under certain conditions iron oxide can be reduced by hydrogen to form iron and water. Is this likely to be an endothermic or exothermic reaction? Explain your answer.
5. Calcium oxide (CaO) has an exothermic reaction with carbon dioxide (CO_2). Write this reaction as a balanced chemical equation. What reaction is this the reverse of?
6. What effect will raising the temperature have on an exothermic reaction and on an endothermic reaction? Explain your answers.
7. In what way does our use of vegetables as food 'reverse photosynthesis'?
8. If you heat ammonium chloride at the base of a long tube it seems to disappear. At the same time, a white powder starts to form at the cool end of the tube. Explain what is happening in terms of both the chemical reaction and the energy changes involved.

A*

If you put a 'used' sodium acetate hand warmer in hot water, the water will cool down as the crystals melt. If you let the hand warmer cool down, then activate it and put it in cold water, the water will warm up as the crystals re-form.

a Design an experiment to show that the same amount of energy is involved on each occasion.

- **i** What are your independent and dependent variables?
- **ii** What other variables might affect your result?
- **iii** How could you control them?

Assess yourself questions

1 A student put a beaker containing 200 cm^3 of 1 M hydrochloric acid, and a watch glass with 1 g of marble chips, onto an electric balance. He then zeroed the display on the balance. After this he tipped the limestone into the acid and put the empty watch glass back on the balance next to the beaker. He recorded the readings from the scale every 20 seconds.

Table 1 The student's results.

Time/s	Balance reading/g
20	−0.11
40	−0.23
60	−0.33
80	−0.40
100	−0.43
120	−0.44
140	−0.44

(a) The student held a drop of limewater on the end of a glass rod over the beaker and it turned milky. What gas was given off? *(1 mark)*

(b) Complete the equation:

$CaCO_3 + 2HCl \longrightarrow CaCl_2 + H_2O +$ ______ *(1 mark)*

(c) Explain why the reading on the balance scale dropped below zero. *(1 mark)*

(d) Plot a graph of the results. Put time along the *x*-axis and mass loss in grams up the *y*-axis. Draw a 'best fit' curve. *(3 marks)*

(e) What would the student have seen happening over the first minute or so? *(1 mark)*

(f) Suggest *two* ways in which the student would have known that the reaction had finished after 2 minutes. *(2 marks)*

2 Another student used the same reaction and apparatus as in question **1** to investigate how the reaction rate changed with temperature. In her experiment she timed how long it took for the mass loss to reach −0.20 g. She repeated the experiment at different temperatures, to see how this affected the reaction.

Table 2 Results.

Temperature/°C	0	12	25	37	50
Time to reach −0.20 g/s	150	72	37	19	10

(a) Suggest some safety precautions that would need to be taken for this experiment. How could the acid be heated safely, and how could it be cooled? *(3 marks)*

(b) Plot a graph of the results, with temperature along the *x*-axis. Draw a 'best fit' curve. *(3 marks)*

(c) Describe the pattern you see in words. How is 'the time it takes' changing as the temperature rises? *(1 mark)*

(d) What does this tell you about the way the speed of the reaction is changing as the temperature rises? *(1 mark)*

(e) Explain this effect in terms of what is happening at the particle level. *(2 marks)*

(f) Add a third row to the table and calculate the rate of the reaction at each temperature. (Rate = 0.20 g/time taken). *(2 marks)*

(g) Plot a new graph of rate against temperature and draw the 'best fit' line. *(3 marks)*

(h) From your 'best fit' line, how much do you need to raise the temperature by to double the rate of this reaction? Explain your reasoning. *(2 marks)*

(i) The student actually repeated each experiment three times at each temperature. The figures shown are the average of the three. Why did she do this? *(2 marks)*

3 **(a)** Sparklers are made from tiny iron filings glued onto an iron wire. Explain why the iron filings burn but the iron wire does not. *(2 marks)*

(b) A student wanted to find out which was more reactive, iron or nickel, so she put samples of each into tubes of 1 M hydrochloric acid, to see which bubbled the most. She had iron filings and nickel granules. Suggest why her results may not be reliable. *(2 marks)*

(c) Iron filings usually fizz faster in acid than iron nails. You might expect very fine iron dust to fizz even faster, but if it has settled out in a layer at the bottom that may not be the case.

(i) Suggest a reason why. *(1 mark)*

(ii) How could you easily get the iron dust to react faster again? Explain your answer. *(2 marks)*

(d) Iron acts as a catalyst in an important industrial reaction that makes ammonia.

(i) Why does the iron have to be in very tiny pieces? *(1 mark)*

(ii) Why does this iron have to be supported on a grid or mesh within the reacting gases? *(1 mark)*

4 A student was investigating the way hydrogen peroxide broke down when metal oxide powders were added. She had a flask containing hydrogen peroxide connected to a 100 cm^3 gas syringe. She added a spoonful of oxide, quickly replaced the bung and took a reading every 20 seconds.

(a) What factors, apart from which metal oxide she chooses, must the student keep constant if this is to be a fair test? *(3 marks)*

(b) Copy and complete the equation for this reaction.
$2H_2O_2 \longrightarrow 2H_2O +$ ______ *(1 mark)*

(c) The metal oxide acts as a catalyst for this reaction. Explain the term catalyst. *(2 marks)*

(d) Figure 1 shows the graph for manganese dioxide, nickel oxide and copper oxide.

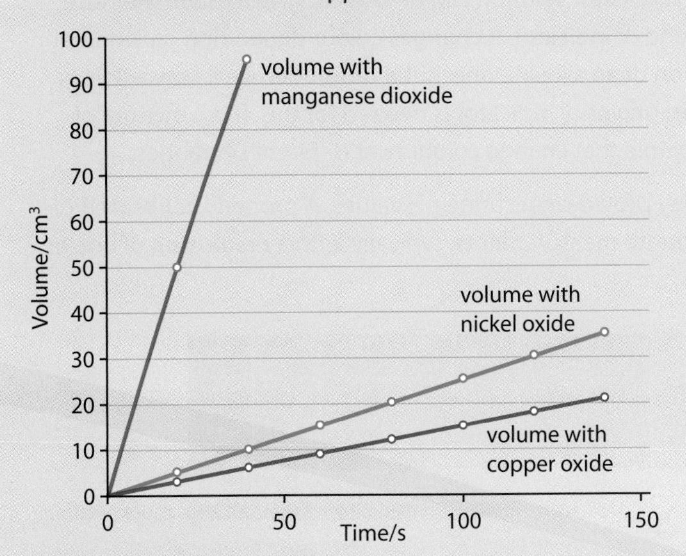

Figure 1 Graph of oxygen production.

Which is the most effective catalyst as shown by these results? *(1 mark)*

(e) Calculate the reaction rate for each of the catalysts. *(3 marks)*

5 For each of the following reactions, state whether it is exothermic or endothermic. If it is exothermic, state whether or not it needs a 'kick start' of energy.

(a) candle wax burning *(2 marks)*
(b) photosynthesis *(1 mark)*
(c) electrolysis of sodium chloride *(1 mark)*
(d) iron displacing copper from copper sulfate solution *(2 marks)*
(e) aluminium displacing iron from iron oxide powder *(2 marks)*
(f) baking soda dissolving in lemon juice *(1 mark)*
(g) caustic soda oven cleaner dissolving grease *(2 marks)*
(h) making sodium chloride from sodium hydroxide and hydrochloric acid *(2 marks)*
(i) gunpowder exploding *(2 marks)*
(j) paraffin burning in a jet engine *(2 marks)*
(k) making copper oxide by heating copper carbonate *(1 mark)*
(l) frying an egg *(1 mark)*
(m) limescale remover removing limescale *(2 marks)*
(n) a lead/acid battery discharging *(2 marks)*
(o) a lead/acid battery charging *(1 mark)*

6 The graphs in Figure 2 were produced by a datalogging set-up using a temperature probe.

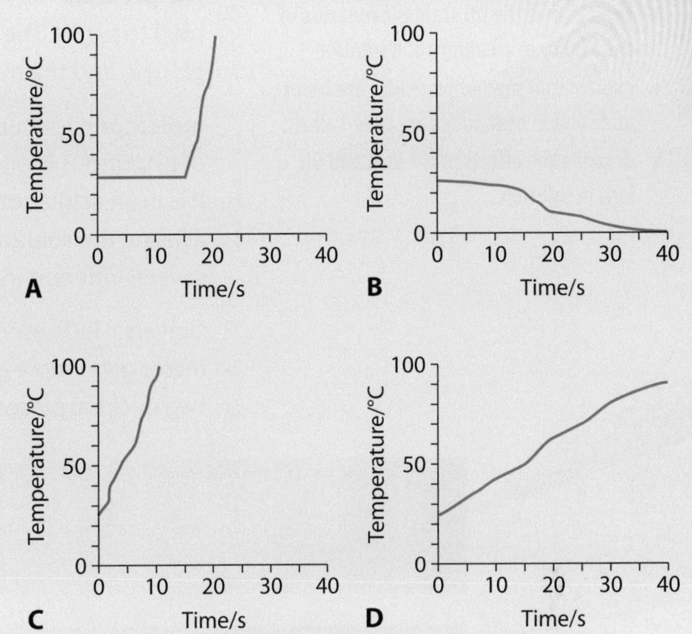

Figure 2 Temperature/time graphs.

Which reaction below could have produced which graph?

(a) solid sodium hydroxide dropped into hydrochloric acid *(1 mark)*
(b) a spoonful of baking soda dropped into a solution of citric acid *(1 mark)*
(c) iron filings dropped into copper sulfate solution *(1 mark)*
(d) a match head lit after 15 seconds *(1 mark)*

7 **(a)** Sodium metal is made by the electrolysis of molten sodium chloride. Is this reaction endothermic or exothermic? *(1 mark)*

(b) Write a balanced chemical equation for this reaction. *(1 mark)*

(c) If you know how much chlorine was produced by a given voltage and current for a given period of time, it is possible to calculate the amount of energy needed to break up a mole of sodium chloride. Explain how you could use this information to predict the energy change that occurs when sodium burns in chlorine. *(2 marks)*

8 Sherbert is a mixture of icing sugar, sodium hydrogen carbonate and citric acid. A student thought that the 'fizzy' feel when you put sherbert in your mouth might be caused by an endothermic reaction. Design an experiment that she could use to test this idea.
In this question you will be assessed on using good English, organising information clearly and using specialist terms where appropriate. *(6 marks)*

Acids and bases

Learning objectives

- explain that the pH scale is a measure of the acidity or alkalinity of a solution
- explain that metal hydroxides are bases and soluble hydroxides are called alkalis
- describe the effects of H^+ ions and OH^- ions in solution.

The pH scale

The **pH scale** is a measure of the acidity or alkalinity of a solution. It runs from pH 0 to pH 14. The most **acidic** solutions have a pH of 0, neutral solutions have a pH of 7, and the most **alkaline** solutions have a pH of 14.

Indicator paper or an indicator solution can be used to give a rough measure of pH. Litmus is one kind of indicator. It changes colour depending on whether it is in an acidic solution or an alkaline one. But it does not reveal how acidic or alkaline the solution is; universal indicator is needed for this. It is a mixture of several different indicators that change colour over different pH ranges.

Universal indicator may provide inaccurate pH values. A properly **calibrated** pH meter gives more accurate measurements, typically with a **resolution** of one or two decimal places.

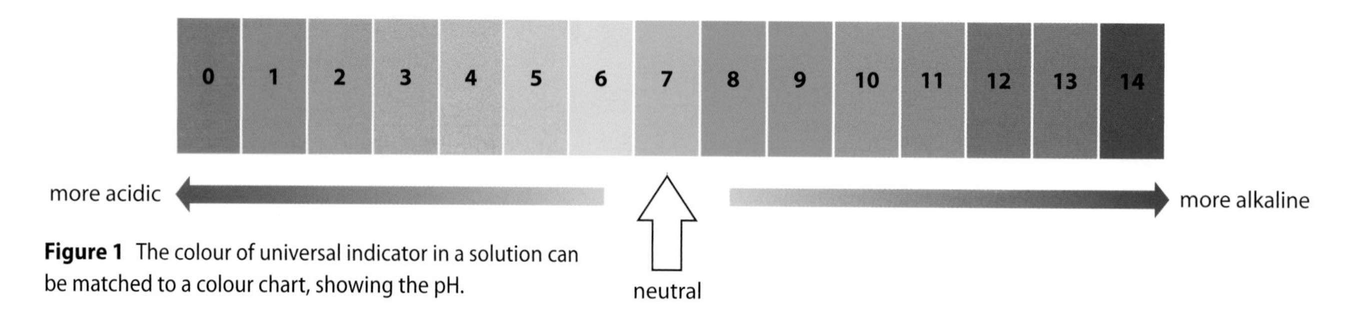

Figure 1 The colour of universal indicator in a solution can be matched to a colour chart, showing the pH.

Science in action

Measuring pH is an important part of monitoring water quality. Pure water is pH 7 at 25 °C but the pH of drinking water varies. The chlorine used to kill harmful microorganisms in drinking water is more effective below pH 7. The World Health Organization recommends that the pH of drinking water should be less than 8.

Bases and alkalis

Copper oxide and other metal oxides are **bases**. They react with acids and neutralise them. Sodium hydroxide and other metal hydroxides are also bases. A **soluble** metal hydroxide is called an **alkali**. Sodium hydroxide is soluble, so it is an alkali as well as a base.

Alkalis have certain properties in common. They:

- turn red litmus paper blue
- form solutions with a pH greater than 7
- release OH^- ions in solution
- neutralise acids to make a salt and water.

Science in action

Material Safety Data Sheets (MSDS) give information about the physical and chemical properties of substances, the hazards they present, and emergency and first aid procedures. For example, oven cleaners contain alkalis. A typical MSDS for an oven cleaner will warn that the substance is corrosive and that gloves should be worn when using it.

Practical

The pH of a solution can be determined using universal indicator paper and a colour chart. It is unwise to dip the paper into the test solution because the dye can leach out and contaminate the test solution. Instead, a clean glass rod is dipped into the test solution. It is touched to the end of a piece of universal indicator paper. The aim is to leave a small spot rather than to soak the paper. The paper should be left for 30 seconds to allow the colour to develop, then the colour of the spot is matched to the nearest colour on the colour chart.

Special 'narrow range' indicator papers are also available. These determine pH values to within 0.5 of a pH unit or better. Typical pH ranges of these papers include 0–2.5, 3–6, 4–7, 7–10 or 10–13.

Alkalis release **hydroxide ions** (OH^-) when they are dissolved in water. These ions make the solution alkaline. **Strong alkalis**, like sodium hydroxide, NaOH, and calcium hydroxide, $Ca(OH)_2$, produce solutions containing high concentrations of hydroxide ions. **Weak alkalis**, like ammonia solution, NH_4OH, produce solutions with lower concentrations of hydroxide ions.

Alkalis are corrosive and will damage skin and eyes, so they must be handled carefully. They feel soapy to the touch because they react with oils and fats to produce water-soluble products. This makes them useful ingredients in household cleaning products.

Examiner feedback

Knowing about hydrogen ion and hydroxide ion concentrations will help you to understand what makes a solution acidic or alkaline. However, you do not need to know about this for the examination.

Acids

Acids release **hydrogen ions** (H^+) when they are dissolved in water. These ions make the solution acidic.

Acids have certain properties in common. They:

- turn blue litmus paper red
- release H^+ ions in solution
- form solutions with a pH less than 7
- react with bases to make a salt and water
- react with many metals to form a salt and hydrogen.

Hydrochloric acid (HCl) and sulfuric acid (H_2SO_4) are **strong acids**. They produce solutions containing high concentrations of hydrogen ions. They are corrosive and will damage skin and eyes, just like strong alkalis. **Weak acids**, such as citric acid, produce solutions with lower concentrations of hydrogen ions. Weak acids have a sharp taste. Think of the taste of lemons, which contain citric acid. Laboratory acids must never be tasted or swallowed.

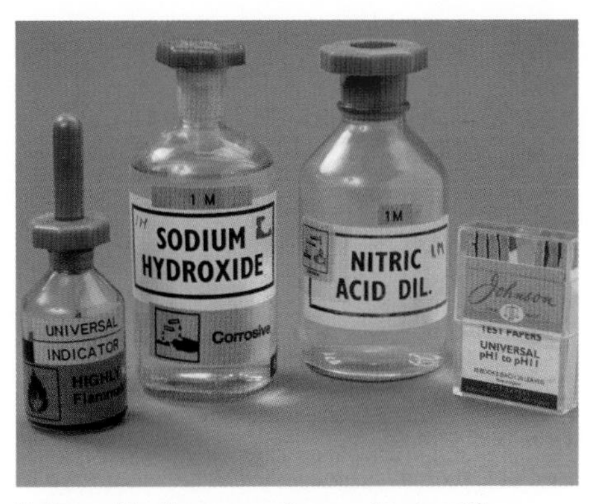

Universal indicator can be used to test if a solution is acidic, alkaline or neutral.

Questions

1. What is the pH scale?
2. Explain why copper oxide is a base but not an alkali.
3. Which ions do acids and alkalis produce in solution?
4. Compare the properties of acids with those of alkalis.
5. Describe the hazards presented by strong acids and strong alkalis.
6. Write balanced chemical equations for the production of H^+ and Cl^- ions by hydrochloric acid, and for the production of H^+ and SO_4^{2-} ions by sulfuric acid.
7. About 1 in 56 000 ethanoic acid molecules release hydrogen ions in solution, whereas almost all nitric acid molecules do so. **(a)** Explain which of these acids is a weak acid. **(b)** Explain which acid will have the lower pH, if both are at the same concentration.
8. Answer these questions in terms of pH and ions. **(a)** The pH of some citric acid is 4 and the pH of some hydrochloric acid is 1. Which is the most strongly acidic, and why? **(b)** The pH of some ammonia solution is 10 and the pH of some sodium hydroxide solution is 14. Which is the most strongly alkaline, and why?

A*

Science skills

The pH scale was proposed in 1909 by a Danish chemist called Søren Sørenson. He knew that the acidity of a solution depended upon the concentration of hydrogen ions. He also knew that these concentrations varied enormously. They could be as low as 1×10^{-14} moles of H^+ ions in each litre of solution to at least as high as 1 mole of H^+ ions in each litre of solution. Sørenson's scale uses 'logarithms' so that an increase of one pH unit is due to an increase in the H^+ ion concentration by 10 times. This idea made the scale much more manageable.

C2 6.2

Neutralisation

Learning objectives

- use state symbols in balanced equations
- describe neutralisation reactions using an ionic equation
- describe how soluble salts can be made from acids by reacting them with alkalis.

State symbols

Consider these two equations:

sodium + water ⟶ sodium hydroxide + hydrogen

$2Na + 2H_2O \longrightarrow 2NaOH + H_2$

The balanced chemical equation gives a lot more information than the word equation. It shows the formula of each substance, and the ratio in which they react or form. **State symbols** provide even more information. Here is the balanced chemical equation again, this time with its state symbols:

$2Na(s) + 2H_2O(l) \longrightarrow 2NaOH(aq) + H_2(g)$

State symbols tell you what state each substance is in: solid (s), liquid (l), gas (g) and dissolved in water – aqueous solution (aq).

The neutralisation reaction

Neutralisation reactions happen when acids and alkalis react together to produce a salt and water. For example:

sulfuric acid + sodium hydroxide ⟶ sodium sulfate + water

$H_2SO_4(aq) + 2NaOH(aq) \longrightarrow Na_2SO_4(aq) + 2H_2O(l)$

In aqueous solution, acids produce hydrogen ions (H^+) and alkalis produce hydroxide ions (OH^-). In neutralisation, the hydrogen ions and hydroxide ions react together to produce water:

$H^+(aq) + OH^-(aq) \longrightarrow H_2O(l)$

This is called an **ionic equation** because it shows what happens to ions in a reaction. The same equation describes neutralisation, whatever acid or alkali is used.

Science in action

Neutralisation reactions are used to relieve the symptoms of indigestion. Stomach acid contains hydrochloric acid, which is a strong acid. The stomach contains a mucus lining that protects against the acid but the oesophagus does not. A sharp pain called 'heartburn' is caused if some of the stomach contents get squeezed into the oesophagus. Antacids are tablets or liquids containing weak alkalis. These react with excess acid and neutralise it.

Examiner feedback

Make sure that you know the ionic equation for neutralisation, including the state symbols for the H^+ ions, OH^- ions and water.

Science skills

Neutralisation is an exothermic reaction. Table 1 shows the amount of energy transferred by heating to the surroundings for three different neutralisation reactions. Each one involves a different strong acid but the same strong alkali.

Table 1 Energy transferred by heating.

Neutralisation reaction	Energy transferred in kJ/mole
$HCl(aq) + NaOH(aq) \longrightarrow NaCl(aq) + H_2O(l)$	57.9
$HBr(aq) + NaOH(aq) \longrightarrow NaBr(aq) + H_2O(l)$	57.6
$HNO_3(aq) + NaOH(aq) \longrightarrow NaNO_3(aq) + H_2O(l)$	57.6

a The table provides evidence in support of the idea that neutralisation is the reaction between H^+ and OH^- ions to form water, rather than the reaction between ions to produce a particular salt. Explain why.

Making soluble salts

Acids and alkalis react together to make a soluble salt and water. When enough acid has been added to completely neutralise the alkali, the water can be evaporated from the salt solution to leave dry crystals of salt.

The **end-point** of the neutralisation reaction, when the acid and alkali have completely reacted, can be found using an indicator. Universal indicator's range of colours makes it difficult to tell when the end-point has been reached, so other indicators are used instead. For example, phenolphthalein is pink in alkaline solutions and colourless in acidic solutions. It is added to an alkaline solution in a conical flask. Acid is added drop-by-drop, from a teat pipette, and the flask swirled after each drop is added, until the pink colour has just disappeared at the end-point.

A white tile makes the colour change at the end-point easier to see.

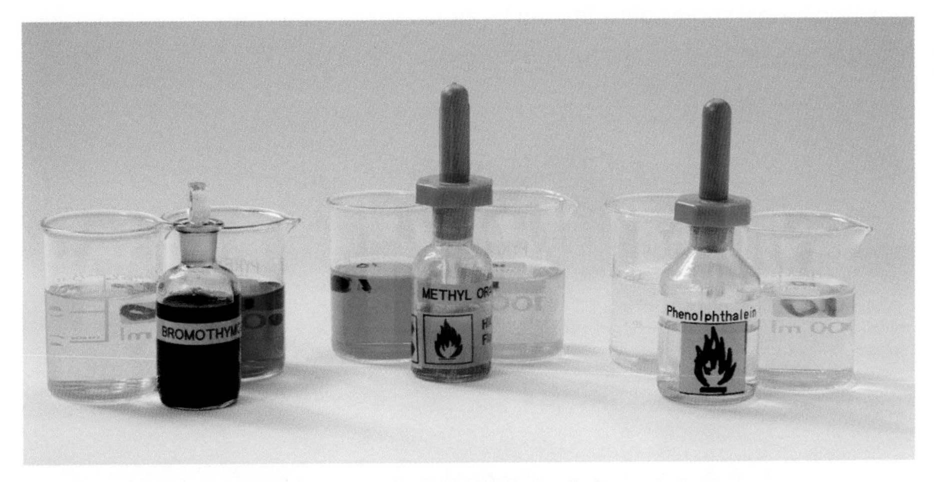

Bromothymol blue is yellow in acidic solutions and blue in alkaline solutions.
Methyl orange is red in acidic solutions and yellow in alkaline solutions.
Phenolphthalein is pink in alkaline solutions and colourless in acidic solutions.

Examiner feedback

The pH of a reaction mixture can also be followed using a pH meter, or by removing drops of the reaction mixture to test its pH with an indicator.

'Activated charcoal' – carbon powder that has been treated to give it a large surface area – is added to remove the phenolphthalein after it has done its job. The mixture is then filtered to remove the charcoal with the indicator bound to it. Water in the filtrate is evaporated to leave crystals of salt.

Questions

1 What information do state symbols provide?

2 What is an ionic equation?

3 Give the names and formulae of the ions produced by acids and alkalis in solution.

4 Describe, with the help of an equation, what happens in neutralisation reactions.

5 Explain why the end-point might be determined using phenolphthalein, instead of using universal indicator solution.

6 Table 1 shows energy data. Explain why it would be useful to include data for reactions involving KOH(aq), rather than just reactions involving NaOH(aq).

7 **(a)** How can an indicator be removed from a salt solution? **(b)** Describe what will happen to the pH of an alkaline solution when excess acid is added to it gradually.

8 Describe how you could produce neutral potassium chloride solution using hydrochloric acid and potassium hydroxide solution.

A*

Taking it further

A piece of glassware called a **burette** can be used to add precise volumes of acid to an alkali. A known volume of alkali is placed in a conical flask. An indicator is added to the alkali, and acid is added to the flask from the burette, until the end-point is reached. The **titre** is the volume of acid needed to exactly neutralise the alkali. Once the titre is known, the titration is repeated without the indicator. The water in the salt solution is then evaporated, as before, to leave crystals of salt.

C2 6.3

More ways to make soluble salts

Learning objectives

- explain how soluble salts can be made by reacting acids with metals or with insoluble bases
- describe how salt solutions can be crystallised to produce a solid salt
- explain that the salt formed depends on the combination of acid and base used.

Examiner feedback

You do not need to recall the reactivity series of metals for the chemistry examinations, as the Data Sheet contains a copy of it.

most reactive

potassium
sodium
calcium
magnesium
aluminium
carbon
zinc
iron
tin
lead
hydrogen
copper
silver
gold
platinum

least reactive

Figure 1 Metals above hydrogen in the **reactivity series** usually react with acids, but those below hydrogen usually do not.

Salts

Sodium chloride, common table salt, is the salt used to flavour and preserve food. It is soluble, i.e. it dissolves in water. Copper sulfate is another soluble salt. Copper sulfate solution is used as a **fungicide** spray for grapevines. The copper ions stop fungal spores germinating by affecting their enzymes, so the fungicide has to be applied before the fungus appears.

Copper sulfate solution is used to prevent fungal disease on grapevines.

Metal or metal oxide?

The reaction of an acid with a metal, or with a metal oxide, is a common way to make soluble salts like sodium chloride and copper sulfate. In general:

metal + acid ⟶ salt + hydrogen

metal oxide + acid ⟶ salt + water

Different acids produce different salts:

- Hydrochloric acid makes chlorides.
- Sulfuric acid makes sulfates.
- Nitric acid makes nitrates.

Sodium chloride might be made by reacting sodium with hydrochloric acid, and copper sulfate by reacting copper with sulfuric acid. However, sodium is a very reactive metal, so it would be dangerous to react it with acid. In contrast, copper is not reactive enough and would not form copper sulfate. To avoid problems like these, metal oxides are often used instead of metals.

Metal oxides are bases. They react with acids and neutralise them. The salt formed depends upon the combination of acid and base used. For example:

- copper oxide reacts with sulfuric acid to make copper sulfate and water
- nickel oxide reacts with sulfuric acid to make nickel sulfate and water.

Practical

Many metal oxides are **insoluble**. When they react with acids, you know that the reaction is complete when there is some metal oxide left over.

The diagrams show how copper sulfate is made by reacting copper oxide with sulfuric acid. Here is the equation for the reaction:

copper oxide + sulfuric acid ⟶ copper sulfate + water

$CuO(s) + H_2SO_4(aq) \longrightarrow CuSO_4(aq) + H_2O(l)$

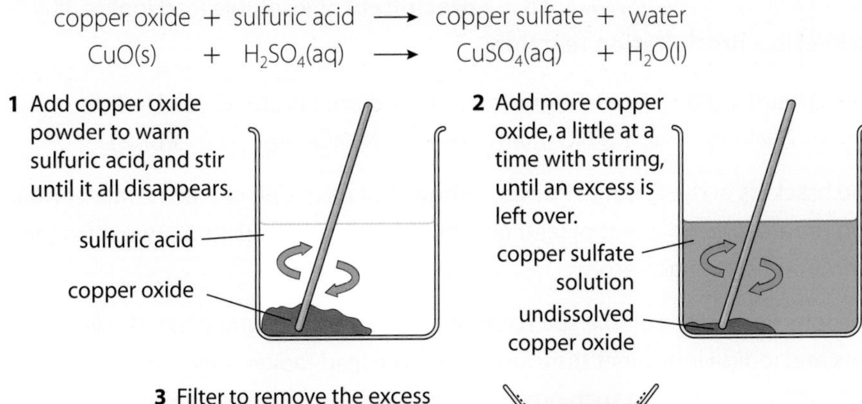

3 Filter to remove the excess copper oxide. Evaporate the water from the solution, leaving blue copper sulfate crystals behind. The more slowly this is done, the larger the crystals will be.

copper sulfate solution

evaporating dish

Figure 2 Making copper sulfate from copper oxide and sulfuric acid.

Science skills

It is important that scientists can communicate their ideas and discoveries to each other easily and clearly. The French chemist Antoine Lavoisier invented a system, published in 1787, to name chemicals. It led directly to the systems used today. Before Lavoisier's work, it was very difficult to work out what a substance was, or what it contained, from its name.

For example, copper sulfate had several names including 'Roman vitriol', 'blue stone' and 'super-vitriolated copper'. Sulfuric acid had even more names. 'Acid of sulphur' was probably easy enough to understand, but other names were not. Chemistry would be much more difficult to understand if you still had to remember names like vitriolic acid, oil of vitriol and spirit of vitriol.

Questions

1. Name two soluble salts and give one use of each.
2. Name the salts formed when nitric acid reacts with: **(a)** magnesium **(b)** copper oxide.
3. Explain why you should make potassium chloride using potassium hydroxide, not potassium.
4. Explain why you could make silver nitrate using silver oxide, but not using silver.
5. Describe how you could make copper sulfate using copper oxide and sulfuric acid.
6. Calcium chloride can be made from hydrochloric acid by reaction either with calcium or with calcium oxide. **(a)** Write a word equation for each reaction. **(b)** Describe what you would expect to observe in each reaction. **(c)** Give a reason, other than the reactivity of calcium, why using calcium oxide may be safer.
7. Zinc nitrate ($Zn(NO_3)_2$) can be made by reacting nitric acid (HNO_3) with zinc, or with insoluble zinc oxide (ZnO). Write balanced chemical equations for the two reactions involved.
8. Magnesium oxide, MgO, is only sparingly soluble in water. Magnesium sulfate, $MgSO_4$, is also known as Epsom salts. It is used in bath water to ease aches and pains. Describe how you could prepare dry magnesium sulfate from magnesium oxide. Include a balanced chemical equation in your answer.

A*

Taking it further

Copper sulfate is white when anhydrous, but blue when hydrated. This is because a copper ion forms a 'complex ion' in solution. Each copper ion becomes chemically bonded to water molecules by 'coordinate bonds', a type of covalent bond.

Route to A*

Transition metals, such as copper and iron, often form more than one ion. The difference is shown using Roman numbers in brackets: Cu^+ is copper(I) and Cu^{2+} is copper(II). The numbers are not used where a metal commonly forms just one ion.

Making insoluble salts

Learning objectives

- explain how insoluble salts can be made by precipitation reactions
- identify the substances needed to make a particular insoluble salt
- describe how precipitation reactions can be used to remove unwanted ions from solutions.

A yellow precipitate of lead iodide forms when sodium iodide solution is mixed with lead nitrate solution.

Examiner feedback

To make an insoluble salt *XY*, a combination of *X* nitrate solution and sodium *Y* solution will always work.

Precipitation reactions

Insoluble salts can be made by reacting together solutions of two soluble salts. Lead nitrate and sodium iodide are soluble salts. They both dissolve in water to form clear, colourless solutions, but clouds of yellow solid appear when they are mixed together. The yellow solid is a **precipitate** of insoluble lead iodide. It is formed in a **precipitation reaction**:

sodium iodide + lead nitrate ⟶ sodium nitrate + lead iodide

$$2NaI(aq) + Pb(NO_3)_2(aq) \longrightarrow 2NaNO_3(aq) + PbI_2(s)$$

The brackets in the formula $Pb(NO_3)_2$ show that each unit of lead nitrate contains two nitrate ions. So a unit of lead nitrate contains one lead atom, two nitrogen atoms and six oxygen atoms.

The ions from each soluble salt have 'swapped partners', and when the lead ions and iodide ions meet they form insoluble lead iodide. Any insoluble salt can be made using such precipitation reactions, provided appropriate solutions of soluble salts are available.

Table 1 The main soluble and insoluble salts.

Soluble	Insoluble
all sodium, potassium and ammonium salts	
all nitrates	
most chlorides, bromides and iodides	lead chloride, bromide and iodide silver chloride, bromide and iodide
most sulfates	barium, calcium and lead sulfates
sodium, potassium and ammonium carbonates	most carbonates
sodium, potassium and ammonium hydroxides	most hydroxides

Preparing an insoluble salt

Most chlorides are soluble, but silver chloride is not. Two soluble salts are needed to make it. One must contain silver ions and the other chloride ions. From the table, you can see that all nitrates are soluble, so silver nitrate will be soluble. In addition all sodium, potassium and ammonium salts are soluble, so sodium chloride will be soluble. This is what happens when they are mixed:

silver nitrate + sodium chloride ⟶ silver chloride + sodium nitrate

$$AgNO_3(aq) + NaCl(aq) \longrightarrow AgCl(s) + NaNO_3(aq)$$

A white precipitate of silver chloride (AgCl) forms. This can be separated by filtration, washed with water and dried in a warm oven.

Practical

Silver chloride crystals can be prepared by mixing silver nitrate solution with sodium chloride solution. The silver chloride precipitate is filtered and washed with distilled water to remove any unreacted substances.

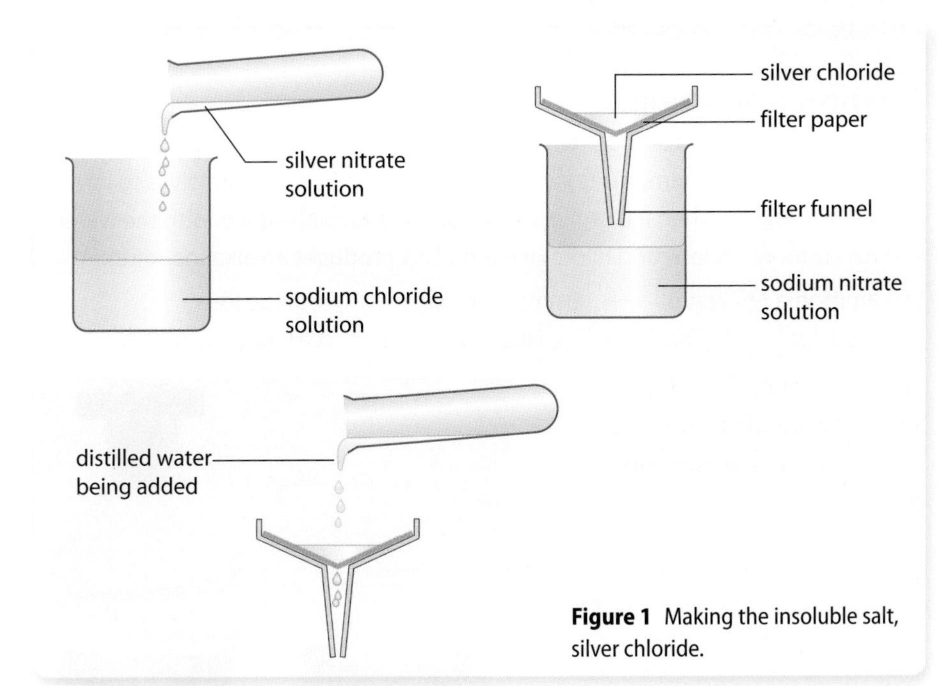

Figure 1 Making the insoluble salt, silver chloride.

Science in action

Silver chloride, silver bromide and silver iodide are light sensitive. The energy supplied by visible light or X-rays is enough to decompose them, forming black crystals of silver. This property makes them useful for traditional photographic film. The increasing use of digital photography, including for X-ray imaging in hospitals, has led to a large decrease in the use of silver for photography.

Science in action

It is important to use the correct amounts of the chemicals when treating water. In 1988 there was an accident at the water works supplying drinking water to Camelford, a town in Cornwall. A tanker driver who was unfamiliar with the site delivered 20 tonnes of aluminium sulfate into the wrong tank by mistake. The dissolved aluminium ions reduced the pH of the water, releasing toxic metals from the pipes. Around 20 000 homes were affected for several weeks afterwards. Health problems included green hair from copper compounds, mouth ulcers, vomiting and rashes. Studies later showed that some people had suffered brain damage due to the aluminium, causing symptoms such as short-term memory loss.

Water treatment

Precipitation reactions can be used to clean up waste **effluent** from factories. This water may contain dissolved transition metal ions such as copper, chromium, cadmium or mercury ions. Transition metal carbonates are insoluble but sodium carbonate is soluble. The effluent can be treated by adding sodium carbonate. Insoluble metal carbonates form precipitates that can be removed by filtering. For example, to treat water contaminated with cadmium:

cadmium ions + carbonate ions ⟶ cadmium carbonate

$$Cd^{2+}(aq) + CO_3^{2-}(aq) \longrightarrow CdCO_3(s)$$

Water from a reservoir must be treated so that it is suitable for drinking. Large pieces of solid such as leaves are removed by sieving. But smaller solids remain suspended in the water, making it cloudy. They stay apart because they are negatively charged and are too small to settle out. Aluminium sulfate is added to clump these particles together. Its positively charged aluminium ions attract the tiny particles, forming a precipitate that settles out faster.

Questions

1. Name two soluble salts and two insoluble salts.
2. Explain what a precipitate is.
3. **(a)** Write a word equation for the reaction between ammonium sulfate and lead nitrate. **(b)** Name the precipitate formed in the reaction.
4. Describe how you would make dry lead chromate from lead nitrate solution and potassium chromate solution.
5. Describe three practical uses of precipitation reactions.
6. Explain, with the aid of a suitable example, what happens in a precipitation reaction involving two soluble salts.
7. Mercury sulfate ($HgSO_4$) is insoluble. Explain how waste water from a factory might be treated with sodium sulfate solution to remove soluble Hg^{2+} ions. Include a balanced chemical equation in your answer.
8. Name two soluble salts that could react together to make silver iodide. Write a word equation and a balanced chemical equation for the precipitation reaction.

Making fertilisers and salts

Learning objectives

- explain that ammonia dissolves in water to produce an alkaline solution
- describe how ammonia is used to produce ammonium salts, which are important as fertilisers
- select an appropriate method for making a particular salt when given appropriate information.

Examiner feedback

Take care not to muddle up ammonia NH_3 with the ammonium ion NH_4^+.

Ammonia solution

Ammonia (NH_3) is a gas at room temperature. It has a very sharp smell and irritates the eyes and throat. Ammonia is very soluble in water. It readily dissolves to form ammonia solution. When it dissolves, some of the ammonia reacts with the water to form ammonium ions and hydroxide ions. This produces an alkaline solution:

ammonia + water ⟶ ammonium ions + hydroxide ions

$$NH_3(g) + H_2O(l) \longrightarrow NH_4^+(aq) + OH^-(aq)$$

Ammonia solution is alkaline. It must be handled with care because it can damage skin and eyes.

Ammonia solution iturns universal indicator solution blue-green.

Ammonium salts

It is often convenient to show ammonia solution as $NH_4OH(aq)$, the compound that would be formed if ammonium ions and hydroxide ions reacted together. Remember, though, that these ions are separate in solution and it is not possible to make solid ammonium hydroxide. Ammonia solution can neutralise acids, just as the other alkalis can. For example:

ammonia solution + sulfuric acid ⟶ ammonium sulfate + water

$$2NH_4OH(aq) + H_2SO_4(aq) \longrightarrow (NH_4)_2SO_4(aq) + 2H_2O(l)$$

The volume of hydrochloric acid needed to neutralise the ammonia solution can be determined using a titration (see lesson C3 4.3). It is usual to add a slight excess of ammonia solution to ensure that the reaction is complete. The ammonia is lost as the water is evaporated from the solution.

Examiner feedback

Take care with brackets in ammonium salts. Ammonium chloride is NH_4Cl, not $(NH_4)Cl$, because there is only one ammonium ion in the formula.

Ammonium salts as fertilisers

As plants grow, they absorb minerals and water through their root hair cells from the soil. Over time, the soil can become deficient in minerals. These must be replaced to allow the plants to grow well. **Fertilisers** are substances added to soil to replace minerals used up by plants. Ammonium salts are very important artificial fertilisers.

Ammonium salts are water soluble, so they are readily absorbed by roots. The ammonium ion also supplies the plant with nitrogen in a usable form. Plants cannot absorb or use nitrogen gas, but nitrogen is essential for making proteins. The growth of plants in nitrogen-deficient soil is stunted.

'Nitrogenous' fertilisers contain ammonium salts such as ammonium nitrate, ammonium sulfate and ammonium phosphate. Of these three, ammonium nitrate NH_4NO_3 contains the largest percentage of nitrogen, making it the best fertiliser for providing nitrogen.

The crop on the left was grown using nitrogen fertiliser. The crop on the right was grown without it. What differences can you see?

Which method for which salt?

Insoluble salts are made using precipitation reactions involving two solutions. There are three ways to make soluble salts. These involve reactions between an acid and:

- a soluble base (alkali)
- an insoluble base
- a metal.

The flowchart summarises how an appropriate method could be chosen. Remember to correctly identify the acid needed:

- hydrochloric acid for chlorides
- sulfuric acid for sulfates
- nitric acid for nitrates.

Examiner feedback

You must be able to choose an appropriate method to make a named salt, given relevant information. You do not need to recall the flowchart below. This is provided to help you practise choosing appropriate methods when you prepare for the examination.

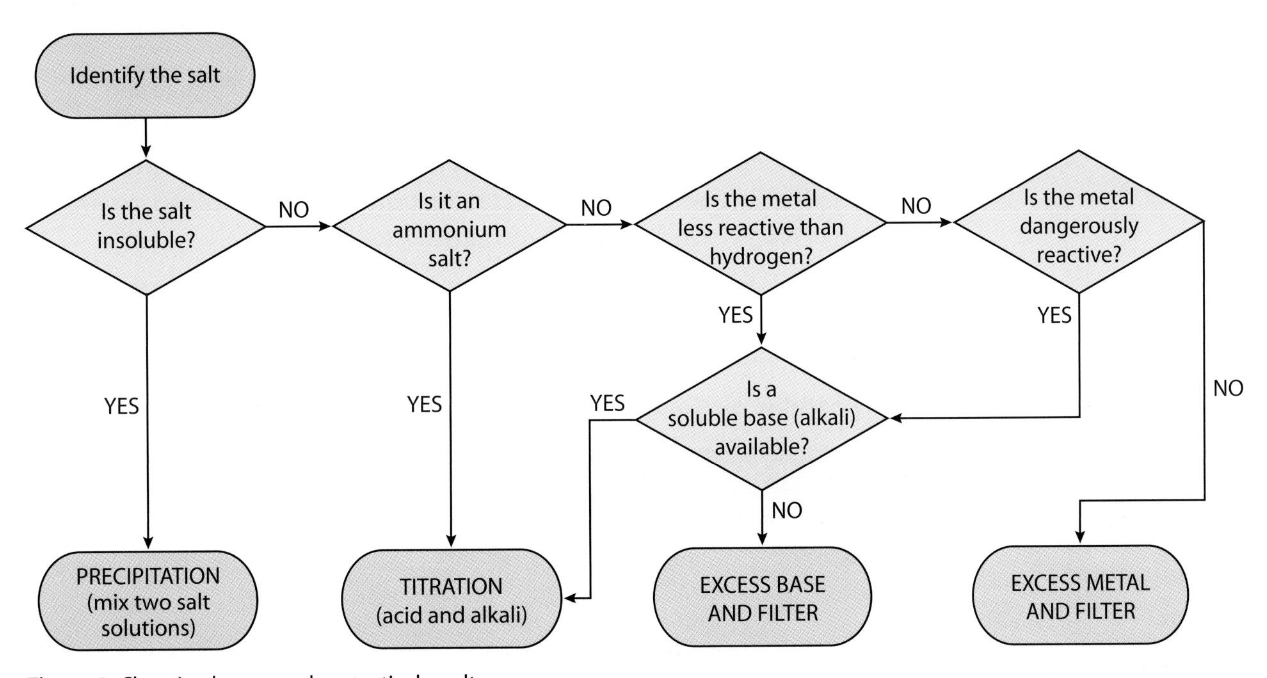

Figure 1 Choosing how to make a particular salt.

Questions

1. Explain, with the help of an equation, why ammonia solution is alkaline.
2. Explain why ammonium salts are useful as fertilisers.
3. Describe how you could prepare solid zinc sulfate using a metal.
4. Explain which method you would use to make: **(a)** potassium chloride **(b)** copper chloride.
5. Ammonia solution itself may be injected into the soil, where it acts as a fertiliser. Describe a benefit and a drawback of doing this.
6. Soluble barium salts such as barium chloride are toxic. However, barium sulfate is useful in medical imaging because it is insoluble and it absorbs X-rays. It allows doctors to examine the structure of their patients' intestines. Explain how you would make dry barium sulfate.
7. Ammonium nitrate NH_4NO_3 and ammonium sulfate $(NH_4)_2SO_4$ are both used as nitrogenous fertilisers. **(a)** Calculate the percentage of nitrogen by mass in each substance. **(b)** Use your answer to part **(a)** to explain which salt is likely to be most useful as a nitrogenous fertiliser.
8. Ammonium phosphate, made using phosphoric acid, H_3PO_4, is useful as a fertiliser. Explain how you could prepare dry ammonium phosphate. Include a balanced chemical equation in your answer.

Electrolysis

Learning objectives

- describe and explain electrolysis of molten ionic substances
- explain oxidation and reduction in terms of losing or gaining electrons
- represent reactions at electrodes using half equations.

Splitting water

Alessandro Volta invented the electric battery in 1800. Scientists were immediately curious to investigate the effect of electricity on different substances. An English chemist called William Nicholson discovered that electricity broke down acidified water into its elements, hydrogen and oxygen. This process is called electrolysis. It also works with other substances.

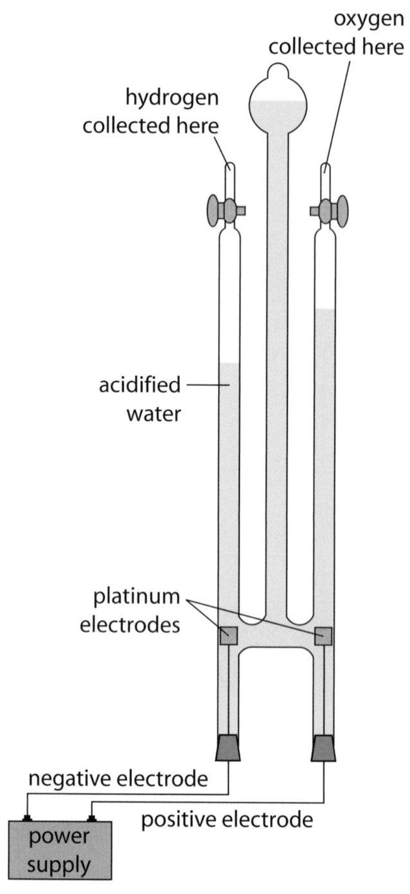

Figure 1 The Hoffman voltameter is often used to demonstrate the electrolysis of water, acidified to increase its **conductivity**.

What is electrolysis?

Electrolysis is the breaking down of an ionic substance into simpler substances using an electric current. The ions in the ionic substance must be free to move for it to work. In an ionic solid, strong electrostatic forces keep the ions in place. However, when the substance is in aqueous solution or is molten, the ions are able to move.

The molten substance or its solution is called the **electrolyte**. Positive and negative **electrodes** carry the electric current into the electrolyte. Negatively charged ions move to the positive electrode and positively charged ions move to the negative electrode. **Oxidation** and **reduction** are often described in terms of gaining or losing oxygen, but they can also involve electrons:

- oxidation is loss of electrons, and it happens to the negatively charged ions at the positive electrode
- reduction is gain of electrons, and it happens to the positively charged ions at the negative electrode.

Examiner feedback

It helps to remember OIL RIG: *o*xidation *i*s *l*oss of electrons, and *r*eduction *i*s *g*ain of electrons. For example, chloride ions lose electrons and are oxidised to chlorine. Zinc ions gain electrons and are reduced to zinc atoms.

Science in action

Very reactive metals like aluminium and potassium are difficult to extract from their compounds. Electrolysis is used to produce such metals on industrial scales. Before the discovery of electrolysis, their existence was suspected but not proven.

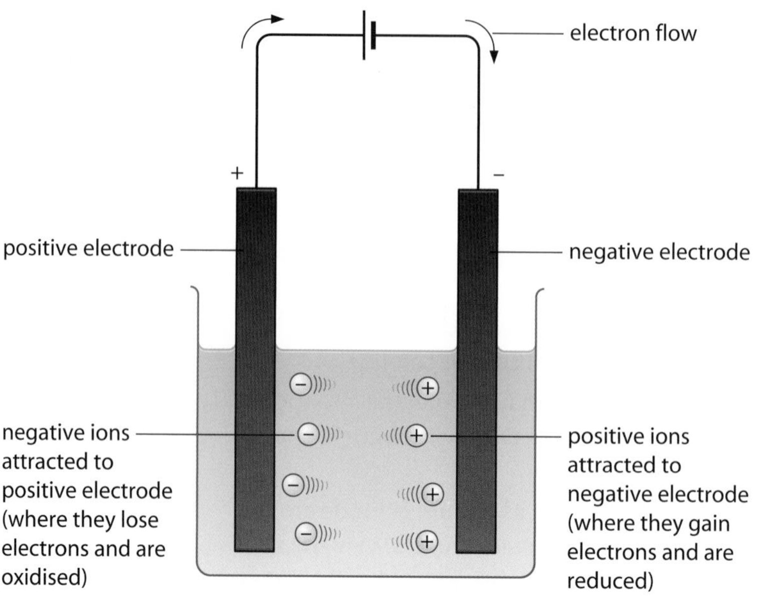

Figure 2 An overview of electrolysis.

Molten ionic substances

Zinc chloride is often used to demonstrate the electrolysis of a molten ionic substance, as it melts at a fairly low temperature (283 °C). Zinc ions move to the negative electrode, where they gain electrons and are **discharged** as zinc metal. Chloride ions move to the positive electrode, where they lose electrons and are discharged as molecules of chlorine gas (Cl_2).

Examiner feedback

Electrolysis is an endothermic process. The energy source is electrical energy rather than thermal energy.

Zinc metal is extracted from zinc compounds by electrolysis. Zinc is coated onto steelwork to help stop it rusting.

Half equations

The reactions at the positive and negative electrodes (**electrode reactions**) can be represented by **half equations**:

At the negative electrode: $Zn^{2+} + 2e^- \longrightarrow Zn$

At the positive electrode: $2Cl^- - 2e^- \longrightarrow Cl_2$

Electrons are shown as e^- in half equations. Notice that zinc ions gain electrons and are reduced, while chloride ions lose electrons and are oxidised. The number of electrons in a balanced half equation must be enough to make the total charge on each side zero.

Taking it further

Half equations can be recombined to give the full balanced equation, if you make sure both half equations have the same number of electrons. If you add both left-hand sides together in the example here, and then both right-hand sides, you get:

$Zn^{2+} + 2e^- + 2Cl^- - 2e^- \longrightarrow Zn + Cl_2$

The electrons cancel out to give:

$Zn^{2+} + 2Cl^- \longrightarrow Zn + Cl_2$ or

$ZnCl_2 \longrightarrow Zn + Cl_2$

Questions

1 In chemistry, what is electrolysis?

2 Explain why electrolysis works if an ionic substance is molten or dissolved in water, but not if it is solid.

3 Explain which products form at each electrode during the electrolysis of molten sodium chloride.

4 Molten lead bromide is sometimes used to demonstrate electrolysis. **(a)** Correctly balance these half equations:

$Pb^{2+} + e^- \longrightarrow Pb$ $\quad Br^- - e^- \longrightarrow Br_2$

(b) Explain what you would see at each electrode.

5 During the electrolysis of water, hydrogen forms at the negative electrode. Balance this half equation:
$H^+ + e^- \longrightarrow H_2$

6 Balance this half equation for the reaction at the positive electrode during the electrolysis of water:
$OH^- - e^- \longrightarrow H_2O + O_2$

7 Potassium was first isolated by Sir Humphry Davy in 1807. He passed electricity through molten potassium hydroxide, KOH. **(a)** Write a balanced half equation for the reaction at the negative electrode. **(b)** Explain whether oxidation or reduction happens at this electrode.

8 Magnesium is produced commercially by the electrolysis of magnesium chloride, ($MgCl_2$), which is abundant and melts at 714 °C. **(a)** Write a balanced half equation for the reaction at the negative electrode. **(b)** The reaction at the positive electrode can be shown as $2Cl^- - 2e^- \longrightarrow Cl_2$. Explain why it may be shown instead as $2Cl^- \longrightarrow Cl_2 + 2e^-$. **(c)** Suggest and explain two reasons why the production of magnesium is expensive.

Electroplating

Learning objectives

- describe and explain electroplating
- predict the products when electrolysing solutions of ions.

A thin layer of metal

Electroplating involves using electrolysis to create a thin layer of metal over the surface of an object. This may be done to improve an object's appearance or to make it more resistant to corrosion or abrasion. It may also be done to reduce costs: for example, inexpensive jewellery is sometimes made from a fairly cheap metal such as nickel, and then plated with gold or silver.

The object to be electroplated is used as the negative electrode. The positive electrode is a piece of the electroplating metal. The electrolyte is a solution containing ions of the electroplating metal. During electrolysis, the metal ions are attracted to the negatively charged object. They gain electrons and are discharged as a thin layer of metal atoms on its surface.

High-end electrical and computer connectors are gold-plated. Gold is corrosion-resistant and a very good conductor of electricity.

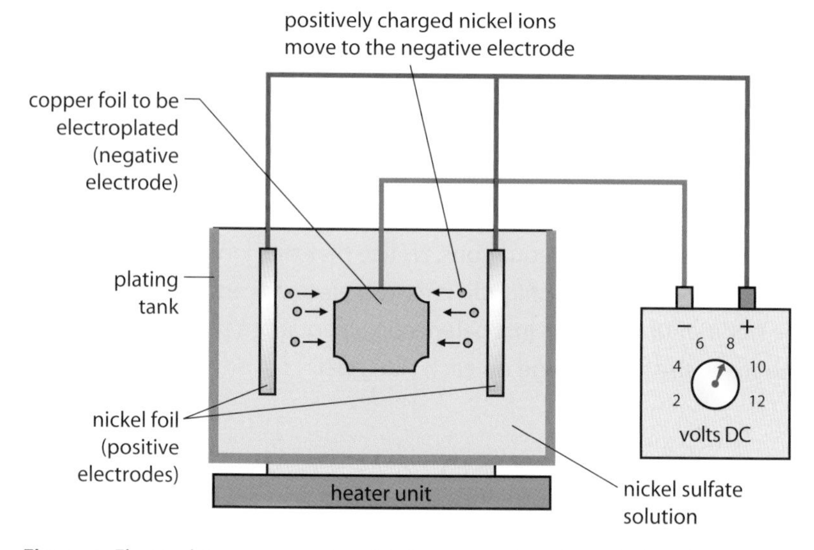

Figure 1 Electroplating copper with nickel. The negative electrode is a piece of copper foil and the positive electrode is a piece of nickel. The electrolyte is a solution of nickel sulfate.

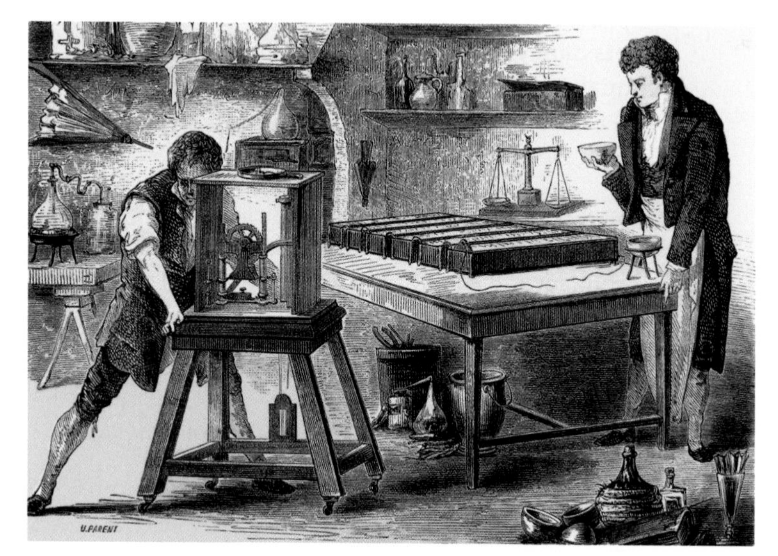

In 1807 Humphry Davy (on the right) became the first person to isolate potassium, using electrolysis. He produced only hydrogen at the negative electrode until he changed the electrolyte from potassium hydroxide solution to molten potassium hydroxide.

Electrolysis of solutions

Electrolysis of a solution of metal ions does not always produce pure metal at the negative electrode. In the electrolysis of copper chloride solution, copper forms at the negative electrode and chlorine at the positive electrode. However, if potassium chloride is used instead, hydrogen forms at the negative electrode rather than potassium. Why is this?

The reason concerns the reactivity series (see lesson C2 6.3). A solution of a metal chloride contains two different positive ions: the metal ions themselves, and hydrogen ions H^+ from the water. Copper is less reactive than hydrogen. When the positive ions reach the negative electrode, copper ions are discharged in preference to hydrogen ions. On the other hand, hydrogen is less reactive than potassium. Hydrogen ions are discharged at the negative electrode in preference to potassium ions.

Examiner feedback

The Data Sheet provided in the examination shows the reactivity series. Hydrogen is discharged at the negative electrode during the electrolysis of a solution of ions, unless the metal is less reactive than hydrogen.

Which non-metal?

Non-metals are produced at the positive electrode during electrolysis. In the electrolysis of copper chloride solution, chlorine is produced at the positive electrode. However, when copper sulfate or copper nitrate solution are used, oxygen is produced instead.

This happens because of differences in how easily different negative ions are discharged in solution. Copper chloride solution contains chloride ions Cl^-, plus hydroxide ions OH^- from the water. Chloride ions are more easily discharged than hydroxide ions. The other halide ions, bromide Br^- and iodide I^-, are also more easily discharged. On the other hand, hydroxide ions are more easily discharged than sulfate ions SO_4^{2-} or nitrate ions NO_3^-. So oxygen is produced during the electrolysis of copper sulfate solution or copper nitrate solution.

Questions

1 Explain why hydrogen is produced when sodium chloride solution is electrolysed.

2 Explain why oxygen is produced when sodium nitrate solution is electrolysed.

3 Name the ions present in the following solutions: **(a)** lead chloride **(b)** silver nitrate **(c)** potassium hydroxide.

4 Name the products formed at each electrode during the electrolysis of each solution in question **3**.

5 Describe how a steel knife could be electroplated with silver using silver nitrate solution.

6 Explain why the electrolysis of dilute hydrochloric acid HCl(aq) produces hydrogen and chlorine, but the electrolysis of dilute sulfuric acid H_2SO_4(aq) produces hydrogen and oxygen.

7 Copper nitrate dissolves in water to produce a blue solution of ions. It can be electrolysed using graphite electrodes. **(a)** Identify the ions in the solution. **(b)** Write a balanced half equation for the reaction that takes place at the negative electrode. **(c)** Suggest why the blue colour fades.

8 Food cans are made from steel electroplated with tin. The tin forms a protective layer, preventing food reacting with the steel. **(a)** Suggest, with reasons, suitable substances for each electrode and for the electrolyte. **(b)** Describe what you would see during the electroplating process. **(c)** Write a balanced half equation for the discharge of tin ions, Sn^{2+}.

A*

Science in action

In the industrial electroplating of metals, it is most important to obtain a clean surface first. Oily contaminants prevent the layer of electroplating metal sticking firmly to the metal below. The metal to be electroplated is usually cleaned using solvents and hot alkaline solutions. The surface is tested to see how clean it is by rinsing the metal in water and holding the surface in the vertical position. On a contaminated surface, water forms beads that run off. It stays as a thin film on clean surfaces.

Examiner feedback

Remember that sulfate solutions do not produce sulfur during electrolysis, nor do nitrate solutions produce nitrogen.

Science in action

Steel rusts when it is exposed to air and water. In the past, cars rusted very quickly because there was little protection for the steel body parts, apart from the paint. If this was damaged, the steel would rust and even more paint would come off as the rust spread underneath the paint. Nowadays great care is taken to protect the steel. For example, it is 'galvanised' by electroplating a layer of zinc on top. Zinc is more reactive than the iron in steel. It continues to protect the steel underneath, even when the zinc layer is damaged.

Taking it further

'Electrode potentials' are used at A level to predict the outcomes of electrochemical reactions. If the metals are arranged in order of their electrode potentials, the list is very similar to the reactivity series. However, it is not identical, because several factors are involved in determining the reactivity of a metal.

Aluminium manufacture

Learning objectives

- describe and explain how aluminium is manufactured by electrolysis.

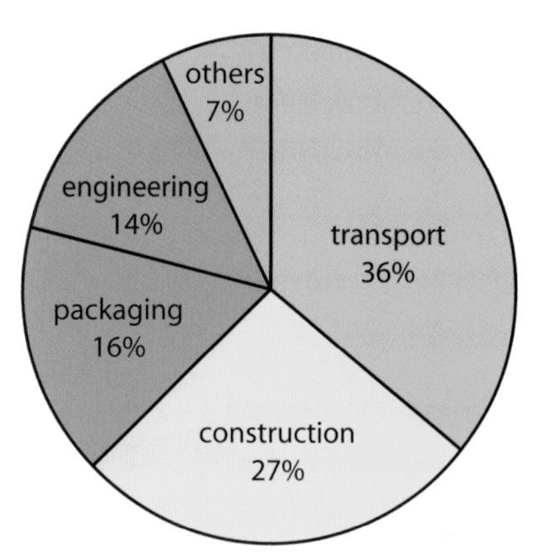

Figure 1 Aluminium is strong, lightweight and resists corrosion. It has many uses.

Science in action

Aluminium is a reactive metal but, unlike iron and steel, it is resistant to corrosion. It is protected from corrosion by a natural microscopically thin layer of aluminium oxide. This stops air and water reaching the metal below and reacting with it. The ability of aluminium to resist corrosion, combined with its strength, low density and ability to conduct electricity and heat well, makes it a very useful metal. For example, aluminium is used to make aircraft and trains, window frames, cooking foil and drinks cans, television aerials and overhead electricity cables.

Examiner feedback

Make sure you understand why cryolite is used in the manufacture of aluminium, including its economic and environmental benefits.

Making aluminium by electrolysis

The American chemist Charles Hall and the French chemist Paul Héroult were both born in 1863 and both died in 1914. In 1886, working on opposite sides of the Atlantic, the two men independently invented a way to produce aluminium by electrolysis. Aluminium is still manufactured today using the Hall–Héroult process. Around 39 million tonnes of aluminium are manufactured in the world each year.

Aluminium smelting uses huge amounts of electricity. This smelting plant was built next to a hydroelectric power plant to provide cheap, plentiful electricity.

Before Hall and Héroult

Aluminium is the most abundant metal in the Earth's crust. It is strong, has a low density and is resistant to corrosion. However, it is high in the reactivity series and is naturally found combined with other elements. For example, **bauxite** is the most abundant aluminium **ore**. The aluminium is present as aluminium hydroxide, which can be processed to make aluminium oxide (Al_2O_3).

Aluminium is too reactive to be easily extracted from aluminium oxide so, for many years, the existence of aluminium was suspected but could not be confirmed. Pure aluminium was first isolated in 1825 by heating aluminium chloride with potassium, an even more reactive metal. Potassium has to be extracted using electrolysis, so the whole process was very expensive. Aluminium was more expensive at that time than silver. The work of Hall and Héroult led to a big reduction in the cost of aluminium manufacture and it became economic to produce aluminium on a large scale.

The Hall–Héroult process

Aluminium oxide is insoluble in water, so it must be molten to allow an electric current to pass through it. Unfortunately, aluminium oxide has a very high melting point, 2054 °C. Heating the oxide to such a high temperature would be very costly.

Hall and Héroult discovered that aluminium oxide dissolves in molten **cryolite** at 1012 °C, a much lower temperature than the melting point of aluminium oxide. Using a molten mixture of aluminium oxide and cryolite for the electrolyte involves less heating. Since the electrolyte is heated electrically, this reduces the electricity costs of manufacturing aluminium. Lower electricity use also means reduced emissions of waste gases to the environment, if the electricity is generated using fossil fuels.

Electrolysis takes place in a steel container, known as a cell. This is lined with carbon, which acts as the negative electrode. The positive electrodes are also made from carbon. They dip into the electrolyte, the molten mixture of aluminium oxide and cryolite. An electric current of around 200 000 A is passed through the electrolyte. This is equivalent to the current drawn by around 15 000 domestic kettles at once.

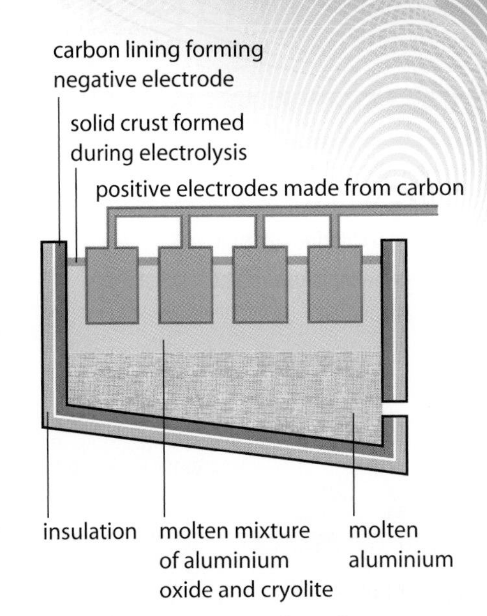

Figure 2 A cross-section through an aluminium cell.

Balanced half equations for the aluminium cell

Aluminium ions are positively charged and move to the negative electrode. Here they gain electrons and are reduced to molten aluminium:

$$Al^{3+} + 3e^- \longrightarrow Al$$

Oxide ions are negatively charged and move to the positive electrode. Here they lose electrons and are oxidised to oxygen:

$$2O^{2-} \longrightarrow O_2 + 4e^-$$

The oxygen reacts with the hot carbon electrodes to produce carbon dioxide. This reaction gradually wears away the electrodes, so they must be replaced regularly.

Examiner feedback

You do not need to recall these half equations for the examination. However, you are expected to be able to complete, and balance, given half equations for the reactions occurring at the electrodes during electrolysis.

Questions

1. Explain why aluminium is manufactured using electrolysis.
2. Describe what happens at each electrode during aluminium manufacture.
3. Suggest why aluminium costs around three times as much per tonne as steel, even though it is the most abundant metal in the Earth's crust.
4. Explain why cryolite is used in the manufacture of aluminium.
5. Write a balanced half equation for the reaction at each electrode, and an overall balanced chemical equation for the production of aluminium from aluminium oxide.
6. Explain why aluminium smelters are often sited near sources of hydroelectricity.
7. **(a)** Explain why, in aluminium manufacture, reduction happens at the negative electrode and oxidation happens at the positive electrode. **(b)** Describe an environmental problem associated with a product of aluminium manufacture by electrolysis.
8. Titanium is another abundant metal that is difficult to extract from its ore. A new method of manufacture involves electrolysing a mixture of titanium dioxide and molten calcium chloride. **(a)** Balance this half equation and explain at which electrode it takes place: $TiO_2 + e^- \longrightarrow Ti + O^{2-}$ **(b)** Explain why oxide ions move through molten calcium chloride. **(c)** Explain, with the help of two equations, why carbon dioxide is produced at the other electrode, which is made from carbon. A*

Science skills

Table 1 compares some of the properties of copper and aluminium.

Table 1 Properties of copper and aluminium.

Property	Copper	Aluminium
density in g/cm^3	8.9	2.7
cost in £/tonne	4400	1320
electrical conductivity	1.6	1.0

a Use the information to explain why aluminium is used for overhead electricity cables, rather than copper.

The chlor-alkali industry

Learning objectives

- describe and explain the production of useful substances by the electrolysis of sodium chloride solution
- describe some uses of these substances.

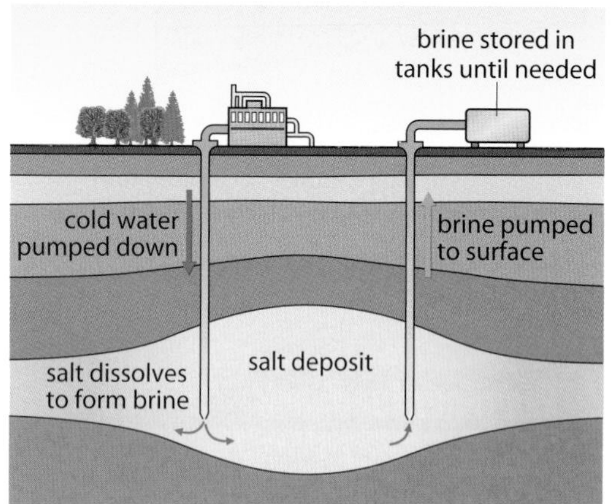

Figure 1 The concentrated sodium chloride solution produced by solution mining is called brine.

Mining solutions

Deep beneath the fields of Cheshire the buried remains of an ancient dried-out sea are being mined. Water is forced through boreholes in underground deposits of rock salt. The water dissolves the salt, forming a saturated salt solution. This is pumped to the surface where it becomes the raw material for the chlor-alkali industry.

Electrolysing sodium chloride solution

Sodium chloride solution contains a mixture of ions: positively charged sodium and hydrogen ions (Na^+, H^+) and negatively charged chloride and hydroxide ions (Cl^-, OH^-). During electrolysis, the positively charged ions move to the negative electrode. Since hydrogen is less reactive than sodium, the hydrogen ions are discharged and hydrogen gas is produced:

$$2H^+ + 2e^- \longrightarrow H_2$$

The negatively charged ions move to the positive electrode. Chloride ions are discharged rather than the hydroxide ions, and chlorine gas is produced:

$$2Cl^- \longrightarrow Cl_2 + 2e^-$$

The remaining sodium ions and hydroxide ions form sodium hydroxide solution. This is the alkali in the 'chlor-alkali' name. The following equation summarises the overall process:

$$\text{sodium chloride} + \text{water} \longrightarrow \text{hydrogen} + \text{chlorine} + \text{sodium hydroxide}$$
$$2NaCl(aq) + 2H_2O(l) \longrightarrow H_2(g) + Cl_2(g) + 2NaOH(aq)$$

Examiner feedback

Remember that sodium metal is too reactive to be produced from sodium chloride solution, but it would be produced if molten sodium chloride were electrolysed instead.

Practical

The electrolysis of sodium chloride solution is easily demonstrated in the laboratory. The presence of hydrogen is confirmed using a lighted splint, which ignites the gas with a 'pop'. The presence of chlorine is confirmed using damp starch iodide paper (or by using damp litmus paper that is bleached), which turns black. The alkaline solution turns universal indicator solution purple.

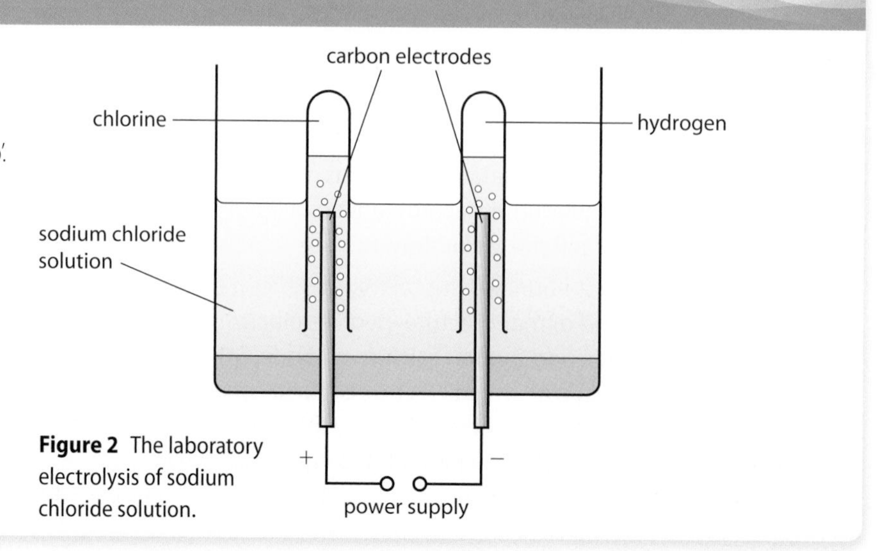

Figure 2 The laboratory electrolysis of sodium chloride solution.

The industrial process

The three products of the chlor-alkali process should not be allowed to mix; otherwise unwanted **by-products** such as sodium chlorate are made. There are different solutions to the problem. One design is called the membrane cell. It contains a polymer sheet, the membrane, which allows positively charged ions to pass through, but not negatively charged ions. As a result, the chlorine and sodium hydroxide cannot react with each other.

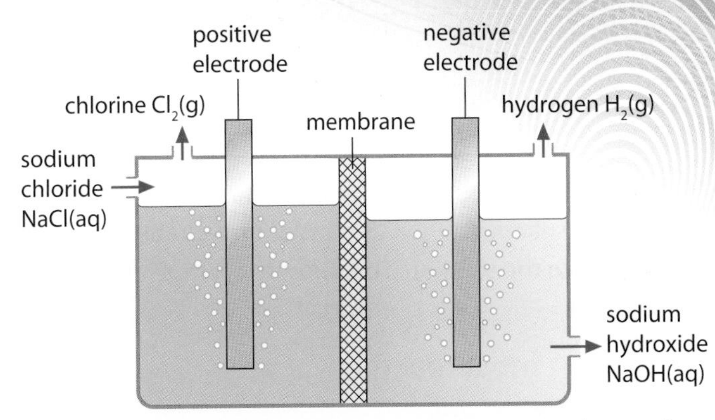

Figure 3 Electrolysis of sodium chloride solution in a membrane cell.

Examiner feedback

You will not be asked to recall any details of the membrane cell. However, you may have to identify the products in a labelled diagram and explain the function of the membrane.

Using the three products

Hydrogen is used in the manufacture of ammonia. It is also increasingly important as a fuel. Chlorine is used to kill bacteria in drinking water. It is used in the manufacture of chloroethene, the monomer for poly(chloroethene) or PVC. Chlorine and sodium hydroxide make sodium chlorate, used as bleach. Sodium hydroxide is used in the manufacture of paper and soap.

Chlorine is used to kill bacteria in swimming pool water.

Questions

1. Write a word equation for the overall process of electrolysing sodium chloride solution.
2. **(a)** Describe two uses for each product from the electrolysis of sodium chloride solution. **(b)** Explain why chlorine should not be allowed to mix with the other two products.
3. Write a balanced half equation for the reaction at each electrode during the electrolysis of sodium chloride solution, and an overall balanced chemical equation.
4. Explain why sodium hydroxide is formed by the electrolysis of sodium chloride solution, but not by the electrolysis of molten sodium chloride.
5. The three products of the chlor-alkali process are formed in a fixed ratio, with equal amounts of the two gases and double the amount of the alkali. Suggest, giving your reasons, two problems that this might cause the chlor-alkali industry.
6. Hydrogen produces only water vapour when it burns. Suggest another reason why it may be viewed as an 'environmentally friendly' fuel.
7. Worldwide, only 20% of electricity is generated using renewable resources. The rest is generated using fossil fuels or nuclear fuels. The chlor-alkali process uses around one per cent of the world's electricity. To what extent might this information affect the view of hydrogen as an 'environmentally friendly' fuel?
8. An older design of cell, the mercury cell, uses a layer of flowing mercury as one of its electrodes. Sodium is produced and dissolves in the mercury, forming a mixture called an amalgam. **(a)** Write a balanced half equation for the production of sodium from sodium ions. **(b)** Explain which electrode, positive or negative, the mercury is. **(c)** The amalgam is mixed with water, letting the sodium react with water to form sodium hydroxide and hydrogen. Write a balanced chemical equation for this reaction. **(d)** Write a balanced chemical equation for the production of sodium chlorate, NaClO, from sodium hydroxide and chlorine.

ISA practice: the best metal for hand-warmers

Many metals in the reactivity series react with 1 mole per dm^3 copper sulfate solution to transfer measurable amounts of energy into the solution. This raises the temperature of the solution.

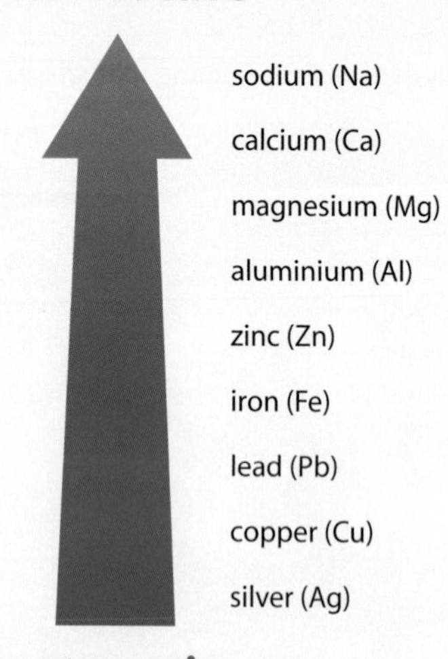

Figure 1 The metal reactivity series.

Section 1

1 Write a hypothesis about the position of a metal in the reactivity series and how it may affect the amount of energy transferred to copper sulfate solution. Use information from your knowledge of displacement reactions to explain why you made this hypothesis. *(3 marks)*

2 Describe how you could carry out an investigation into this factor.

You should include:

- the equipment that you would use
- how you would use the equipment
- the measurements that you would make
- a risk assessment
- how you would make it a fair test.

You may include a labelled diagram to help you to explain the method.

In this question you will be assessed on using good English, organising information clearly and using specialist terms where appropriate. *(9 marks)*

3 Design a table that you could use to record all the data you would obtain during the planned investigation. *(2 marks)*

Total for Section 1: 14 marks

Section 2

Two students, Study Group 1, carried out an investigation into the metals. They investigated the first two metals in the first lesson, and two days later tested the other three. Their results are shown in Figure 1.

Magnesium raised temperature of solution by 11°C
Iron raised temperature of solution by 2°C
Zinc raised temperature of solution by 3°C
Copper did not raise temperature of solution at all
Aluminium raised temperature of solution by 7°C

Figure 1 Study Group 1's results.

4 **(a)** Plot a graph of these results. *(4 marks)*

(b) What conclusion can you draw from the investigation about a link between the temperature change and the reactivity series? You should use any pattern that you can see in the results to support your conclusion. *(3 marks)*

(c) Do the results support the hypothesis you wrote in answer to question 1? Explain your answer. You should quote some figures from the data in your explanation. *(3 marks)*

Opposite are the results from three more study groups.

Table 1 shows the results of Study Group 2, two other students who investigated the hypothesis. They used five metals and recorded the temperature change of their solutions after 10 minutes.

Table 1 Study Group 2's results.

Metal	Temperature change of solution 10 minutes after adding metal/°C
magnesium	6
iron	0
zinc	1
copper	0
aluminium	3

Study Group 3 was a third group of students. Their results are given in Table 2. They used 50 cm^3 of solution and 5 g of metal.

Table 2 The results obtained by Study Group 3.

Metal	Temperature change of the solution/°C			
	Test 1	Test 2	Test 3	Mean of tests
copper	0	1	0	0
silver	1	0	1	2
zinc	6	5	6	6
magnesium	22	24	21	22
calcium	5	4	4	5

Study Group 4 were researchers for the hand-warmer manufacturers looked on the internet and found out how much each of the metals would cost to buy. This information is shown in Table 3.

Table 3 Cost of each of the metals used in the investigations.

Metal	Cost of 25 g of the metal/£
magnesium	0.04
iron	0.02
zinc	0.04
copper	0.15
aluminium	0.04
calcium	2.58
silver	0.43

Study Group 4 also measured the energy released by each metal when 10 g of metal is displaced, and when 1 mole of the metal (A_r) is displaced. Table 4 shows the results.

Table 4 Energy released by each metal.

Metal	Energy released/kJ	
	10 g displaced	1 mole of metal (A_r) displaced
magnesium	461	1105.4
iron	131.5	736.4
zinc	122.7	797.5
copper	n/a	n/a
aluminium	430.6	1162.6
calcium	n/a	n/a
silver	n/a	n/a

5 **(a)** Draw a sketch graph of the results from Study Group 2. *(3 marks)*

(b) Look at the results from Study Groups 2 and 3. Does the data support the conclusion you reached about the investigation in question 5(a)? Give reasons for your answer. *(3 marks)*

(c) The data contain only a limited amount of information. What other information or data would you need in order to be more certain whether the hypothesis is correct or not?
Explain the reason for your answer. *(3 marks)*

(d) Look at Study Group 4's results. Compare them with the data from Study Group 1. Explain how far the data supports or does not support your answer to question 5(b). You should use examples from Study Group 4's results and from Study Group 1. *(3 marks)*

6 **(a)** Compare the results of Study Group 1 with Study Group 2. Do you think that the results for Study Group 1 are *reproducible*?
Explain the reason for your answer. *(3 marks)*

(b) Explain how Study Group 1 could use results from other groups in the class to obtain a more *accurate* answer. *(3 marks)*

7 Applying the results of the investigation to a context.

Suggest how ideas from the original investigation and the other studies could be used by the manufacturers to decide which metal they should use to obtain a temperature rise of 20 °C using only 25 g of metal. *(3 marks)*

Total for Section 2: 31 marks

Total for the ISA: 45 marks

Assess yourself questions

1 Explain what the symbols (s), (l), (g) and (aq) are, and what they mean. *(5 marks)*

2 Figure 1 shows how copper sulfate crystals can be made.

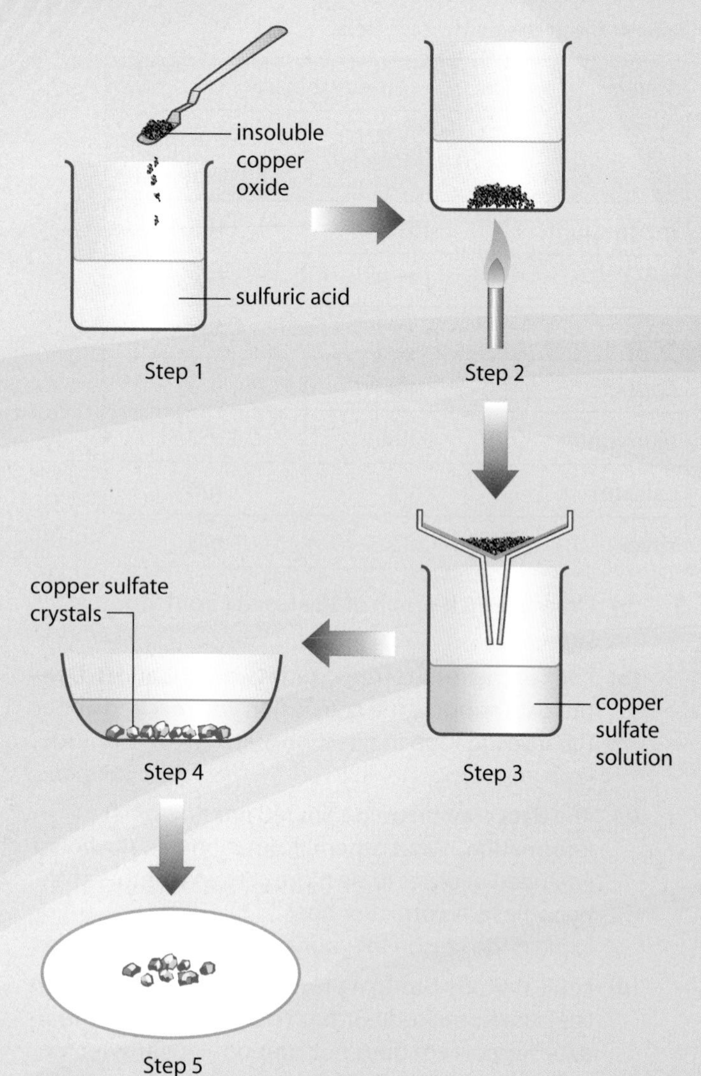

Figure 1 Making copper sulfate crystals.

(a) Explain why the mixture of sulfuric acid and copper oxide is warmed. *(1 mark)*

(b) Write a word equation for the reaction. *(2 marks)*

(c) Why is the copper sulfate solution filtered? *(1 mark)*

(d) Describe how you could make copper sulfate crystals form quickly from the solution. *(1 mark)*

3 The information below describes how silver chloride, an insoluble salt, can be made.

Mix silver nitrate solution with an excess of sodium chloride solution. Filter the precipitate of silver chloride from the mixture. Wash it several times with de-ionised water. Dry the silver chloride and store it in a dark bottle.

(a) What is a precipitate? *(1 mark)*

(b) Why was an *excess* of sodium chloride used in the preparation? *(1 mark)*

(c) How do you know, from the information in the paragraph above, that silver chloride is insoluble rather than soluble in water? *(1 mark)*

(d) Describe one method that could be used to dry the silver chloride precipitate. *(1 mark)*

(e) Write a word equation for the reaction between silver nitrate and sodium chloride. *(2 marks)*

4 Table 1 shows which substances are likely to be soluble and which are likely to be insoluble.

Table 1 Soluble and insoluble.

Soluble	Insoluble
all sodium salts	
all nitrates	
most other chlorides	lead chloride and silver chloride
most other sulfates	lead sulfate and barium sulfate
sodium hydroxide	most other hydroxides

Use the information in the table to help you answer the following questions.

(a) Name *two* insoluble lead salts. *(2 marks)*

(b) Which substance, barium nitrate or barium hydroxide, would be suitable to mix with sodium sulfate solution to make barium sulfate? Give reasons for your answer. *(2 marks)*

(c) Name *two* soluble substances that could be mixed together to make lead chloride. *(2 marks)*

5 Soluble salts can be made from acids by reacting them with metals, with insoluble bases, or with alkalis. Explain, giving reasons, the most suitable method to prepare the following salts from hydrochloric acid.

(a) potassium chloride *(3 marks)*

(b) copper chloride *(3 marks)*

6 Different acids produce different salts.

(a) Name the salt produced by each of the following pairs of reacting substances.

(i) sodium hydroxide and hydrochloric acid *(1 mark)*

(ii) potassium hydroxide and sulfuric acid *(1 mark)*

(iii) zinc oxide and sulfuric acid *(1 mark)*

(b) Name the acid needed to produce sodium nitrate. *(1 mark)*

7 Ammonium salts are important as fertilisers.

(a) What is a fertiliser? *(1 mark)*

(b) Name the salt produced by reacting ammonia solution with the following acids.

(i) sulfuric acid *(1 mark)*

(ii) nitric acid *(1 mark)*

(c) Ammonium chloride can be made from ammonia solution.

(i) Name the acid needed. *(1 mark)*

(ii) Describe what would happen to the pH of the ammonia solution if excess hydrochloric acid was added to it. *(1 mark)*

8 The presence of certain ions makes solutions acidic or alkaline.

(a) Give the name or formula of the following ions.

(i) The ion that makes solutions acidic. *(1 mark)*

(ii) The ion that makes solutions alkaline. *(1 mark)*

(b) Write a balanced chemical equation, involving these ions, to represent neutralisation. Include the correct state symbols. *(3 marks)*

9 Figure 2 shows the electrolysis of molten lead iodide.

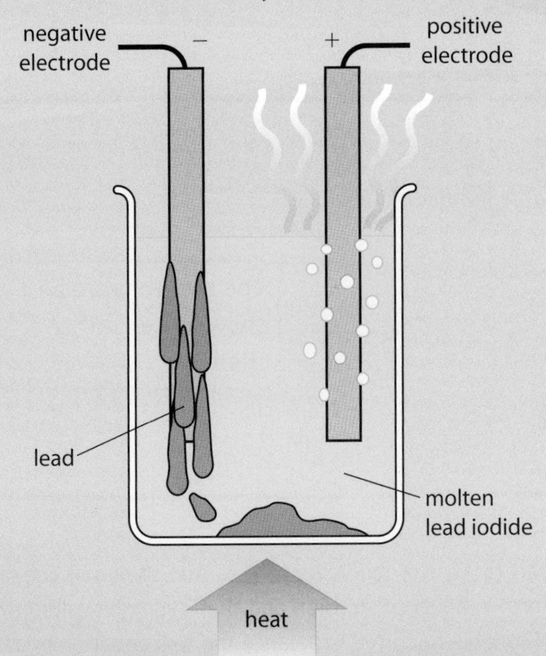

Figure 2 Molten lead iodide can be electrolysed using carbon electrodes.

(a) What is meant by the word *electrolysis*? *(3 marks)*

(b) Explain why the positively charged ions move to the negative electrode. *(1 mark)*

(c) Name the product formed at the positive electrode. *(1 mark)*

(d) Explain why an electric current will pass through molten lead iodide, but not through solid lead iodide. *(2 marks)*

10 Aluminium is manufactured by the electrolysis of a molten mixture of aluminium oxide and cryolite, as seen in Figure 3.

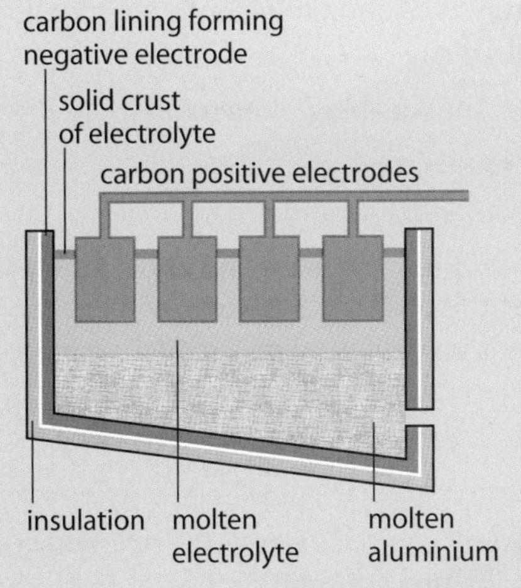

Figure 3 Aluminium is extracted from aluminium oxide by electrolysis.

(a) Why is molten cryolite used? *(2 marks)*

(b) Aluminium forms at the negative electrode.

(i) What does this tell you about aluminium ions? *(1 mark)*

(ii) Explain, in terms of electrons, whether the aluminium ions are oxidised or reduced. *(2 marks)*

(c) Explain why carbon dioxide is produced at the positive electrode. *(2 marks)*

11 When Humphry Davy first attempted to produce potassium by electrolysis in the 1800s, he used potassium hydroxide solution. He was unsuccessful until he used molten potassium hydroxide.

(a) Explain why an electric current can pass through potassium hydroxide solution. *(2 marks)*

(b) Explain why hydrogen, rather than potassium, forms at the negative electrode during the electrolysis of potassium hydroxide solution. *(2 marks)*

(c) Name the gas that would form at the positive electrode during both of Humphry Davy's experiments. *(1 mark)*

(d) Name the gas that forms at the positive electrode during the electrolysis of:

(i) copper chloride solution *(1 mark)*

(ii) copper sulfate solution *(1 mark)*

12 The electrolysis of sodium chloride solution is an important industrial process.

(a) Name the gas formed at the negative electrode. *(1 mark)*

(b) Complete and balance the half equation below.

$Cl^- \longrightarrow Cl_2$ *(2 marks)*

(c) Name the alkali produced during the electrolysis of sodium chloride solution, and give one important use for it. *(2 marks)*

GradeStudio Route to A*

Here are three students' answers to the following question:

The electrolysis of sodium chloride solution is very useful industrially.

(a) Explain the meaning of *electrolysis*. *(2 marks)*

(b) **(i)** Name Gas A.

(ii) Name the alkali formed.

(iii) Complete the half equation for the reaction at the positive electrode, and explain whether it is an oxidation or a reduction reaction:

$Cl^- \longrightarrow Cl_2$ *(4 marks)*

In this question you will be assessed on using good English, organising information clearly and using specialist terms where appropriate.

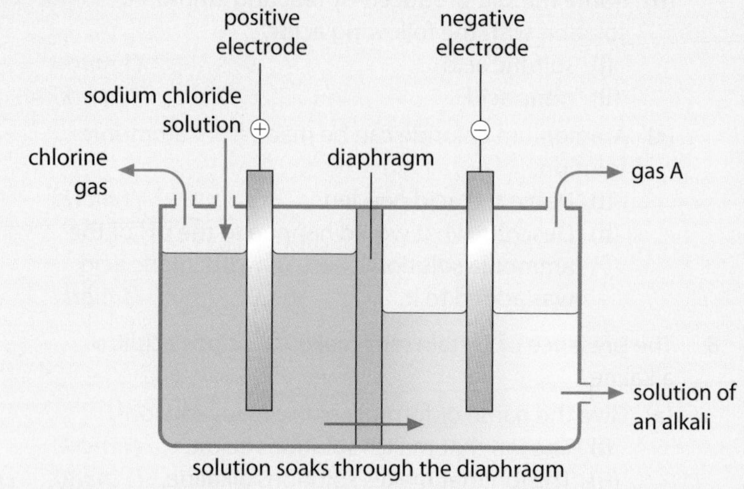

Figure 1 A membrane cell for the electrolysis of salt solution.

Read the answers together with the examiner comments. Then check what you have learnt and try putting it into practice in any further questions you answer.

B Grade answer

Student 1

(a) Electricity passes through a solution and new substances are made.

(b) (i) Sodium.

(ii) Sodium hydroxide.

(iii) $Cl^- - 2e \longrightarrow Cl_2$

It is an oxidation reaction.

This is too vague – 'elements are released' would be better.

The candidate ignores the 'gas' clue in the question.

The electrons are best shown as e^- rather than as e.

Examiner comment

The candidate has not answered the question carefully enough. In part (a) he knows that electricity is involved and gains one mark. However, the remainder of his answer is too vague to gain the second mark. He fails to mention that the electric current passes through a molten or aqueous ionic compound.

In part (b) (i), the candidate forgets that sodium is too reactive to form in aqueous solution and that hydrogen will form instead. He is unable to correctly balance the half equation in part (b) (iii) but correctly states that it is an oxidation reaction. However, he forgets to explain why it is.

Read the whole question carefully.

- Look at the number of marks available for each section so you know how many different points to make.
- Use both your knowledge *and* the information given to you in the question.
- Make sure you use scientific terminology correctly.
- Take care not to write a correct and an incorrect answer to the same question: give one answer and move on.

A Grade answer

Student 2

(a) This is when an electric current passes through an ionic compound in solution.

(b) (i) Hydrogen.
(ii) Hydroxide.
(iii) $2Cl^- \longrightarrow Cl_2 - 2e^-$
It is an oxidation reaction because it loses electrons.

This is one of the ions in solution not the alkali itself.

The candidate shows the correct number of electrons but on the wrong side.

Take care with the use of the word 'it', which means two different things here.

Examiner comment

The candidate has answered part (a) well. She correctly states that an electric current must pass through, and she links this to an ionic compound in solution. She might have added that the ionic compound could also be molten, or that decomposition takes place. Nevertheless, she gets full marks here. She correctly answers part (b) (i) but then starts to go wrong. Hydroxide ions are produced by alkalis in solution, but they are not alkalis. The candidate should have identified the alkali itself, sodium hydroxide. In part (b) (ii) an almost correct answer was spoiled by a silly mistake, as the electrons are shown on the wrong side of the equation. She does correctly explain that the reaction is oxidation. However, the answer would have been clearer if the second 'it' had been 'chloride ions' instead.

A* Grade answer

Student 3

(a) Electrolysis is when an electric current is passed through a molten or aqueous ionic compound. This causes the ionic compound to decompose.

(b) (i) Hydrogen.
(ii) Sodium hydroxide, NaOH.
(iii) $2Cl^- \longrightarrow Cl_2 + 2e^-$
Oxidation is loss of electrons, so it is an oxidation reaction because the chloride ions lose electrons.

This is a more precise term than 'break down' or 'split up'.

The candidate gave the correct formula but was only asked to name the alkali.

The candidate begins her answer with what she knows.

Examiner comment

The candidate has answered in great detail, taking care to use scientific terms correctly. Her answer to part (a) covers three marking parts but of course she can only gain a maximum of two marks. This is alright when the answer is short, like this one, but she could run out of time if she wrote too much in longer answers. She makes good use of words such as molten, aqueous, ionic and decompose.

The candidate gains full marks in part (b), too. She correctly identifies the gas in part (b) (i) and the alkali in part (b) (ii). However, she was only asked to name the alkali but also gave its formula. Luckily this formula was correct, otherwise she would have lost the mark because the examiner would have to choose between a correct and an incorrect answer. In part (b) (iii) the candidate gives a more sophisticated balanced half equation in which the electrons are shown on the right-hand side. The Specification shows $2Cl^- - 2e^- \longrightarrow Cl_2$ as one of its examples, and this would have been acceptable too. The candidate gives a clear and complete explanation of why the reaction is an oxidation reaction.

Examination-style questions

1 Hydrocarbons are molecular substances containing carbon and hydrogen only.

(a) Butane is a hydrocarbon with the molecular formula C_4H_{10}. The atoms in a butane molecule are held together by covalent bonds. A diagram of a butane molecule is shown below.

```
    H   H   H   H
    |   |   |   |
H — C — C — C — C — H
    |   |   |   |
    H   H   H   H
```

Butane has a low boiling point and is a gas at room temperature. Explain why butane has a low boiling point. *(2 marks)*

(b) Propane is a hydrocarbon with the molecular formula C_3H_8. Calculate the percentage by mass of carbon in this compound. *(1 mark)*

(c) Cyclohexane is another hydrocarbon.

(i) Analysis showed that it contains 85.7% carbon by mass. Calculate the empirical formula of cyclohexane. *(4 marks)*

(ii) The mass spectrum of cyclohexane is shown below. Use the data from the molecular ion to find the relative formula mass of cyclohexane. *(1 mark)*

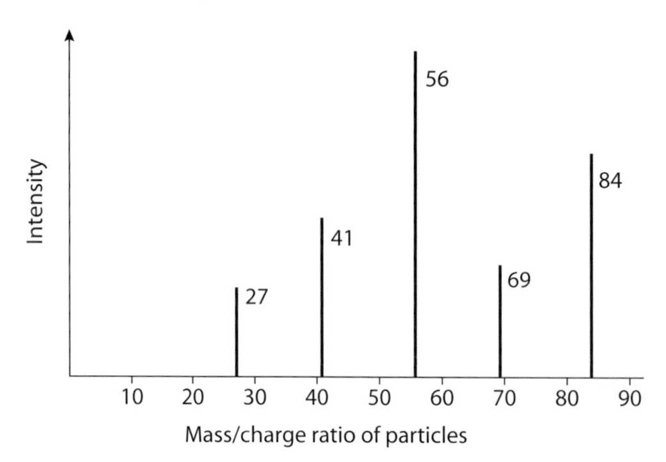

2 Tungsten (W) is a metal. It has many uses, including use in the production of X-rays, as the filament in filament light bulbs and in some electrical contacts. It has a very high melting point and conducts electricity.

(a) Describe the structure and bonding in the metal tungsten, and use it to explain why tungsten conducts electricity and has a high melting point. *(6 marks)*

Tungsten metal is often extracted from the compound tungsten oxide (WO_3) produced from the ore wolframite. In the final stage of the extraction process, WO_3 is reacted with hydrogen.

$$WO_3 + 3H_2 \longrightarrow W + 3H_2O$$

(b) Calculate the mass of hydrogen needed to react with 1 kg of tungsten oxide (WO_3). (Relative atomic masses: H = 1, O = 16, W = 184) *(3 marks)*

(c) The maximum theoretical yield for the extraction of tungsten from 200 g of tungsten oxide is 159 g. In an extraction process from 200 g of tungsten oxide, only 140 g of tungsten was formed.

(i) Calculate the percentage yield for this reaction. *(1 mark)*

(ii) Give one reason why the yield is less than 100%. *(1 mark)*

3 **(a)** Potassium bromide is an ionic compound made from the reaction between potassium, a group 1 metal, and bromine, a group 7 non-metal. It is used by vets to treat epilepsy in dogs. Potassium bromide has a high melting point and conducts electricity when molten or dissolved in water.

(i) Which of the following is the correct formula of potassium bromide?

KBr_2, K_2Br, KBr, K_2Br_2 *(1 mark)*

(ii) Explain why potassium bromide has a high melting point. *(2 marks)*

(b) Diamond and graphite are forms of the element carbon. They both have a giant covalent structure. Describe and explain one similarity in, and one difference between, the physical properties of diamond and graphite by a consideration of their structure and bonding.

In this question you will be assessed on using good English, organising information clearly and using specialist terms where appropriate. *(6 marks)*

(c) Poly(ethene) is a thermosoftening polymer used to make plastic bags and bottles. Poly(ethene) softens and melts on heating. Melamine is a thermosetting polymer that is used in the manufacture of kitchen worktops. Melamine does not soften or melt on heating. Explain why poly(ethene) softens on heating, but melamine does not, by considering their structure and bonding. *(2 marks)*

4 A student investigated how the rate of reaction between zinc granules and sulfuric acid varied with temperature. The reaction can be represented by the equation:

$Zn(s) + H_2SO_4(aq) \longrightarrow ZnSO_4(aq) + H_2(g)$

She used a gas syringe and a stopclock to time how long it took to produce 20 cm^3 of gas at different temperatures.

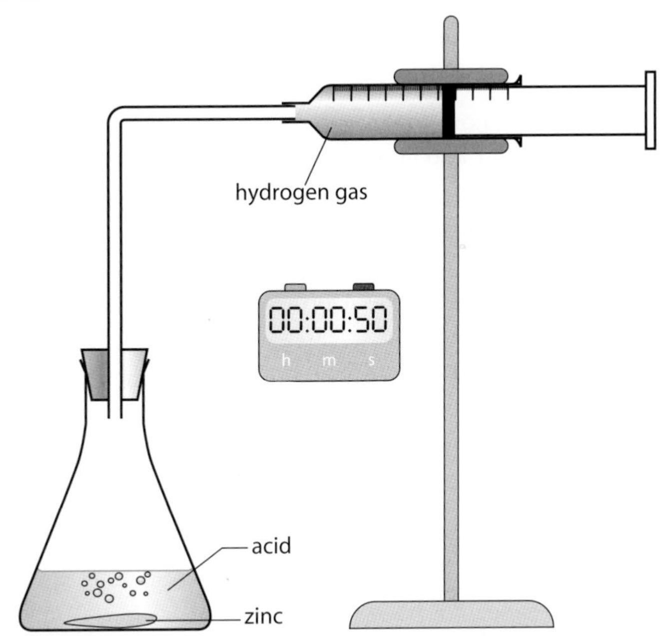

(a) Suggest three variables that the student needed to keep constant in this reaction. *(3 marks)*

Here are the results of her experiment:

Temperature (°C)	0	25	37	50	63	76
Time taken (s)	799	197	101	49	25	12

(i) Plot a graph of her results, adding a line of best fit. *(3 marks)*

(ii) Use your line to predict the time to produce 20cm^3 of gas at 12°C in this experiment. *(1 mark)*

(c) Use your graph to describe, in detail, how the time taken to produce 20 cm^3 of gas in this reaction changes with temperature. *(2 marks)*

5 Read the advertisement below and then answer the question.

Whizzo re-usable hand-warmers

Fed up with wasting money on those 'use once – throw them away' handwarmers? Our liquid sodium acetate-filled hand-warmer can be used over and over again. When you need it, press the button and watch the crystals grow as the pouch steadily warms up. Then, to 'refill' it with energy to produce heat for next time, simply pop it into a bowl of hot water. The crystals will take in energy as they melt, storing it up for future use. The sodium acetate will stay liquid, even when it cools down. That is, until you press that button again! Use as often as you like and save money!

Use information in the advertisement and your knowledge of the processes involved to answer this question.

Explain, as fully as you can, how energy is stored in this device, how it can be released by starting the process of crystallisation, and how and why it may be re-used over and over again.

In this question you will be assessed on using good English, organising information clearly and using specialist terms where appropriate. *(6 marks)*

6 Potassium phosphate, K_3PO_4, is used as a fertiliser. It can be made by reacting phosphoric acid, H_3PO_4, with an alkali.

(a) Name an alkali that could react with phosphoric acid to make potassium phosphate and water. *(1 mark)*

(b) Write a balanced symbol equation for the reaction between this alkali and phosphoric acid. *(2 marks)*

(c) Identify the ions that make solutions acidic or alkaline, and write a balanced equation for the reaction between them. *(3 marks)*

7 Read the information in the box and then answer the question.

Sodium chloride solution contains sodium ions, Na^{+}, hydrogen ions, H^{+}, chloride ions, Cl^{-}, and hydroxide ions, OH^{-}. The electrolysis of sodium chloride solution produces three important reagents for the chemical industry. The diagram below shows the apparatus used to electrolyse sodium chloride solution.

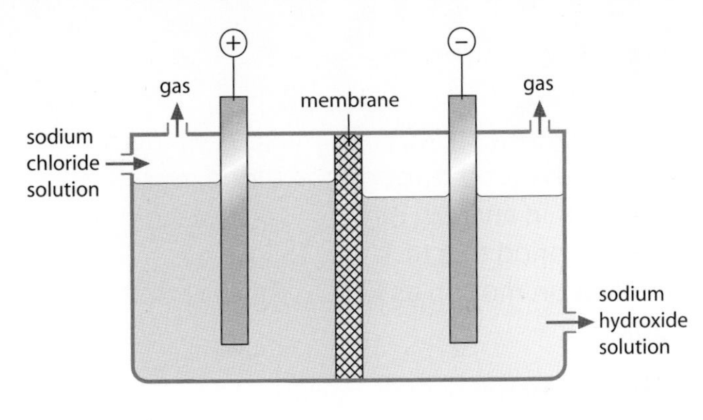

(a) Use information in the box and your knowledge of this process to answer this question.

Explain, as fully as you can, how chlorine, hydrogen and sodium hydroxide solution are formed in this process.

In this question you will be assessed on using good English, organising information clearly and using specialist terms where appropriate. *(6 marks)*

(b) Explain why sodium chloride solution can undergo electrolysis, but solid sodium chloride cannot. *(2 marks)*

(c) Describe an industrial use of sodium hydroxide. *(1 mark)*

P2 Forces and motion

Every day you use forces. You need them to walk, talk and make any kind of movement. Forces inside your body push blood through the blood vessels, move food through the gut and force air in and out of your lungs. Machines and engines increase the forces we can bring to a task. The following chapters explore forces and their effect on motion as well as the way that forces transfer energy between objects.

The first section looks at how a force can bend, stretch, compress or change the shape of an object. It explains how we can use forces to hold an object still, keep it moving at a constant velocity or even accelerate it. Frictional forces and reaction forces in a range of everyday situations are also examined.

These forces make objects move and we go on to look at the measurement of speed, velocity and acceleration and how they are linked to the size and direction of forces. This leads us to examine how scientists and engineers use their understanding of motion to design safety features in cars, and how the motion of cars is governed by the forces that act on them.

Understanding energy and the way we control it is extremely important. The second section looks at the transfer of energy from one form to another, including the concepts of 'energy loss' and 'work done'.

Finally, we explore the concept of the conservation of momentum, which includes the physics of collisions and explosions.

Test yourself

1 Explain what a Sankey diagram is used for.

2 Explain why energy is never really lost and what usually happens to waste energy.

3 List as many different forms of energy as you can.

4 Describe how energy might be transferred from one form to another and explain how this can occur.

Objectives

By the end of this unit you should be able to:

- describe the effects of forces on springs and other objects
- describe a range of factors that affect the way in which a vehicle stops
- explain the effects of forces on motion in a variety of situations
- explain how safety systems in cars work to protect passengers
- calculate the momentum changes that occur as objects interact
- calculate the amount of energy an object has in a range of situations.

P2 1.1

Introduction to forces

Learning objectives

- describe the effects of forces on objects
- predict the way that a resultant force will affect an object
- calculate resultant forces on an object
- represent the size and direction of forces using arrows on diagrams
- explain how a reaction force behaves.

The Space Shuttle's engines thrust down with a force of over 3 million newtons. The reaction force pushes the rocket upwards.

The Earth pulls the Moon towards it. The Moon pulls the Earth towards it with exactly the same force, but in the opposite direction.

The effects of forces

Pushing, pulling, squeezing and stretching are all descriptions of the effects of **forces** on objects. Forces can change the shape of an object, make it start or stop moving, change its speed or change the direction it moves in. Forces are measured in newtons (N).

When two objects interact, the force on one object (the **action force**) is matched by an equal and opposite force (the **reaction force**) on the other. For example, when a rocket moves through space, the rocket engines push out hot gas backwards at high speed. The reaction force makes the rocket move forwards. The action is expelling the hot gas. The reaction is the forward force on the rocket.

There is always an action and a reaction force, whether the object moves or stays still. A kitten sitting on the table has weight; its weight pushes down on the table, and the table pushes up against this weight.

Resultant forces

When a number of forces act at a point, they may be replaced by a single force that has the same effect on the motion as the original forces all acting together. This single force is called the **resultant force**. Each individual force has a size, measured in newtons, but also a direction. If two forces act in the same direction they add together. If they act in opposite directions they subtract.

If the resultant force on an object is zero, it will either stay still or keep on moving at the same speed in the same direction as before. Skaters gliding along on the ice have virtually no resultant force on them – just a small amount of friction (see lesson P2 1.5). They keep on moving in a straight line at an almost constant speed until they apply a braking force to stop.

Figure 1(a) shows two equal-sized forces pushing in opposite directions. Because the forces are equal and opposite they counteract each other. The resultant force is zero, so the object will either stay still, or if it is already moving it will continue moving at the same speed and in the same direction.

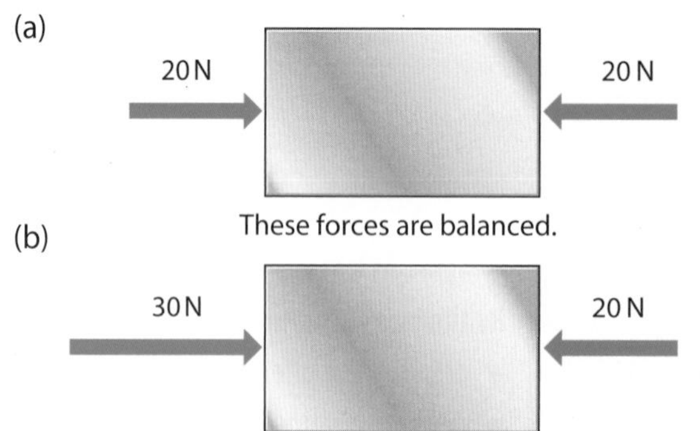

Figure 1 Forces in opposite directions counteract each other.

Figure 1(b) shows two unequal forces pushing in opposite directions. One force subtracts from the other:

$30\,N - 20\,N = 10\,N$

The resultant force is 10 N to the right, so the object **accelerates** to the right.

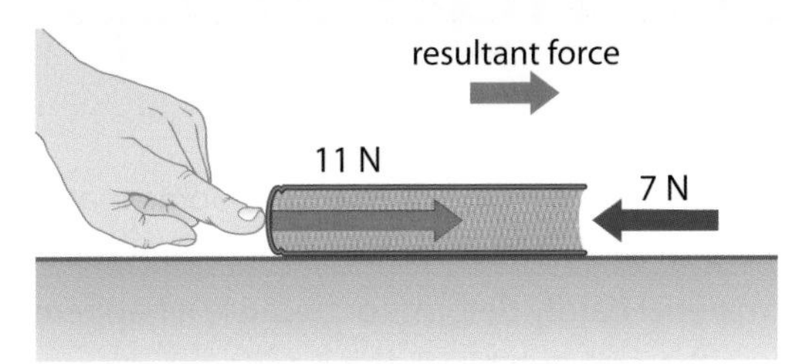

Figure 2 What is the resultant force on the book?

The book in Figure 2 has two forces on it pushing in opposite directions. If we think of the force to the right as being positive, then the resultant force is:

$11\,N - 7\,N = 4\,N$

Since the value is positive, the force is to the right.

In Figure 3 there are three forces acting. The sum of the forces is a resultant of 20 N to the right.

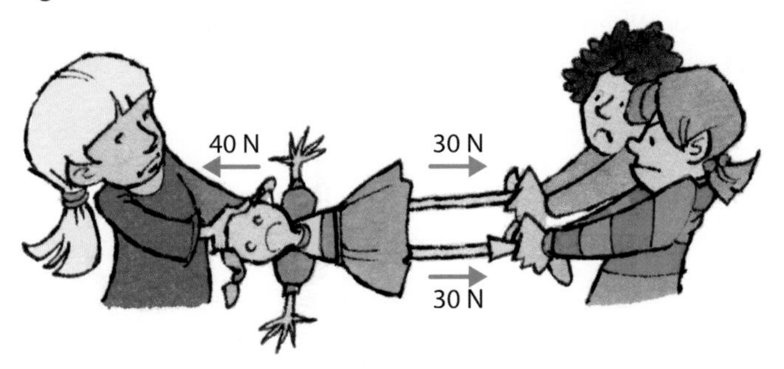

Figure 3 Who will win the tug of war?

Route to A*

A child weighing 400 N sitting on a bench is pushed sideways with a force of 40 N. The friction between the child and the bench is only 38 N. Draw a sketch that shows the forces on the child including the resultant force. Write a short explanation of the way the forces act on the child.

Taking it further

When forces act in a straight line they are quite easy to add or subtract, but when a force pushes an object at an angle to the direction it is moving, the calculations become more complicated. Imagine an aircraft coming in to land that suddenly gets hit by a gust of wind blowing across the runway. Try to describe what would happen to the motion of the aircraft if the pilot did nothing to correct the situation.

Questions

1. Describe the forces involved if a cook tries to close a kitchen drawer with a force of 4 N, but it will not close. Use a diagram to help you.
2. If an object has no resultant force on it, what will it do if it is not already moving?
3. What force would you expect a chair to push upwards with if you sit on it?

   Draw a diagram and include arrows showing the forces.
4. Why is it difficult for scientists to experiment on masses to demonstrate that if there is no resultant force on an object it will either stay still or carry on moving in a straight line?
5. If two objects were floating near each other in space but not moving away or towards each other, describe what would eventually happen and why.
6. You may have noticed that the force you need to start something moving from rest is greater than the force you need to keep it moving. This is not your imagination. Use the image in Figure 4 to help explain why this would be the case.

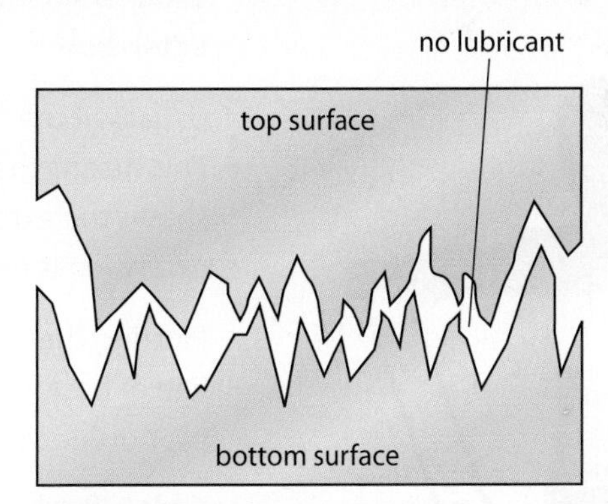

Figure 4 Magnified view of two surfaces in contact with each other.

7. Oil is often used to reduce friction between two surfaces. Use diagrams to explain how this works.
8. If you look at Figure 2 showing two forces on the book, you will notice that the weight and reaction force from the table are not shown. Imagine the book had several other books piled on top of it. Describe what would happen to the various forces on the book and explain the probable effect on the sizes of the forces shown.

Forces that change shapes

Learning objectives

- describe how forces can change the shape of an object
- explain how force and extension are related for a stretched spring
- calculate the spring constant of a spring being stretched.

Changing shapes

Forces can deform or change the shape of an object. Some materials, such as the foam in a cushion, deform easily. However, we often cannot see a change in shape, either because the material hardly deforms or because it happens too quickly. A golf ball flattens when the golf club hits it, but it springs back again very fast.

The force on a golf ball can be as high as 9000 N when the club hits it. This makes the ball deform.

Elastic deformation

Objects that are **elastic** can be pushed or pulled out of shape but will regain their original shape when released. Elastic bands are an obvious example. If a force squashes or stretches an elastic object, then work is done. Energy is stored in the object as elastic potential energy. When an archer pulls the string on a bow, chemical energy in the archer's body is transferred into elastic potential energy in the bow and string. Work is done by the archer to change the shape of the string and bow. When the string is released, that energy is transferred again, this time into kinetic energy as the arrow flies through the air.

Springs

To study how elastic objects behave we often use springs, because we can easily see and measure how they change.

From Figure 1 you can see that when a force is applied to a spring its length changes. The extra length in the spring is called **extension**.

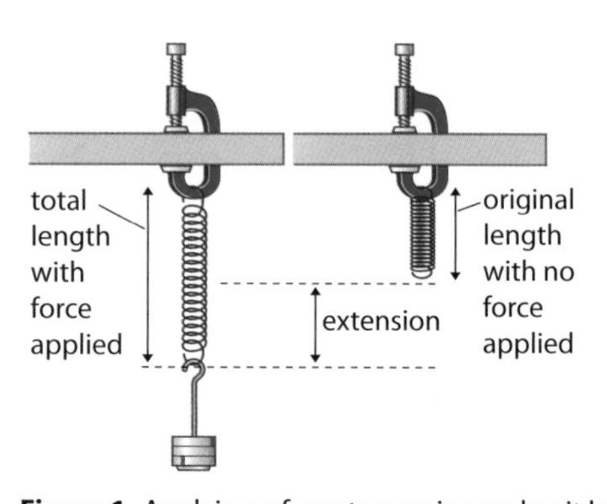

Figure 1 Applying a force to a spring makes it longer.

Usually, if the force on a spring doubles, then the amount it stretches is doubled. This means that the extension is **proportional** to the force applied. Table 1 shows the extension of the spring with different forces. With a force of 10 N the extension is double the extension with a force of 5 N.

Plotting the data from the table produces Figure 2. You can see that the line of best fit is straight and passes through the **origin**. Graphs that show proportionality are always like this.

Table 1 Results from a spring experiment.

Force/N	0	1	2	3	4	5	6	7	8	9	10
Spring length/cm	2.0	2.8	3.7	4.5	5.2	6.0	6.7	7.6	8.4	9.3	10.0
Spring extension/cm	0	0.8	1.7	2.5	3.2	4.0	4.7	5.6	6.4	7.3	8.0

Pulling on the resistance bands stores elastic potential energy.

The spring constant

There is a simple equation that links force and extension:

$F = k \times x$

where F = force in newtons (N), k = the spring constant in newtons per metre (N/m) and x = extension in metres (m).

If you plot a graph as in Figure 2, the spring constant is given by the slope of the graph. When measuring the spring constant, make sure you convert to the correct units. To get the correct value for the spring constant, the extension must be in metres.

Example 1

From Table 1, we find that with a force of 5 N the total length of the spring is 6.0 cm.

The extension is: total length − original length

6.0 cm − 2.0 cm = 4.0 cm

The extension must be converted to metres for the calculation = 0.04 m.

We can use this extension value to work out the spring constant.

force = spring constant × extension ($F = kx$)

Rearranging, we get:

$$k = \frac{F}{x}$$

$$k = \frac{5}{0.04}$$

$$k = 125\ \text{N/m}$$

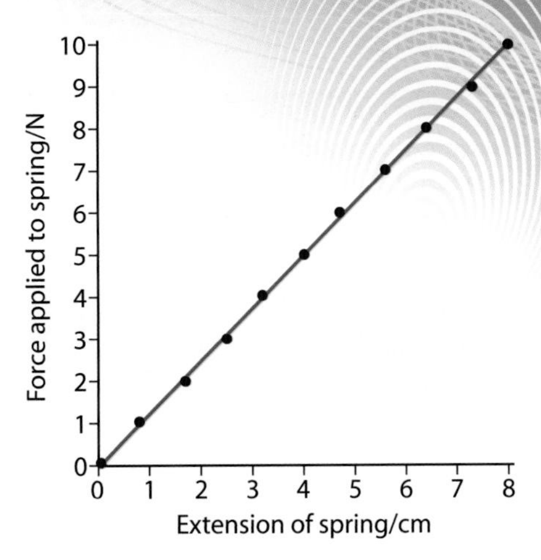

Figure 2 The spring constant for this experiment is worked out from the slope of the graph. Notice extension is in metres in the calculation.

Examiner feedback

Be careful: extension is how much longer a spring becomes, not just the length of the spring.

Questions

1 By looking at the photograph of the golf ball estimate how much the side is pushed in when it is hit.

2 Describe how you can tell that two variables are proportional from looking at a graph.

3 Describe what Figure 2 would look like for a stiffer spring.

4 Glass is a very hard material. Does it behave elastically when a glass marble is dropped on a tiled floor? Justify your answer.

5 **Table 2** The data collected by a student experimenting on a spring.

Force/N	Length/m
0	0.05
1	0.053
2	0.056
3	0.059
4	0.062
5	0.065
6	0.068
7	0.071
8	0.079
9	0.077
10	0.080

(a) Draw a graph from the data in Table 2 and calculate the stiffness of the spring. **(b)** There is a problem with the data. Describe the problem and what a scientist using the data would do when plotting the graph. What should the student have done when he plotted the graph and found the problem?

6 When a spring is pulled with too large a force, it deforms permanently and, although it does get shorter again, it does not go back to its original length when the force is removed. Sketch the shape of the graph of force against extension that you might expect in this situation.

7 A spring has a stiffness of 500 N/m and an original length of 4 cm. Draw a graph of the length of the spring as forces up to 8 N are placed on it.

8 All solid materials behave elastically to some extent. Explain why a bell is a particularly good example of an object that behaves elastically.

A*

Distance, speed and velocity

Learning objectives

- draw and interpret graphs of distance and speed against time
- calculate speeds
- describe the difference between speed and velocity.

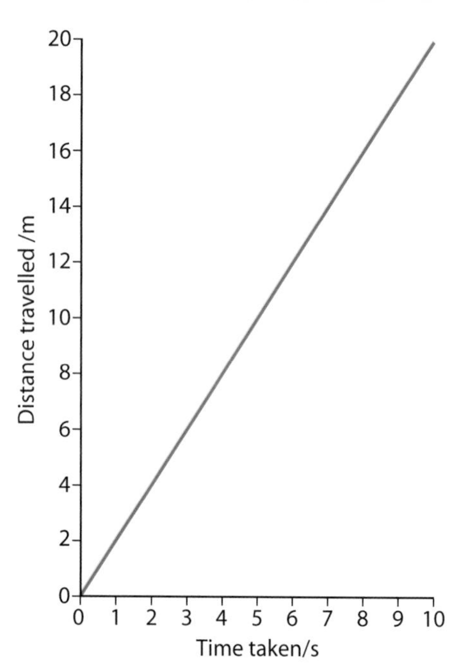

Figure 1 This object moves 2 metres every second.

Route to A*

A particle in a nuclear research laboratory is measured as travelling 0.3 mm in 0.06 microseconds. What speed is it travelling at? (A micro second is one millionth of a second.)

Examiner feedback

The gradient of a distance–time graph represents speed.

Examiner feedback

In questions check each graph carefully: is it distance–time or velocity–time?

Calculating speed

Speed is a measure of how far something moves in a known time:

$$v = \frac{d}{t}$$

where v = speed (metres per second, m/s), d = distance (metres, m) and t = time (seconds, s).

For example, suppose a dog runs 100 metres in 5 seconds:

$$\text{speed} = \frac{100\text{ m}}{5\text{ s}} = 20\text{ m/s}$$

Graphs of distance and time

It is quite easy to represent the way an object moves by plotting a graph of distance moved against time, as in Figure 1.

Table 1 Data for Figure 1.

Time taken/s	0	1	2	3	4	5	6	7	8	9	10
Distance moved/m	0	2	4	6	8	10	12	14	16	18	20

To calculate the speed from Figure 1 you must find out how far the object travels and how long it takes. This will give you the average speed.

In this case the object travels 20 metres in 10 seconds.

$$v = \frac{d}{t} = \frac{20}{10} = 2\text{ m/s}$$

Notice how the graph in Figure 1 is a straight line. This means that the speed is constant.

The **gradient** of a graph shows the way that one variable is changing compared with another. The gradient of a distance–time graph shows how distance is changing with time – this represents the speed. If the gradient is steeper, the speed is higher.

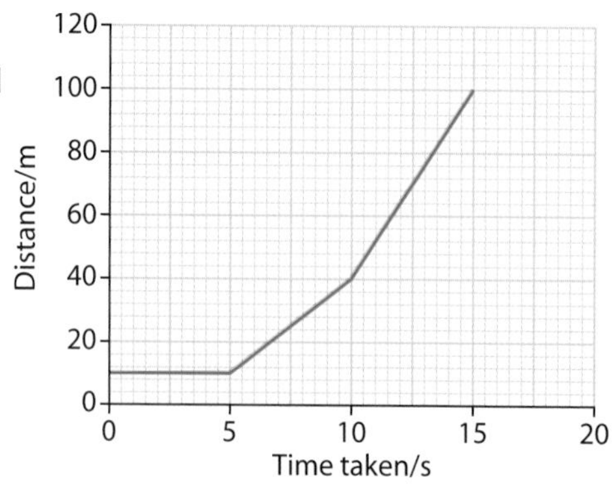

Figure 2 A distance–time graph for the first 15 seconds of a cyclist's journey.

Example 1

Figure 2 shows changes in a cyclist's speed during the first 15 seconds of a journey. The graph is made up of three straight-line sections. We can work out the cyclist's speed for each of these sections.

First section: speed $= \frac{\text{distance}}{\text{time}} = \frac{0}{5}\text{ m/s} = 0\text{ m/s}$

Second section: speed $= \frac{\text{distance}}{\text{time}} = \frac{30}{5}\text{ m/s} = 6\text{ m/s}$

Third section: speed $= \frac{\text{distance}}{\text{time}} = \frac{60}{5}\text{ m/s} = 12\text{ m/s}$

The steeper parts of the graph show higher speeds.

Velocity versus time graphs

Speed is a measure of how fast something is going. Velocity is a measure of an object's speed in a given direction. If athletes running around the bend of a track travel 8 metres every second, they are running at a constant speed of 8 m/s. However, their velocity changes because their direction is constantly changing.

Plots of velocity against time look similar to distance–time graphs. However, what they show is very different.

The graph in Figure 3 is a velocity–time graph for a train leaving a station. The first part of the graph shows the train moving faster (accelerating). In the second section the graph levels out because the train is travelling at a constant velocity.

A really important part of this graph is the area under the line.

The area is velocity $\times$ time, which equals distance.

The velocity of athletes on a bend constantly changes.

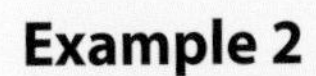

Example 2

In Figure 3, how far does the train travel?

We can work this out from the area under the graph.

The area under the first part of the line is half a rectangle, so we can work out the area of the rectangle and divide by two.

So, area $= 60 \times 30 \div 2 = 900$

For the next 80 seconds, the area under the graph is the area of a rectangle. This is $80 \times 30 = 2400$

So for the whole period shown on the graph, the train travels a total of $900 + 2400 = 3300$ m or 3.3 km

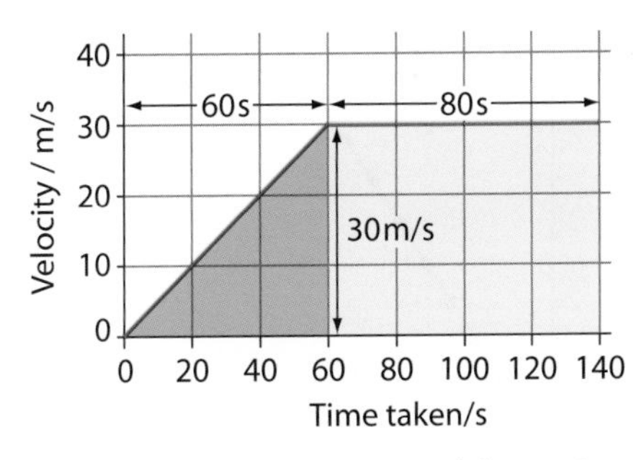

Figure 3 The velocity–time graph for a train leaving a station.

Questions

1. Calculate the speed of a bullet moving 625 metres in 2.8 seconds.
2. A golf ball rolls down a slight slope at a constant speed in a straight line, until it falls into the hole. Sketch the shape of a distance–time graph for this movement.
3. Sketch a velocity–time graph for the movement of the golf ball in question 2.
4. Write down an equation for converting m/s to km/h.
5. In 2008 a marathon runner covered a distance of 42.195 km in a time of 2 hours, 3 minutes and 59 seconds. What was his average speed in metres per second?
6. Put these in order of which travels furthest:
   - a cycle travelling at a constant speed of 2 m/s for 35 seconds
   - a car that steadily speeds up from 0 to 15 m/s in 8 seconds
   - a motorcyclist who slows down from 30 m/s to 10 m/s in 3.2 seconds.

   Show your working for each vehicle.
7. Draw the velocity–time and distance–time graphs for a woman running at a constant speed of 2.3 m/s for 7 seconds.
8. A car travels at a constant speed of 10 m/s for 24 seconds and then slows down to 4 m/s over the next 8 seconds. Draw a graph of this motion and calculate the distance the car moves in the 32 seconds described.

P2 1.4 Acceleration

Learning objectives

- describe acceleration from analysing distance– and velocity–time graphs
- explain how acceleration is linked to mass and force
- calculate acceleration from data and from graphs.

Acceleration

Acceleration is a measure of how quickly something gets faster. As an equation this is:

$$a = \frac{v - u}{t}$$

where a = acceleration in metres per second squared (m/s^2), v = final velocity in metres per second (m/s), u = initial velocity in m/s, and t = time in seconds.

If a dog starts to chase a ball, it might go from rest, i.e. not moving, to 10 m/s in 2 seconds. This means it has got 5 m/s faster each second. The dog's acceleration is therefore 5 m/s^2.

If we plot a graph of the dog's movement using a velocity–time graph, the gradient shows the acceleration. In this case, the gradient, and therefore the acceleration, is constant.

Remember, the area under the line for a velocity–time graph equals the distance travelled. The area under the line in Figure 1 is half of 2 × 10, i.e. 10. So the dog has travelled 10 metres in 2 seconds.

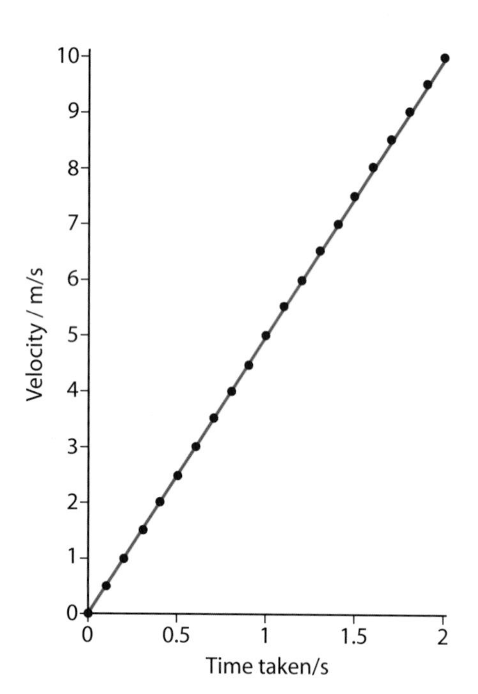

Figure 1 Graph of velocity versus time for an accelerating dog.

Force, mass and acceleration

If you push or pull something that was not moving and it moves, you have accelerated it. How this works is actually quite simple.

- The harder you push or pull something, the more it accelerates.
- The heavier an object is, the less it accelerates, given the same force.

If a team of dogs pulls a sled it accelerates more quickly than with only one dog, because the pull is greater. If you load the sled with equipment it will accelerate less, assuming the team pulling it, i.e. the force, stays the same.

Dogs pulling a sled.

Mass, force and acceleration are linked by a simple formula:

$$F = ma$$

where F = force in newtons (N), m = mass in kilograms (kg) and a = acceleration in metres per second squared (m/s^2).

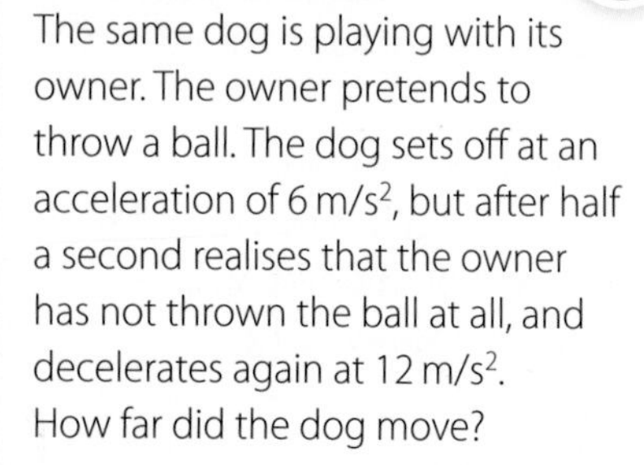

Route to A*

The same dog is playing with its owner. The owner pretends to throw a ball. The dog sets off at an acceleration of 6 m/s^2, but after half a second realises that the owner has not thrown the ball at all, and decelerates again at 12 m/s^2. How far did the dog move?

Forces on a moving object

A resultant force acting on an object that is already moving can have several effects, depending on the direction of the force. A force in the direction of movement will cause it to accelerate. A force from another direction could change the direction the object is moving in, or slow it down. Both of these are kinds of acceleration too. Slowing something down is negative acceleration, or **deceleration**.

Science skills

A student makes a toy rocket and performs an investigation using light gates to find how quickly it moves. The mass of the rocket is 100 g and it goes from rest to 20 m/s in 0.5 s. She works out that the acceleration of the rocket was 40 m/s^2. This means that the accelerating force was 4 N.

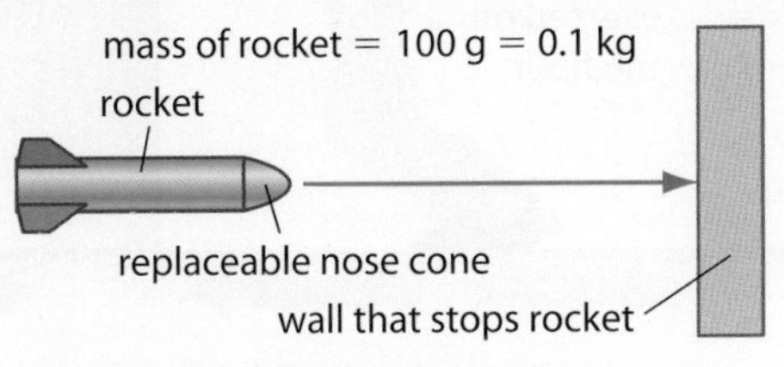

Figure 2 The rocket accelerates due to a force of 4 N.

The rocket then hits the wall and stops. It does not stop instantly but takes 0.02 seconds.

a Calculate the deceleration of the rocket as it hits the wall.

When the student repeats the experiment, she uses rockets with different masses.

b Describe what effect this will have and justify your answer.

Science skills

A pair of students have been set the challenge of measuring the acceleration of a dog. They use a stopwatch and a tennis ball for it to chase. They measure the time the dog takes to run 25 metres from a standing start. It takes the dog 5 s to cover the 25 metres. They calculated the average speed as 5 m/s and then calculated the acceleration as 1 m/s^2.

c Comment on the mistakes they have made in their calculations and advise the students on how they can use their data to estimate the acceleration of the dog and the assumptions they might have to make.

Questions

1 A student found that increasing the length of the disposable nose cone decreased the deceleration when the rocket in Figure 2 hit the wall. This stopped the rocket getting damaged so much. Use the relationship between force, mass and acceleration to suggest why this makes sense.

2 The student repeated the toy rocket experiments (see Science skills box) with the rocket moving vertically. She found that accelerations were smaller. Explain why.

3 The student then conducted experiments with the rocket over greater distances on playing fields. She found that the acceleration was not constant. Acceleration decreased as the rocket got faster. Look back to the previous work on forces and explain why this might be.

4 A force of 15 N pushes a ball with a mass of 3 kg. What is its acceleration?

5 A force of 25 N is applied to a trolley. It accelerates at 2 m/s^2. What is its mass?

6 Calculate the acceleration for each of the graphs in Figure 3.

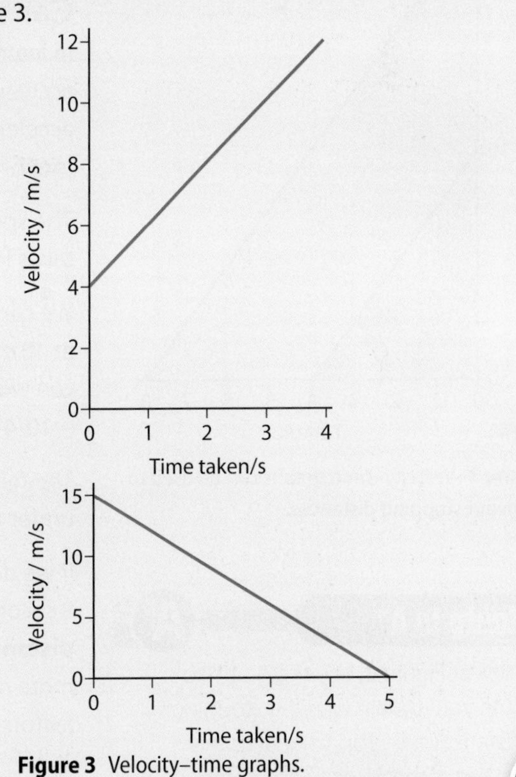

Figure 3 Velocity–time graphs.

A*

P2 1.5

Slowing down and stopping

Learning objectives

- describe how friction acts in opposition to movement
- describe how resultant forces affect motion
- explain the factors affecting stopping distances
- describe the effect of alcohol and drugs on stopping distances.

Friction is a force too

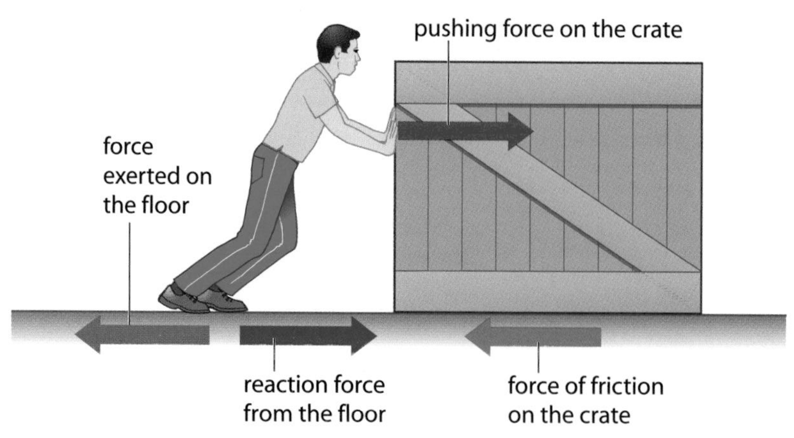

Figure 1 Friction forces oppose the movement of the box.

If an object is moving, friction acts in the opposite direction to the movement. If you push something, friction will act in the opposite direction to your push.

Figure 2 At constant speed, resistive and driving forces balance and the resultant force is zero.

The forces resisting the motion of a car include air resistance or drag, tyre rolling resistance and friction in the bearings. When a car is moving at a constant speed, the engine produces just enough forward force to counteract these resistive forces. The resultant force on the car is zero, so its speed and direction do not change. Friction can also be useful. It is the friction between the tyres and the road that allows the car to move forwards, and without friction the car would not stop.

Car stopping distances

In lesson P2 1.4 we showed that the bigger the force on an object, the more it accelerates. The same rules apply to stopping a car, except that in this case the acceleration is negative – the car decelerates. The better a car's brakes, the more frictional force they apply, and the faster the car stops.

The speed of the car also affects how quickly it stops. At higher speeds a car takes longer to stop, given the same braking force, than at lower speeds.

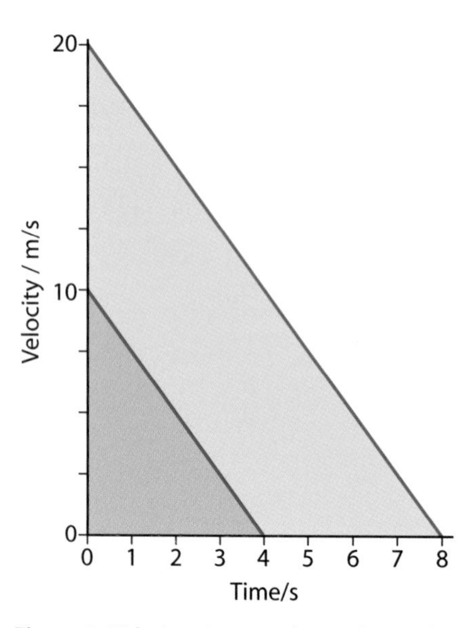

Figure 3 Velocity–time graphs can be used to work out stopping distances.

We can see how this works by looking at an example. Supposing a car travelling at 10 m/s stops in 4 seconds. Figure 3 shows a velocity–time graph. From this, we can work out that the acceleration is the gradient of the graph: $-10/4 = -2.5\ m/s^2$

The minus sign shows that this is deceleration. The stopping distance is the area under the line, which is 20 m.

If we double the speed but keep the same acceleration, the car now takes 8 seconds to stop. The area under the graph is now 80 metres – the **braking distance** has quadrupled. As speed increases, the braking distance increases more rapidly than the speed. Braking distances are increased even more by factors such as worn car tyres or faulty brakes.

Route to A*

Look at Figure 3. Use the graph to help you explain why the stopping distances are proportional to the square of the velocity of the car.

Thinking distance

In a real situation, a car's stopping distance will be greater than the braking distance because the driver will take a small amount of time to respond to a hazard and put their foot on the brake. This extra distance is usually called the **thinking distance**. The **stopping distance** is the thinking distance added to the braking distance. The thinking distance increases with speed, although not as much as the braking distance.

Many things can affect the thinking distance. If a driver is tired or distracted, or if they have been drinking alcohol or taking drugs, their reactions will be slower and so the thinking distance will be longer. In most countries drivers are forbidden by law from having more than a certain level of alcohol in the blood when driving.

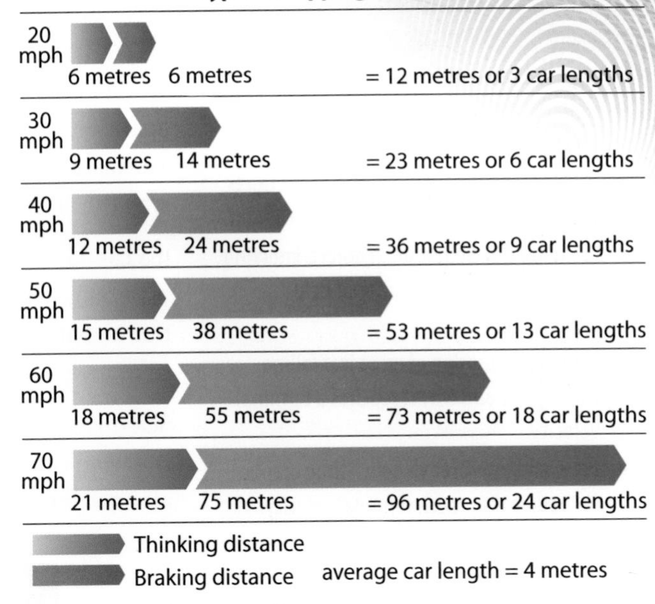

Figure 4 Typical stopping distances.

Making roads safer

Braking distances can change dramatically with the road conditions. If there is water or ice on the road or if the road surface is poor, vehicles will need much larger braking distances. To take account of this, many motorways have displays that make it possible to vary the speed limit. If weather conditions are poor or there are problems with the road surface, the speed limit is reduced. Variable speed limits can also be used to keep traffic flowing more effectively. This helps to reduce collisions because drivers are not frequently changing from high to low speeds.

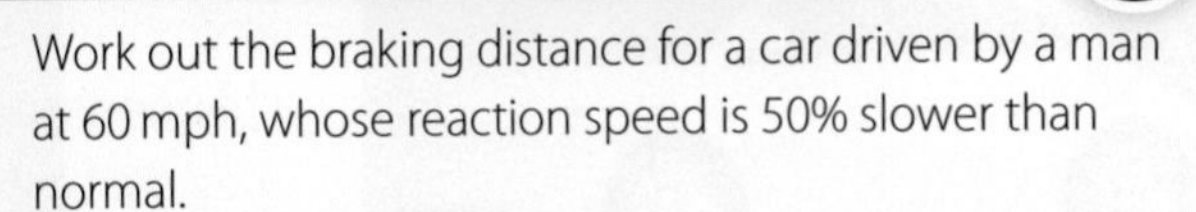

Route to A*

Work out the braking distance for a car driven by a man at 60 mph, whose reaction speed is 50% slower than normal.

Questions

1. In a test, the reaction time of a driver was found to be 1.5 seconds. How far would a car moving at 45 m/s move before the driver could start braking?
2. What force would be needed to stop a lorry that weighed 15 000 kg from 20 m/s in 8 seconds?
3. A cyclist uses a force of 60 N to move a bicycle at a constant speed of 5 m/s. What is the total resistive force on the bicycle?
4. If the maximum braking force that the tyres of a car can exert is 2000 N and the car has a mass of 800 kg, calculate the maximum deceleration of the car and the distance it would travel before it stopped if it was travelling at 15 m/s.
5. Tyre treads are designed to remove water from the surface of the road immediately beneath the tyre. Explain why smaller tread depths will be less effective at removing water from the surface of the road.
6. In a scientific experiment, the time it took for people to press a button when a light came on was about 0.2 seconds. Describe what factors might make a driver's reaction time longer than this.
7. In Formula 1 car racing, manufacturers have found that tyres made from relatively soft rubber give more grip, so cars can move around corners faster. However, harder tyres last longer. Describe the conflicting issues that a tyre manufacturer faces when marketing a tyre.
8. As fuel becomes more expensive, some people have suggested that the speed limit on motorways be reduced to 60 miles per hour. Explain why this would reduce the cost of motoring and improve road safety.

P2 1.6

Terminal velocity

Learning objectives

- describe the way that forces change when an object moves through a fluid
- explain how the shape of an object affects the forces on it as it moves through a fluid
- calculate the weight of an object, knowing its mass and the gravitational field strength
- explain the motion of an object falling through air and why it reaches a terminal velocity
- draw and interpret velocity–time graphs for bodies that reach terminal velocity.

Streamlining

When an object moves though a fluid such as air or water, resistive forces act against it. The resistive forces on a car include air resistance, see lesson P2 1.5. The faster a car moves, the greater the air resistance trying to slow it down. This is why the amount of fuel used increases as a car goes faster.

If a car has a streamlined shape this reduces the forces due to air resistance, especially at higher speeds. High-speed cars tend to be rounded at the front and pointed at the back. This is similar to fast-moving fish, which are also streamlined to reduce resistive forces.

Scientists investigate the way that air flows over the body of a new car design or the wing of an aeroplane using wind tunnels. The most streamlined shapes allow air or water to flow over them smoothly, with very little **turbulence**.

The smooth lines of this car reduce the frictional forces on it at speed.

Testing airflow over a car in a wind tunnel. Smoke trails show the general airflow. What do the woollen strings stuck to the car show?

Taking it further

When an astronaut is orbiting in space they are often described as being weightless. The astronaut still has mass and is still in the Earth's gravitational field, although this reduces with distance away from the Earth. The weightlessness feeling is because the astronaut is, in fact, falling. Therefore, there is no force from the floor of the spacecraft pushing the astronaut upwards, because that is falling too. The weight is accelerating the astronaut towards the Earth.

Mass and weight

The weight of an object is the force pulling it towards the ground. On Earth, this force is approximately 10 N for every kilogram of mass. This can be written as an equation:

$$W = m \times g$$

where W = weight in newtons (N), m = mass in kilograms (kg) and g is a constant called the gravitational field strength (N/kg).

So on Earth, a mass of 3 kg has a weight of approximately 30 N.

Terminal velocity

To help understand terminal velocity, we will look at a parachutist jumping from a plane. Figure 1 is a velocity–time graph of what happens. Section A is just after the parachutist jumps. His weight pulls him down, and he accelerates towards the Earth.

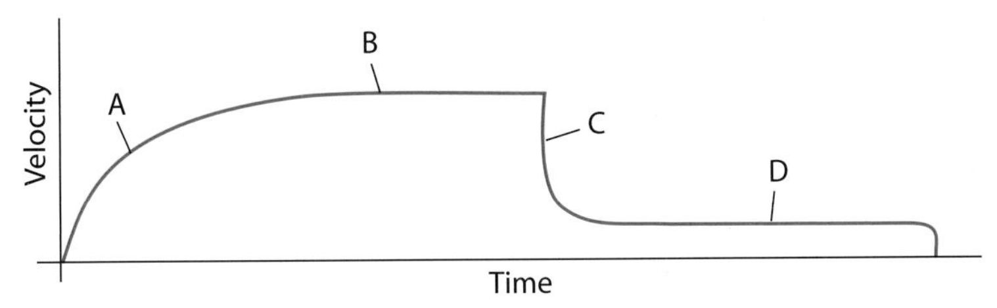

Figure 1 How the velocity of a parachutist changes during a jump.

As the parachutist moves faster, the upward force of the air resistance increases until it exactly balances his weight. At this point, the resultant force on the parachutist is zero and there is no further acceleration. This is called **terminal velocity** – section B in Figure 1.

Terminal velocity varies, depending on the size and shape of the falling object. If the parachutist opens his parachute, the large area of the parachute experiences a large amount of extra air resistance. The parachutist slows quickly, as in section C, then reaches a slower terminal velocity (section D).

Questions

1 What is the weight in newtons of a 5.5 kg mass?

2 An object has a weight of 2.3 N. What is its mass in kg?

3 What is the gravitational field strength on the Moon if an object of mass 10 kg weighs 16 N?

4 A teacher challenged students to get identical lumps of clay to fall fastest through a liquid. One student decided to cheat by putting small lumps of metal inside the clay. Their friend said that this would not work as it would make the object bigger and this would slow it down. Explain which student is right and why.

5 Use the data below to draw a graph of a pencil falling out of an office window several floors up. Explain why the last few values for velocity are the same.

Time/s	0.0	0.2	0.4	0.6	0.8	1.0	1.2	1.4	1.6	1.8	2.0
Velocity / m/s	0	1.8	2.4	2.6	2.7	2.8	2.8	2.8	2.8	2.8	2.8

6 A skydiver falls from a plane and reaches a top speed of 40 m/s after 20 seconds. She then falls for 30 seconds until she opens her parachute. After opening the parachute her downward speed is 6 m/s and she falls for a further 40 seconds. Draw a velocity–time graph for this descent.

7 A falcon folds its wings to dive at high speed onto its prey below. Explain how folding the wings helps the falcon to increase its speed. Include comments about weight and drag.

8 When a parachutist falls out of a plane, he quickly gains speed and reaches his terminal velocity. While falling, he takes out of his pocket a small lead sphere and a wooden ball of about the same size. The lead ball falls faster than the parachutist, while the wooden ball seems to move upwards compared with the parachutist. The lead ball is heavier than the wooden ball. Describe the forces on the two objects as each changes speed and reaches its own terminal velocity.

Skydivers use a parachute to reduce their terminal velocity.

Science skills

A student investigating how the shape of an object affects its terminal velocity in a liquid tries three different shapes in three different liquids. The results are shown in the table.

Table 1 Time taken for an object to fall 10 cm through a liquid.

	Time/s		
	liquid 1	**liquid 2**	**liquid 3**
shape a	5	20	25
shape b	7	30	35
shape c	8	45	50

a From the data, state which is the thickest liquid and explain why you think this.

b The student decided to start the stop-clock when the objects passed a line about 8 cm below the surface of the liquid. Explain why this was a good idea.

c Describe how to make the experiment a fair test and also improve the reliability of the results.

Route to A*

Many people think that when a skydiver opens their parachute they move upwards suddenly. Explain why there is this misconception and what is really going on.

Force and energy

Learning objectives

- describe how energy is transferred when a force moves an object through a distance
- calculate the work done when a force acts through a distance
- describe how kinetic energy is reduced and the temperature of the brakes increases when a vehicle brakes.

Forces that transfer energy

In almost every situation there are forces at work. Even the weight of an ant on a leaf is balanced by the leaf pushing back to support the tiny weight of the ant. If an object stays still, then no energy is **transferred**. When an object moves in the same direction as a force that is pushing or pulling the object, then energy is transferred.

In Figure 1, because the boy does not move, no energy is transferred. In Figure 2, the woman moves the suitcase through a distance and so energy is transferred from the woman to the suitcase. There is **work done** by the woman: the harder she pulls and the further she moves, the more energy she transfers.

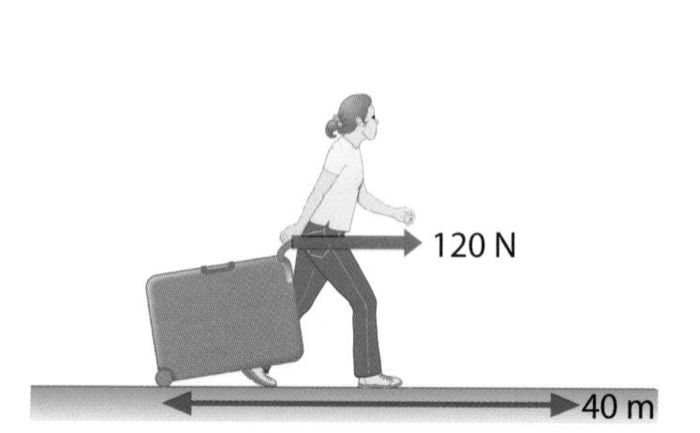

Figure 2 Here, the force moves the suitcase.

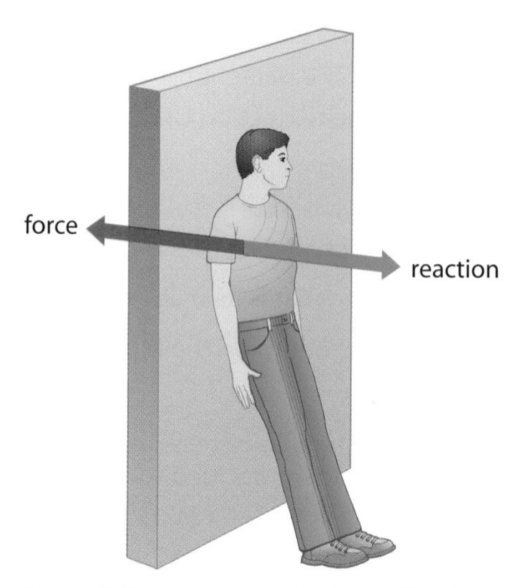

Figure 1 Here, a force pushes the wall and a force reacts to the push.

Work done = energy transferred

It is easy to calculate work done, using the formula:

$W = F \times d$

where W = work done or energy transferred in joules (J), F = force applied in newtons (N) and d = distance moved in metres (m).

From this equation we can see that 1 joule = 1 newton metre.

In the case of the woman pulling the suitcase:

$W = 120\,\text{N} \times 40\,\text{m}$

$= 4800\,\text{J}$

Energy loss

Energy can change form, but it is never destroyed or used up. When scientists say that energy is 'lost', they mean that it has changed (transformed) from one form into another and has moved out of the system they are studying. For example, when a car crashes it loses **kinetic energy**. The car no longer has the kinetic energy it had before the crash. Some kinetic energy has become sound, and some has slightly heated up the car and its surroundings.

Examiner feedback

Remember to change the force units into newtons (N) and distance units into metres.

Route to A*

If the woman pulling the suitcase then moved onto a much harder floor surface, the friction would reduce significantly. For half a second she does not realise that the friction has reduced. Describe what happens during this half a second, and justify your answer using estimates for the forces and masses involved.

Friction and energy

If you rub your hands together, they get warmer. What is happening is that the kinetic energy produced by your muscles is transferred to your palms through heating. In fact it is almost always the case that friction between moving objects produces heating. For example, when a car slows down using its brakes, friction between the brake pads and discs leads to the kinetic energy of the car being transferred to the pads and discs, which get hot.

The friction in the brakes on this racing car makes the brakes glow red-hot.

Science in action

Disc brakes for high-performance cars often have air cooling systems. The discs are hollow and air can circulate through the discs. This cools them more effectively because there is a greater surface area for the air to remove the energy.

Taking it further

Some very fast cars have air brakes as part of their braking system. Describe how this improves the slowing capability of the car and why they only work at high speed.

Route to A*

It takes 2000 J of energy to increase the temperature of the brake disc on each wheel by 1 °C. If the car has a mass of 1 tonne and slows from 20 m/s to 15 m/s, by how much will the discs' temperature rise?

Science in action

The glove boxes inside the passenger compartment of some cars are designed to open slowly. To get the glove box to open slowly engineers use 'smart grease'. The grease is a lubricant and allows movement of the glove compartment door, but resists fast movements, so the door opens slowly.

Questions

1. A force of 200 N pushes a car 30 m. Calculate the amount of work done.
2. If a drawer moves 40 cm and the person opening it transfers 1 J of energy against friction forces, what was the friction force in the drawer system?
3. A crane lifts a 2 t load 40 m to the top of a building. How much work has been done against the weight of the load?
   Note: 1 t (tonne) = 1000 kg. Remember that weight = mass × gravitational field strength (10 N/kg on Earth).
4. To keep a bus moving on a flat road the engine needs to be pushing the bus all of the time. Describe the data you would need to collect to calculate the work done by the bus engine against friction.
5. Explain why a lubricant can reduce the energy lost through heating in a car engine.
6. As the Space Shuttle enters the Earth's atmosphere it heats up. Draw a diagram to explain the changes in energy as the Space Shuttle goes from high above the Earth's atmosphere to landing on an airfield.
7. If you slide down a rope, you can get a friction burn. Explain why your hands get hot in this situation in terms of force, distance and work done.
8. Two people, one walking and one riding a bike, move along a path at about the same speed. What similarities and differences can you think of between the forces on the cyclist and those on the walker?

P2 2.2

Power

Learning objectives

- describe how power is related to energy and time
- calculate the power output of a system
- explain how the power rating of a system governs its ability to do work.

Power, energy transferred and time taken

To understand the concept of **power**, it is best to look closely at an **energy transfer** in two different situations.

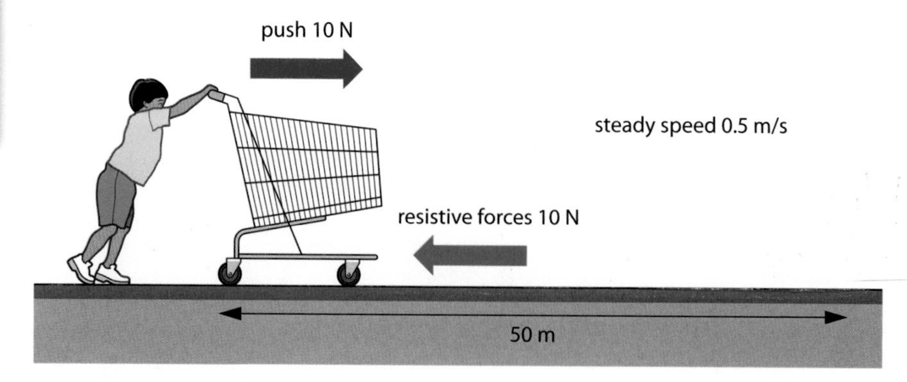

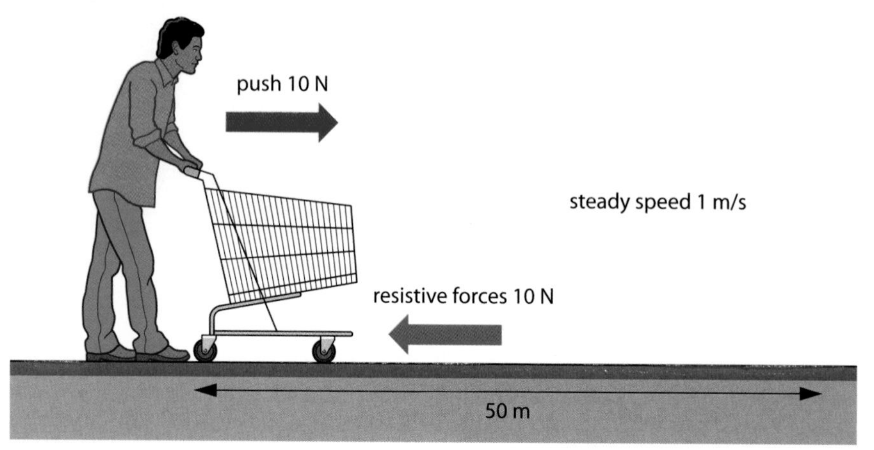

Figure 1 Both the child and the man push at a steady speed, but the man pushes faster.

In Figure 1(a) a small child pushes a trolley. She can keep it moving at a steady speed if she pushes all the time. She needs to keep pushing because resistive forces push in the opposite direction to the movement of the trolley. In Figure 1(b) the father pushes the trolley. He too can keep it going at a steady speed but because he is more powerful than the child, he can make it move faster.

$W = F \times d$

where W = work done in joules (J), F = force applied in newtons (N) and d = distance moved in metres (m).

Both the child and the father push the trolley 50 metres using a force of 10 newtons, so they both do the same amount of work:

$W = 10\,\text{N} \times 50\,\text{m}$

$= 500\,\text{J}$

However, the time taken to push the trolley 50 metres is different. The child took 100 seconds, so she transferred 5 joules of energy every second. The father only took 50 seconds, so he transferred 10 joules of energy every second. Both did the same amount of work, but the child took longer than the father.

Route to A*

If a mass of 40 kg has a resultant force of 10 N on it for 8 s, it will accelerate. Show that the energy transferred by pushing the mass is 80 J.

Taking it further

The principle of work done also applies to electrons. If an electron is in an electric field it will be accelerated. As it gains speed, the electric field is doing work on the electron and it is gaining kinetic energy. In a cathode ray oscilloscope this energy is given out as light when the electron hits the screen.

Science in action

The Victorians were very inventive and understood about work, energy and forces. At Lynmouth and Lynton they designed a funicular railway that made water do the work for them. The same principles of force, distance and work done apply to this antique mechanism as to putting a satellite into orbit.

The rate at which energy is transferred, that is the work done in a given time, is known as power. The unit of power is the **watt (W)**. Because many systems are very powerful, engineers often use kilowatts (kW, 1000 W) or megawatts (MW, 1 000 000 W) instead of watts.

To calculate the power output of a system you can use the formula:

$$P = \frac{E}{t}$$

where P = power in watts (W), E = energy transferred in joules (J) and t = time taken in seconds (s).

When 1 joule of energy is transferred in 1 second, the power is 1 watt.

In the trolley example, the child's power is 5 W (500 J in 100 s). The father's power is 10 W (500 J in 50 s). So the father is more powerful than the child.

Large power stations like this can produce between 500 and 2300 megawatts of power.

Power stations

Many power stations transform chemical energy from fossil fuels into electrical energy, although an increasing proportion of our electricity comes from renewable sources.

An electricity generating station can have a power of 500 megawatts. This means that it transfers 500 million joules of energy every second.

If a power station is running for 10 seconds, it transfers 500 000 000 × 10 = 5 000 000 000 (5 billion) joules of energy.

Questions

1 The energy transferred by a small electric motor is 10 J and it takes 5 seconds to move a lever. What is the power of the motor?

2 Two car engines are working at full capacity. The engine in the red car transfers 50 000 J in 5 seconds and the engine in the blue car transfers 20 000 J in 1 second. Which engine is more powerful? Show your working to calculate the power of each engine.

3 How long will it take a 500 W heater to give out 3000 J of energy?

4 How much light energy does the bulb in Figure 2 give out in 1 minute?

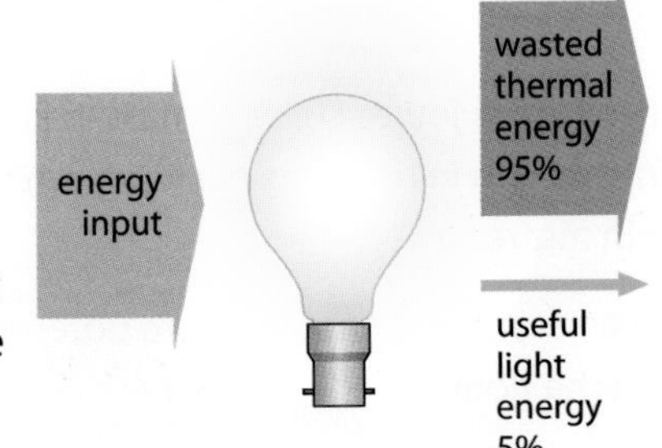

Figure 2 Energy input and output for a 100 W filament light bulb.

5 A new car has an 80 kW engine. It has a top speed of 190 km/h. Explain why it cannot go any faster than this. Calculate how much energy is being transferred and explain where the energy is going.

6 In a typical car engine, the chemical energy in the petrol is about four times greater than the amount of energy transformed into useful work by the engine. Explain why.

7 A small electric motor can output 2 watts of energy. It can be made to do the same job of moving a CD tray as a more powerful 15 W motor. Use the relationship between power, energy and time to describe the advantage of using a more powerful motor.

8 The following wording is taken from an advert for batteries.

'This is the most powerful battery we have ever made! It lasts up to five times longer than an ordinary battery.'

Use your understanding of energy and power to explain to the advertising department of the company what they have got wrong in their advert and construct a scientifically correct advert to replace it.

A*

Potential and kinetic energy

Learning objectives

- describe how energy is transferred as an object rises or falls
- calculate the potential energy of an object
- calculate the kinetic energy of a moving object.

Gravitational pull

Any two masses will attract each other, because all masses exert a gravitational pull on all other masses. The gravitational pull we are most used to is between us and the Earth, so we often use the words 'force of gravity' to describe the attraction between objects on or near the Earth and the Earth itself.

Gravity and energy

In lesson P2 2.1 we used the equation:

$W = F \times d$

where W = work done or energy transferred in joules (J), F = force applied in newtons (N) and d = distance moved in metres (m).

We can use this equation to calculate work done against gravity.

If you lift a tin with a mass of 2 kg from the floor and put it on a shelf 1.5 m above the ground, you have done work against the weight of the object. We can calculate the amount of work done as follows.

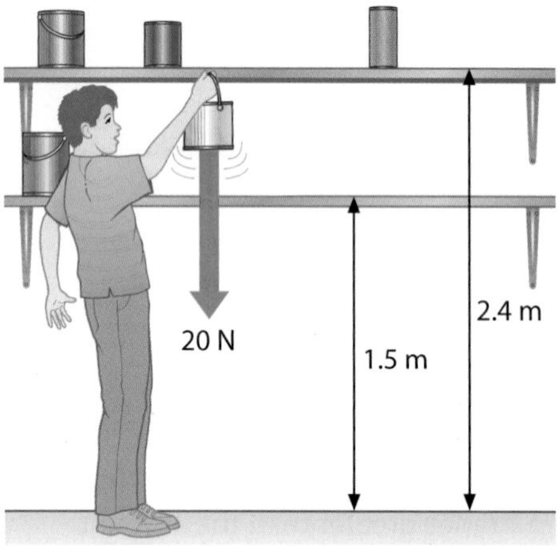

We need to know the weight of the tin rather than its mass. In lesson P2 1.6 we learned that the weight of an object can be calculated by multiplying the mass by the Earth's gravitational field strength (roughly 10 N/kg).

The weight of the tin is therefore 2 kg × 10 N/kg = 20 N.

From the equation above,

work done = downward force (weight of tin) × distance lifted
= 20 N × 1.5 m
= 30 J

The energy you transfer is stored because the mass is in the Earth's gravitational field. It has **potential energy**.

To calculate the potential energy of a mass you use the formula:

$E_p = m \times g \times h$

where E_p = potential energy in joules (J), m = mass in kilograms (kg), g = gravitational field strength in newtons per kilogram (N/kg) and h = change in height in metres (m).

So for our 2 kg tin we can say that it has 30 J more potential energy when it is on the shelf than it had on the floor.

Figure 1 Transferring energy to potential energy by lifting a mass.

Route to A*

If the efficiency of your body when lifting a can of paint is 10%, how much energy does your body convert if you lift four of the yellow cans of paint to the top shelf from the floor?

Kinetic energy

Kinetic energy is energy due to movement. The formula for calculating kinetic energy is:

$E_k = \frac{1}{2}mv^2$

where E_k = kinetic energy in joules (J), m = mass in kilograms (kg) and v = speed in metres per second (m/s).

Examiner feedback

Note that when we calculate the change in potential energy, h is the change in vertical height.

For example, if a cyclist and bike together have a mass of 100 kg and are moving at 3 m/s, their kinetic energy is:

$$E_k = \frac{1}{2} \times 100 \times 3^2$$

$$= 450\ \text{J}$$

In a car the engine transfers some of the chemical energy from the fuel to kinetic energy.

Imagine a tennis ball thrown vertically up into the air. As it leaves the tennis player's hand it is moving upwards, so it has kinetic energy. As it rises it slows down so has less kinetic energy but it is getting higher and so has more potential energy. The kinetic energy is transferring into potential energy. At the top of its movement, it stops for an instant and so has no kinetic energy and all of the energy is stored as potential energy. When it falls again the potential energy is transferred back into kinetic energy.

Science in action

The gravitational field of the Earth is about 9.81 N/kg. The exact strength depends where you are on the Earth. Gravity at the poles is measurably larger than gravity at the Equator.

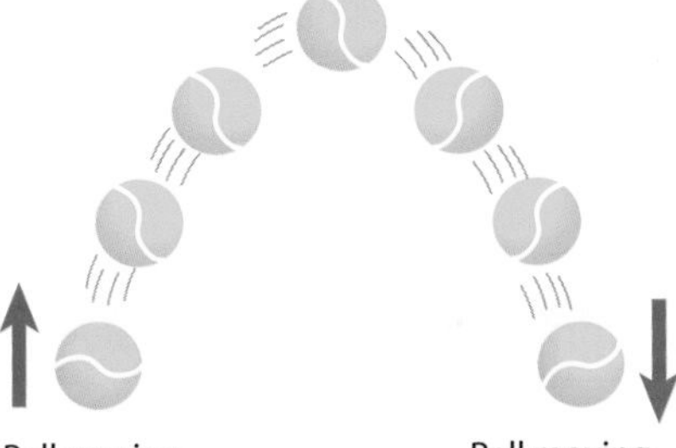

Figure 2 Energy can be transferred between kinetic energy and potential energy.

Questions

1. Calculate the amount of potential energy that you give a 2 kg ball if you throw it 3 metres upwards.
2. Describe the energy transfers that take place when a firework rocket goes off.
3. A meteor hits the Earth's atmosphere at a speed of 50 km/s. If the meteor weighs 40 kg calculate its kinetic energy, and describe what happens to all this energy.
4. Calculate the extra potential energy the paint tin in Figure 1 would gain if it was moved from the lower shelf to the higher one.
5. Calculate the kinetic energy of a 5000 kg bus moving at 20 m/s.
6. The Victoria Falls on the Zambezi River in Africa is one of the largest waterfalls in the world. As the water falls, its potential energy must be transferred. Describe what happens to the potential energy.
7. Calculate the amount of kinetic energy a 30 g mouse would have as it runs at 1.5 m/s across the floor.
8. If you look closely at Figure 2 you will note that the last label says it is 'gaining kinetic energy' as it falls. In fact this kinetic energy is slightly less than the kinetic energy it had when it was thrown upwards. Explain why this is and compare the potential energy with the kinetic energy the ball had at the start and the end of the throw.

A*

Taking it further

If you apply the idea of potential energy = *mgh* to spacecraft, the situation rapidly becomes more complicated. As you move away from the Earth, the gravitational field strength drops rapidly so calculations of potential energy can get quite complicated. NASA scientists working out the forces on a rocket as it climbs through the atmosphere also have to contend with changes in the density of the air as the rocket gets higher, which reduces air drag.

P2 2.4 Momentum

Learning objectives

- calculate the momentum of a mass
- explain why momentum has direction as well as size
- describe how force is related to change in momentum and time.

Examiner feedback

Notice that momentum is calculated from velocity and not speed. It has both size and direction, just like velocity.

Taking it further

In nuclear physics it is also important to understand the momentum of tiny particles. Being able to calculate the momentum of a single electron moving at high speed allows scientists to understand how the basic building blocks of the universe behave.

Route to A*

If a meteorite is moving at 70 km/s and has a mass of 0.005 g, what would be its momentum?

Examiner feedback

When you are calculating momentum, you need to choose which direction is positive and which is negative. It does not matter which direction you choose to be positive as long as you keep to that decision for all of the masses you are studying.

What is momentum?

Momentum is a property of a moving object, calculated by multiplying the object's mass by its velocity:

$p = m \times v$

where p = momentum in kilogram metres per second (kg m/s), m = mass in kilograms (kg) and v = velocity in metres per second (m/s).

Momentum is a useful mathematical way of modelling what happens when masses collide or things explode.

If a large mass is moving fast it has a lot of momentum. If you want to stop it in a hurry, you will need to apply a large force. This is why in a rugby game, a defending team will try to stop an attacking player getting up speed. A fast-moving player has more momentum than one at rest, so they are harder to stop.

Example 1

If a boat has a mass of 150 kg and is moving north at 2 m/s, then its momentum is:

momentum = mass × velocity

= 150 kg × 2 m/s

= 300 kg m/s north

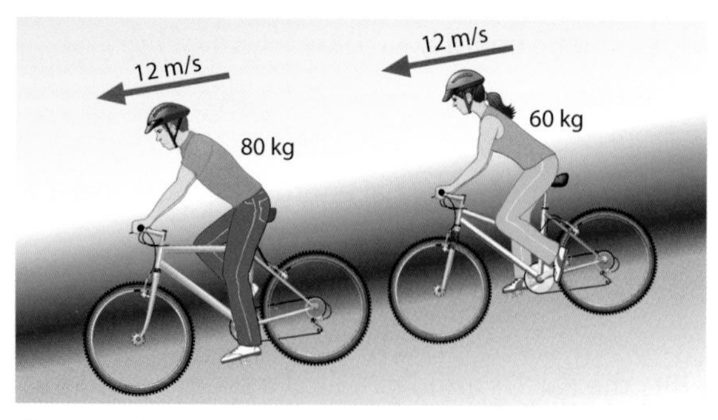

Figure 1 The riders have the same velocity but different momentum.

In Figure 1, the momentum of the boy on the bike is 80 kg × 12 m/s = 960 kg m/s to the left.

The girl has the same velocity but her mass is less: her momentum is 60 kg × 12 m/s = 720 kg m/s to the left.

If a bullet with a mass of 1 g is fired out of a gun at a velocity of 250 m/s, the momentum of the bullet can be found in the same way as for the cyclists above. However, we must be careful to use the correct units.

1 g = 0.001 kg

So, using the equation: momentum = $m \times v$

the momentum of the bullet = 0.001 × 250

= 0.25 kg m/s

At the other end of the scale, a 747 jet flying at 900 km/h with a mass of 400 tonnes has a massive momentum.

400 tonnes = 400 000 kg

$$900 \text{ km/h} = \frac{900\,000}{3600} = 250 \text{ m/s}$$

Again, using momentum = mass × velocity

momentum of the 747 = 400 000 × 250

= 100 000 000 kg m/s

If you calculate the momentum of cars X and Y in Figure 2, you will find that the momentum is the same for both cars, but in opposite directions. Here the positive direction is to the right.

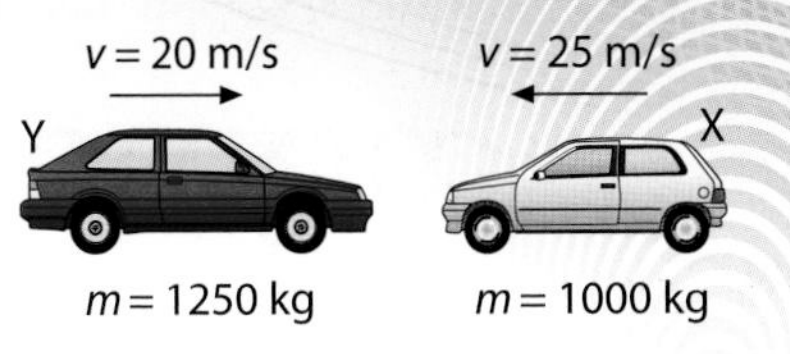

Figure 2 These cars have the same momentum but in opposite directions.

Car X: mass = 1250 kg, velocity = 20 m/s

Momentum = 1250 kg × 20 m/s

= 25 000 kg m/s

Car Y: mass = 1000 kg, velocity = −25 m/s

Momentum = 1000 kg × −25 m/s

= −25 000 kg m/s

Note the minus sign for car Y, showing that the momentum is to the left, in the opposite direction to that of car X.

Taking it further

It is difficult to weigh an object in space because it is 'weightless'. However, you can find its mass by calculation from the effects of its momentum.

Science in action

Some vehicles are fitted with devices that pull the seatbelt tight around the driver and passenger if the car crashes. This means that the passenger does not 'hit' the seatbelt as they move forward relative to the car.

Route to A*

A student suggested that making the seatbelts elastic would help reduce the injury to passengers in a crash. Use your understanding of the safety features in car design to explain whether you think this is a good idea or not.

Questions

1. Calculate the momentum of the following: **(a)** a 15 kg dog running at 10 m/s **(b)** a 5 g bullet moving at 300 m/s **(c)** a 12 t lorry moving at 20 m/s.
2. Describe how the momentum of two identical masses moving at the same speed can be different.
3. What is the change in momentum for a 300 kg vehicle slowing down from 15 m/s to 12 m/s?
4. Use the equation: force = $\frac{\text{change in momentum}}{\text{time}}$ to explain why it takes a greater force to stop a lorry than a car if they are both travelling at the same velocity.
5. Explain, using diagrams and estimates for the mass, speed and momentum of the head of a hammer, why the head of the hammer needs to have quite a lot of mass when it is used to drive in large nails.
6. A canal barge carrying 50 000 kg of stone is moving at 0.5 m/s. Explain why the person in charge needs to start slowing the barge down quite a long time before it arrives at the lock gates.
7. You are on a space station where everything is weightless. Explain why being hit by a heavy lump of metal would still hurt.
8. A pile driver can be used to drive metal rods into the ground. It is a large mass of metal that is slowly lifted and then dropped onto the end of the metal rod. Describe the amount of momentum a pile driver might have as it is lifted, falls and then hits the metal rod. What does your understanding of the idea that for each force there is an equal and opposite force tell you about the motion of the Earth in this situation? A*

The law of conservation of momentum

Learning objectives

- explain what conservation of momentum means
- use calculations to predict the motion of masses after a collision
- describe examples of conservation of momentum in a closed system.

Practical

Use sticky tack to hold two chopsticks on a surface so that a marble will roll along them. Position one marble in the middle of your 'marble run'. Roll another similar marble towards the stationary one. What happens?

Think about what you observe in terms of the conservation of momentum.

before collision 4 m/s → 5 kg

before collision 4 m/s ← 5 kg

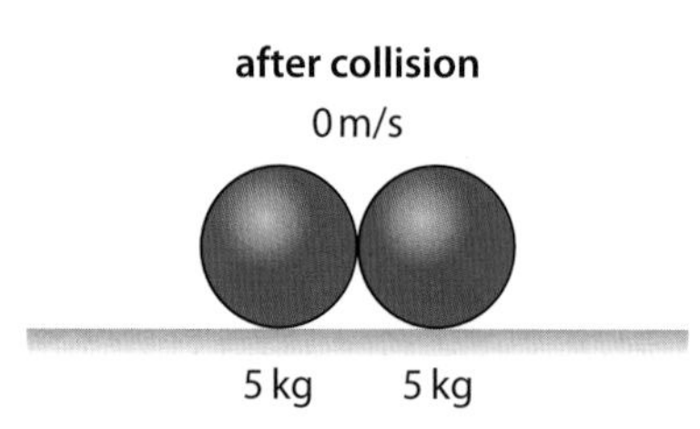

Figure 1 A collision between two balls moving in opposite directions.

Examiner feedback

From explosions of stars to collisions of sub-atomic particles, momentum is always conserved provided no external forces act.

Masses in a closed system

Imagine a person sitting on a surface even more slippery than ice. There is no friction at all between them and the surface. If they try to move they get nowhere. How can they get off the surface?

The answer is for the person to throw something. They will then move in the opposite direction to the object they throw. The momentum of the person and the object before they threw it was zero. If you add up the momentum of the person and the object after it is thrown the total momentum is still zero, but now they are both moving apart. This is an example of a **closed system**, where no forces from the external environment act. In a closed system, total momentum does not change.

Conserving momentum

If two objects collide in a closed system, the total momentum is the same before and after the collision. This idea is known as the **law of conservation of momentum**.

The easiest collision to understand is when two objects of equal mass, moving at equal speeds in opposite directions, hit each other and stick together. Imagine two sticky balls rolled towards each other. If the balls hit and stick together, then after the collision they will be stationary. The law of conservation of momentum helps us to understand why.

Example 1

Using the formula:

$$p = m \times v$$

where p = momentum in kilogram metres per second (kg m/s), m = mass in kilograms (kg) and v = velocity in metres per second (m/s),

the momentum of the left-hand ball in Figure 1 can be calculated as

5 kg × 4 m/s = 20 kg m/s to the right.

The other ball has the same size of momentum but in the opposite direction, i.e. 20 kg m/s to the left, or −20 kg m/s to the right. This is the key point: movement to the left is the same as negative movement to the right.

Adding the momentum of the left-hand and right-hand balls together we get:

20 kg m/s + (−20 kg m/s) = 0.

So even though both balls are moving, their total momentum is zero.

By the law of conservation of momentum, their total momentum after the collision must also be zero. If the balls stick together, then the result is that they stop moving.

Elastic collisions

When balls bounce off each other things are more complicated but the law of conservation of momentum still applies.

Example 2

Suppose a lighter ball moves towards and hits a heavier, stationary ball. After the collision both balls move.

We will make the rule that all values to the right are positive.

Before the collision:

Momentum of red ball = 5 kg × 4 m/s = 20 kg m/s

Momentum of blue ball = 8 kg × 0 m/s = 0 kg m/s

So the total momentum is 20 kg m/s to the right.

After the collision:

Momentum of blue ball = 8 kg × 3 m/s = 24 kg m/s

We already know that the total momentum before the collision was 20 kg m/s.

According to the law of conservation of momentum, the total momentum after the collision must also be 20 kg m/s. Since the blue ball has a momentum of 24 kg m/s, the momentum of the red ball must be:

20 kg m/s − 24 kg m/s = −4 kg m/s (i.e. 4 kg m/s to the left).

We know that $p = m \times v$.

Rearranging, $v = \frac{p}{m}$. This means that after the collision the red ball is moving left at 4 kg m/s ÷ 5 kg = 0.8 m/s.

In a real situation momentum would not be conserved because some energy would be lost to the environment in the collision.

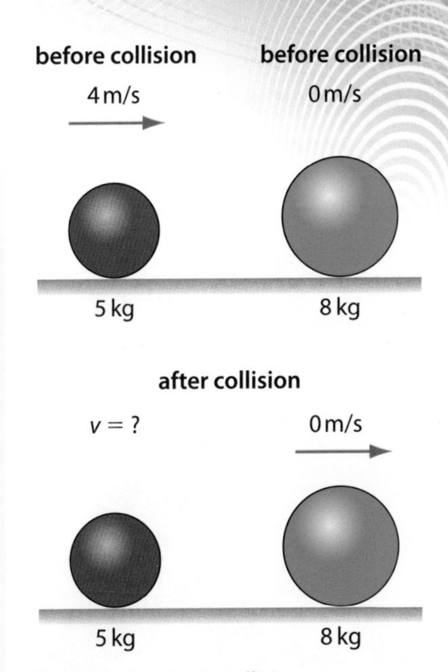

Figure 2 An elastic collision.

Taking it further

If a 1 g bullet travelling at 250 m/s hits a wooden target with a mass of 500 g and gets buried in it, how fast will the wooden target move?

Questions

1. When a snooker ball hits another stationary snooker ball of equal mass, the moving snooker ball stops and the stationary one carries on at the same speed (or very nearly). Describe what is happening in terms of the momentum of the two balls.
2. If an 85 kg basketball player jumps up and throws a basketball horizontally at 10 m/s to another player, he moves backwards because of the throw. The basketball weighs 425 g; how fast will the player move backwards?
3. Explain why a gun will move backwards when fired, but more slowly than the bullet being fired moves forward.
4. When a cricket bat weighing 2 kg and moving at 30 m/s hits a ball travelling towards it at 30 m/s with a mass of 160 g, the ball then moves off backwards and the bat slows to 24 m/s. What is the velocity of the ball?
5. When a tennis racquet hits a tennis ball, the total momentum is not conserved. Why is this?
6. If an object is moving in a circle, its velocity is continually changing but its speed is not. Explain what is happening to the momentum of the object.
7. On a space walk your safety line breaks, and you are left floating in space, just out of reach of the spacecraft but not getting any further away from it. How could you get yourself back to the spacecraft?
8. If a bullet of mass 15 g moving at 300 m/s hits and buries itself in a tree, the tree does not seem to move. Where has the momentum gone? Conservation of momentum demands it must be somewhere. Explain what has happened to the momentum.

P2 2.6 Car safety

Learning objectives

- evaluate the benefits of different car braking systems
- explain how air bags, crumple zones, seatbelts and side-impact bars improve car safety.

Braking systems

We saw in lesson P2 2.1 that when a car brakes, its kinetic energy is transferred through heating to the brake pads and discs. However, if the brakes grip too hard they cause the wheels to lock and the car skids. In this case there is friction between the road and the car tyres, which both heat up. Skidding is dangerous because the driver loses control of the steering.

Many modern cars have an antilock braking system (ABS). When a driver brakes hard, the ABS senses if the car's wheels lock up. It reduces the braking force and the wheels start to turn again. The steering still works, so the driver can control the car, although the braking distance may not actually be any shorter.

Most modern cars have disc brakes. A calliper squeezes the brake pads against the brake disc.

KERS

A kinetic energy recovery system (KERS) is a way of making a vehicle's braking system more energy-efficient. KERS uses an electric generator instead of brakes to slow the car down. Kinetic energy is transformed into electrical energy that is stored in a battery instead of being lost as heat.

KERS braking does not work at very low speeds, and in an emergency stop it may not be able to supply enough braking force. KERS is therefore backed up by conventional brakes.

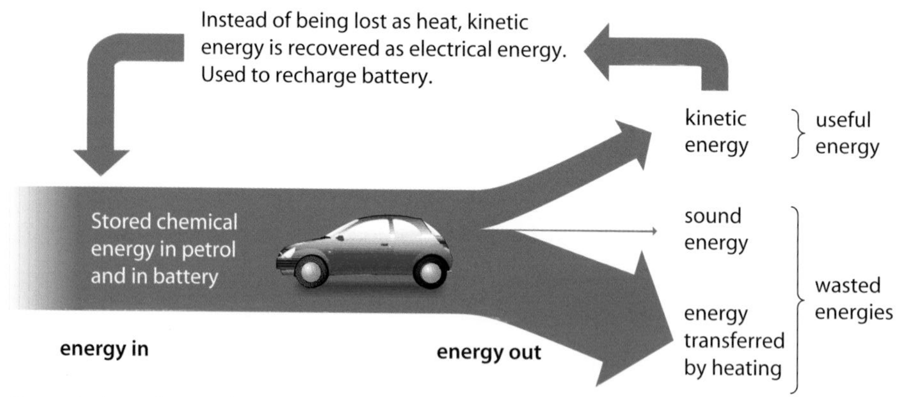

Figure 1 Flow of energy in a car using KERS to recover energy from braking.

Science in action

Some Formula 1 cars may use a KERS that stores energy in a heavy flywheel. This stored energy can then be released as a boost when accelerating.

Car safety devices

Modern cars are fitted with many safety devices, including seatbelts, air bags, crumple zones and side-impact bars. These devices help reduce injury to car occupants by reducing the forces on them during a collision.

The force on a car occupant in a crash can be calculated from:

$F = ma$

where F = force in newtons (N), m = mass in kilograms (kg) and a = acceleration in metres per second squared (m/s^2)

and

$$a = \frac{v - u}{t}$$

where a = acceleration in metres per second squared (m/s^2), v = final velocity in metres per second (m/s), u = initial velocity in m/s, and t = time in seconds.

To reduce the force we need to reduce the deceleration. To reduce the deceleration we need to increase the time the car takes to slow down.

It all comes down to making the crash happen over as long a time as possible.

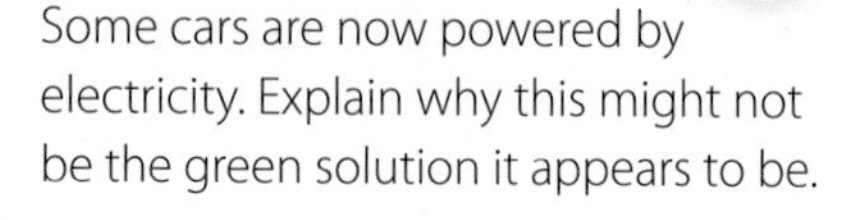

Route to A*

Some cars are now powered by electricity. Explain why this might not be the green solution it appears to be.

Crumple zones and seatbelts

If a person hits a concrete block head-on, with no protection, they come to a stop in a tiny fraction of a second. If the person is strapped into a car and the bonnet section of the car crumples up, the time for the impact to happen is much longer. This makes the force that slows the passenger down much smaller and so they suffer less injury. Air bags also work by increasing the time taken for impact.

Quite a few crashes happen where one car rams the side of another. Manufacturers now build stiff side-impact bars into cars so that the occupants have some protection. Some cars also have side-impact air bags.

The most important safety device of all is the seatbelt. Other safety features have little effect if car occupants are flung about during a crash.

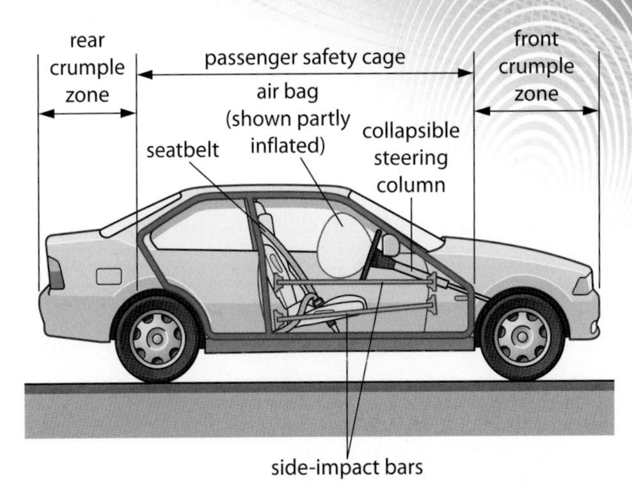

Figure 2 Safety features in modern car design.

Cars are crash-tested to see how well their safety features work.

Questions

1. If a car slows down from 20 m/s to stop in 0.1 seconds, calculate the deceleration of the car.
2. **(a)** A passenger wearing a seatbelt in the car from question 1 has a mass of 80 kg. Calculate the force on the passenger as the car slows down.
   (b) The driver in the same crash was not wearing a seatbelt and hit the windscreen. She took only 0.01 seconds to stop.
   Calculate her deceleration and the force on her during the crash.
3. Explain the difference between deceleration and acceleration.
4. Table 1 contains information about the velocity of two crash-test dummies in a crash. Plot the data for both dummies on the same graph.

Table 1 Crash data.

Time/s	0.00	0.01	0.02	0.03	0.04	0.05	0.06	0.07	0.08	0.09	0.10
Velocity dummy 1 / m/s	20	20	19	18	15	12	9	6	2	1	0
Velocity dummy 2 / m/s	20	20	20	20	20	20	20	20	20	20	0

   Which crash-test dummy do you think was wearing a seatbelt?

5. A 2000 kg electric vehicle is moving at 10 m/s. When it slows down, it uses a KERS. As a result, 40% of the kinetic energy is recovered as electrical energy. Calculate the amount of electrical energy that is recovered.
6. To make use of a KERS, a racing car would need to have a heavy flywheel to store kinetic energy. Why might some engineers suspect that the advantages of KERS are not as great as might be expected?
7. Use the equation: acceleration = change in velocity / time to explain how crumple zones reduce the deceleration of the passenger during a collision.
8. Cars are designed to make the engine slide beneath the passenger compartment in a head-on collision. Explain why this improves the chances of a passenger's survival in a crash.

Route to A*

Some modern cars use an automatic system that cuts the engine when the car pulls up at traffic lights. It uses some fuel to start the car again; this is equivalent to about 4 seconds when stationary with the engine on. The system can save fuel but the car is initially more expensive to buy. Describe the factors that would influence the decision of a buyer when deciding whether or not to buy this more expensive option.

ISA practice: materials for crumple zones

The front and rear ends of motor cars are designed to crumple when involved in collisions. This helps to protect the occupants from injury.

You have been asked to explore the behaviour of different materials that might be used in the design of crumple zones in cars. These materials include steel, copper, aluminium, brass and PVC.

Section 1

1 Write a hypothesis about how the density of a material affects its ability to crumple. Use information from your knowledge of forces and energy to explain why you made this hypothesis. *(3 marks)*

2 Describe how you could carry out an investigation into this factor.

You should include:

- the equipment that you could use
- how you would use the equipment
- the measurements that you would make
- how you would make it a fair test.

You may include a labelled diagram to help you to explain the method.

In this question you will be assessed on using good English, organising information clearly and using specialist terms where appropriate. *(6 marks)*

3 Think about the possible hazards in the investigation.

(a) Describe one hazard that you think may be present in the investigation. *(1 mark)*

(b) Identify the risk associated with this hazard that you have described, and say what control measures you could use to reduce the risk. *(2 marks)*

4 Design a table that you could use to record all the data you would obtain during the planned investigation. *(2 marks)*

Total for Section 1: 14 marks

Section 2

A group of students, Study Group 1, investigated a hypothesis about how the density of a material relates to its ability to crumple. They tested it by dropping a mass on a sample of each material from the same height.

Figure 1 shows the results they obtained. The experiment was repeated three times with each metal.

Steel
Density: 7.8 g/cm3
Bend: 3.5 mm, 3 mm and 3 mm

Aluminium
Density: 2.7 g/cm3 density
Bend: 14.5 mm; 15 mm and 18.5 mm

Brass
Density: 8.5 g/cm3
Bend: 5.5 mm, 6 mm and 5.5 mm

PVC
Density: 1.4 g/cm3
Bend: bent at right angle all three times

Copper
Density: 8.9 g/cm3
Bend: 7.5 mm, 8.5 mm and 7 mm

Figure 2 Study Group 1's results.

5 **(a)** Plot a graph or bar chart of these results. *(4 marks)*

(b) What conclusion can you draw from the investigation about a link between the density of the material and the amount of crumple? You should use any pattern that you can see in the results to support your conclusion. *(3 marks)*

(c) Look at your hypothesis, the answer to question 1. Do the results from Study Group 1 support your hypothesis? Explain your answer. You should quote some figures from the data in your explanation. *(3 marks)*

Below are the results from three other study groups.

Table 1 shows the results from two other students, Study Group 2. They investigated the same hypothesis as Study Group 1, but using thinner materials.

Table 1 Study Group 2's results.

Material	Amount of bend
steel	4.5
copper	17
aluminium	Totally bent
brass	13
PVC	Totally bent

Study Group 3 was a group of researchers working for a car manufacturer. They measured the force on a crash-test dummy during destructive testing of a new car design. The forces were measured during collisions where identical cars travelling at different speeds were crashed into a concrete block (see Figure 2). The results are given in Table 2.

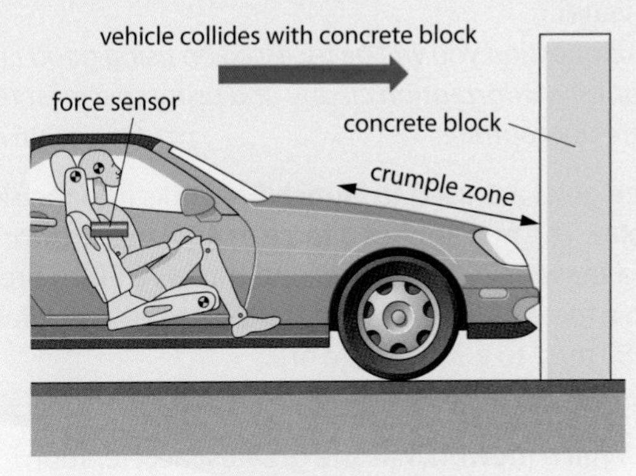

Figure 2 Experimental set-up for testing forces during a crash.

Table 2 Results for Study Group 3.

Speed / m/s	Maximum force on dummy/N			Average force on dummy/N
2	158	165	157	160
4	629	652	660	647
6	1500	1100	1450	1350
8	2400	2420	2440	2420
10	3950	3990	3994	3978

Study Group 4 is a group of scientists who compared the speed of collision with the crumple effects for 1-metre lengths of different materials with the same cross section. Table 3 shows their results.

Table 3 Results for Study Group 4.

Speed of collision / m/s	Amount material crumpled (%)				
	Steel	Copper	Aluminium	Brass	PVC
2	14	8	25	11	60
4	28	15	48	25	75
6	39	23	61	34	90
8	53	31	70	42	90

6 **(a)** Draw a sketch graph of the results from Study Group 2. *(3 marks)*

(b) Look at the results from Study Groups 2 and 3. Does the data support the conclusion you reached about the investigation in answer to question 5(b)? Give reasons for your answer. *(3 marks)*

(c) The data contain only a limited amount of information. What other information or data would you need in order to be more certain whether the hypothesis is correct or not?
Explain the reason for your answer. *(3 marks)*

(d) Look at Study Group 4's results. Compare them with the data from Study Group 1. Explain how far the data supports or does not support your answer to question 5(b). You should use examples from Study Group 4's results and from Study Group 1. *(3 marks)*

7 **(a)** Compare the results of Study Group 1 with Study Group 2. Do you think that the results for Study Group 1 are *reproducible*?
Explain the reason for your answer. *(3 marks)*

(b) Explain how Study Group 1 could use results from other groups in the class to obtain a more *accurate* answer. *(3 marks)*

8 Applying the results of the investigation to a context.
Suggest how ideas from the original investigation and the other studies could be used by the manufacturers to decide on the best material to use for a crumple zone.
(3 marks)

Total for Section 2: 31 marks

Total for the ISA: 45 marks

Assess yourself questions

1 Explain why the following statement is incorrect: 'If you don't push something it won't move.' Use an example to illustrate your answer. *(3 marks)*

2 Use a labelled diagram to describe all of the forces that would be acting on a cyclist riding along a straight road at a steady speed. *(6 marks)*

3 Explain how *three* forces, all of which are larger than 40 N, can produce a resultant force of less than 10 N. *(2 marks)*

4 **(a)** Explain the difference between length and extension for a spring. *(2 marks)*

(b) Use the data in Table 1 to produce a graph of the results. *(4 marks)*

Table 1 Data from an experiment where a student stretched a spring.

Force/N	Length/cm
0	6.0
0.1	6.5
0.2	7.1
0.3	7.5
0.4	7.9
0.5	8.6
0.6	8.9
0.7	9.5
0.8	10
0.9	10.6
1.0	11.0

(c) Describe why this graph supports the suggestion that the force applied is proportional to the extension of the spring. *(2 marks)*

(d) From the data work out the spring constant in N/m. *(2 marks)*

(e) A stiffer spring of the same length was tested using the same forces. Add a line to your graph that shows the relationship between force and extension for this new spring. *(2 marks)*

(f) A weaker spring of the same initial length was also tested. It was found that, after the experiment, its length when the force was removed was 7.2 cm. Explain why this might have happened. *(2 marks)*

5 The following is an extract from a live radio commentary during the launch of a space vehicle.

'This massive man-made machine is slowly inching its way off the launch pad. All 500 tonnes of it are balanced on a pillar of flame, and now it starts to gather speed and climb into the night sky.'

Use the information in this passage to fully describe the forces involved with the take-off of this vehicle. *(5 marks)*

6 When somebody dives off the high board in a swimming pool, they move upwards to start with and then plunge into the water. The high board is further off the surface of the water than the depth of the water. Use your understanding of forces to describe fully the motion of the diver from the time that they leave the board to the moment they reach their deepest point in the water.

In this question you will be assessed on using good English, organising information clearly and using specialist terms where appropriate. *(6 marks)*

7 Three guns are used to launch fireworks into the sky at a display. The first applies a force of 40 N to a firework that has a mass of 200 g. The second throws a 500 g firework with a force of 80 N and the third gives an acceleration of 180 m/s^2 to a 300 g firework.

(a) Which firework has the greatest force on it? *(1 mark)*

(b) Which firework has the greatest acceleration? *(1 mark)*

8 Figure 1 shows a distance–time graph for a car. Calculate its speed at a, b and c. *(3 marks)*

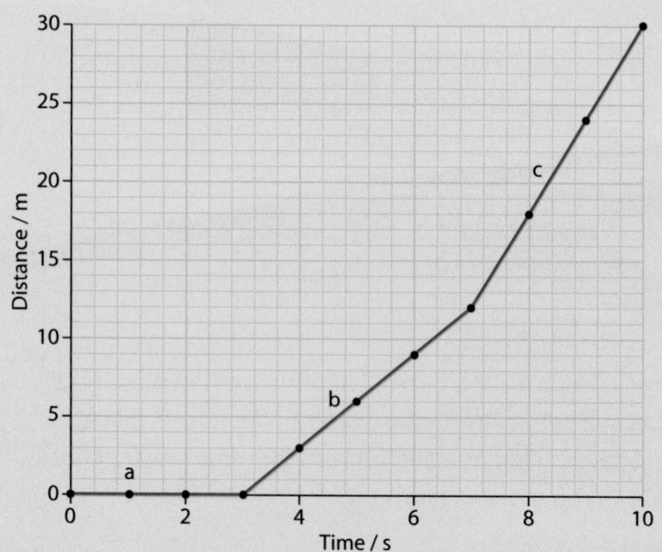

Figure 1 Graph of distance against time.

9 Use the data in Table 2 to draw a graph of velocity against time for a radio-controlled car.

Table 2 Time and velocity.

Time/s	0	1	2	3	4	5	6	7	8	9	10
Velocity / m/s	0	0	2	4	6	6	6	6	6	3	0

(a) Calculate the distance travelled by the car. *(1 mark)*

(b) Calculate the maximum acceleration of the car. *(1 mark)*

(c) If the mass of the car is 2 kg, what is the greatest accelerating force on the car? *(1 mark)*

10 A driver in a car sees a deer run out into the road and puts on the brakes. Unfortunately the car hits the deer. Explain why it is much more likely that the deer will be severely injured if the car was travelling at 40 miles per hour than if it was travelling at 30 miles per hour. *(4 marks)*

11 A car is travelling at 5 m/s when a second car pulls out into the road. The first car decelerates at 2 m/s^2. Draw a velocity–time graph to show the motion of the first car as it stops. From the graph work out the distance the car travels before it stops. *(5 marks)*

12 A car with a mass of 800 kg brakes with a force of 2000 N. Calculate its deceleration. *(2 marks)*

13 A tractor is pulling a plough through soil with a force of 1500 N. The tractor drags the plough a distance of 3.2 km during the day. The tractor has a mass of 3 tonnes and the plough weighs 500 kg. How much work does the tractor do to pull the plough? *(3 marks)*

14 As it moves along, a car engine not only pushes the car forward, but also generates electricity to recharge the battery and run the lights. Two students discussing this issue disagree. One student thinks that the car is moving anyway so the lights do not take any fuel to run. The other student insists that it must use more fuel to have the lights on than to have them off. Choose one student's argument to support and explain which student is correct. *(2 marks)*

15 Below is an excerpt from a magazine article about regenerative braking used in trains in Switzerland. Read the article and then explain why this system would be unlikely to have many applications in this country. *(2 marks)*

'In Switzerland, many of the train lines in the more mountainous areas use overhead electric power lines and regenerative braking. This reduces power consumption on the railway. In some places, one train engine going up a mountain can be powered by the descent of two other trains. The electric motors that are used to push the trains up the mountains turn into generators when they come down again, feeding electricity back into the overhead power lines.'

16 It takes a parent 10 seconds to push a pushchair up a slope. The parent pushes with a force of 80 N and the slope is 5 metres long.

(a) How much work has the parent done? *(1 mark)*

(b) What was their power output? *(1 mark)*

17 A power station's power output is 500 MW. If the power station was only 40% efficient, how much energy would it need, in the form of fuel, to operate for 1 hour? *(3 marks)*

18 If a 100 g ball is thrown up into the air at 5 m/s, and assuming that no energy is lost, calculate the following:

(a) The kinetic energy that the ball has initially. *(1 mark)*

(b) The height it will reach before it stops. *(1 mark)*

(c) The speed it will be travelling at when it comes down again. *(1 mark)*

(d) In reality, as the ball comes down, it will move a little more slowly that when it was thrown up, and it will not get quite as high as you have calculated. Explain why this is so. *(1 mark)*

19 In an experiment using a friction-compensated runway and an 800 g trolley, students measured the velocity of the trolley as it was pulled down the slope by different forces.

(a) Describe what is meant by a friction-compensated slope. *(2 marks)*

(b) Calculate the acceleration of the trolley if a force of 0.4 N pulled it down the slope. Show your working. *(2 marks)*

(c) If the force of 0.4 N continued for 3 s, what would be the change in the velocity of the trolley? Show your working. *(1 mark)*

(d) Using the answer from part (c), calculate the increase of momentum of the trolley after 3 s. Show your working. *(1 mark)*

(e) The trolley is stopped by a brick placed at the end of the runway. To protect the trolley, the teacher asked the pupils to place a piece of sponge in front of the brick. Use your understanding of the change in momentum of colliding masses to explain how the sponge protects the trolley. *(2 marks)*

20 A student trying to model the way that a crumple zone works used a steel tube strapped to an egg. The egg is dropped from a height of 20 cm on to the bench. Another student used a tube made of thin cardboard. Explain why one student ended up with a broken egg and the other ended up with a crumpled tube. *(4 marks)*

21 Explain how the ideas in question 20 relate to the design of crumple zones in cars that are made from steel sections. *(5 marks)*

22 A 400 g toy plane moving at 4 m/s crashes into a wall and is damaged.

(a) How much momentum did the plane have before the crash? *(1 mark)*

(b) How much momentum did it have after the crash? *(1 mark)*

(c) How much kinetic energy did it have before the crash? *(1 mark)*

(d) How much kinetic energy did it have after the crash? *(1 mark)*

(e) Describe where the momentum and kinetic energy have gone. *(1 mark)*

GradeStudio Route to A*

Here are three students' answers to the following question:

When car manufacturers are designing cars, they must give high priority to the safety of the driver and passengers. Seatbelts work with crumple zones to protect the driver and passengers if there is a collision. Explain how engineers use their understanding of force and acceleration in the design of these safety features.

In this question you will be assessed on using good English, organising information clearly and using specialist terms where appropriate. *(6 marks)*

Read the answers together with the examiner comments. Then check what you have learnt and try putting it into practice in any further questions you answer.

Grade answer

Student 1

Speed kills people in cars. A seatbelt holds the driver still when the car hits a wall so all of the impact is taken by the crumpel zone of the car. If the crumpel zone is big, it slows down less too. If no seatbelt, the crumpel zone doesn't work.

The student has misunderstood here: it is the high acceleration that injures the passenger.

The answer should explain that the crumple zone works by reducing acceleration, which leads to a smaller force on the occupant.

The length of the crumple zone does not affect the amount of slowing down; it affects the time it takes to slow down, and this in turn affects the force.

Spelling mistakes and poor sentence construction.

Examiner comment

This candidate seems to have understood that the injury to passengers comes from the fact that the car slows down suddenly but they have not made this clear. They have not mentioned force and have not explained why a crumple zone reduces injury. It is high acceleration (deceleration) that causes injury rather than 'speed'. The candidate fails to explain that the crumple zone increases the time it takes for the collision to happen, and so reduces the acceleration. Although they mention the role of the seatbelt, it is not clear that they have understood how it works with the crumple zone. The candidate needs to have explained that without the seatbelt, the driver would effectively hit a stationary steering wheel and so the crumple zone would have no effect.

- Read the question carefully.
- Put the key points into a logical sequence using a flowchart or mind map.
- Make sure that links between parts of the answer are clearly made.
- Be full in your explanation and extend explanations by using words and phrases like 'because', 'so that', 'as well as', 'and so', 'this also means'.
- Make sure you have used as much science as possible in your answer.
- Check that your answer addresses all parts of the question.

A Grade answer

Student 2

It would be better to say that the force on the chest also causes high acceleration of the internal organs, and that it is the high acceleration that causes internal injuries.

The link between the crumple zone and acceleration is well made but the seatbelt explanation is too superficial.

The designers understand that the thing that causes injury to the driver is the force on their chest when the car slows down. The amount of force depends on the mass of the driver and the acceleration of their body. If the car has a crumple zone, the time it takes to slow down is also bigger, so the acceleration is less. This means that the force on the driver is less. The seatbelt also helps by reducing the impact on the driver during the collision.

Examiner comment

Like student 1, this candidate has shown a good understanding of the way that the crumple zone works and has made clear links between the injuries caused through force and how these are reduced by extending the time of the impact. However, the explanation of the role of the seatbelt is underdeveloped and they should have explained how the crumple zone and the seatbelt work together.

A* Grade answer

Student 3

Good identification of what causes the injury.

Good link between speed change, time and acceleration.

Good extension of what the injuries are.

The injuries caused to the driver during a head-on crash with another car or vehicle are most severe when the acceleration of the driver is high. The designers work to reduce the acceleration of the driver during the impact. One method is to use a crumple zone. The engine compartment and bonnet area are designed to crumple during a collision. This spreads the time of the collision over a longer time than if the car bonnet region was rigid. The passenger compartment is made very rigid and the seatbelt holds the driver in place so that they slow down with the car as the crumple zone crushes. Because the change in speed happens over a longer time the acceleration of the driver is less. This produces less force on the driver's chest so reducing external injuries and also reduces the acceleration of the internal organs so reducing internal injuries too.

Examiner comment

This candidate has covered all the main points. It is the effect of acceleration on the internal organs that causes most injury, rather than the force itself, which causes external injuries. The way that the crumple zone and the seatbelt work together is clear. The contrast with the ability of the front area of the car to crumple as well as the rigidity of the passenger compartment is clearly made. The candidate clearly understands the link between force and acceleration.

P2 Electricity, radiation, atoms and the stars

The following chapters examine four areas: electric currents in circuits, atomic structure and radioactivity, nuclear power, and the life cycles of stars.

Our modern lives depend on controlling the flow of electric current. Whether it is to play music, wash clothes or simply to provide light, electricity has become an essential part of our lives. All electrical circuits are made up of components that have an influence on the way that current flows through them. The first section analyses the behaviour of components in series and parallel circuits, as well as mains electricity in the home. In the second section alternating and direct currents are explored, and we learn how to measure these with a cathode ray oscilloscope.

One of the great achievements of the twentieth century was the understanding of the structure of atoms. Atomic structure is fundamental to much of modern science. The third section looks at the structure of atoms and especially the atomic nucleus. Topics include ionising radiation, the half-lives of isotopes and the dangers of nuclear radiation. This leads into an exploration of how electricity is generated using nuclear fission, and how nuclear fusion may be an important energy source in the future.

The final section explores the life of stars and how stars and planets are formed. All stars are formed in a similar way, when a cloud of hydrogen is pulled together by gravity and heats up. Stars eventually run out of fuel and change, but how they change depends on their mass. We also explore the formation of elements heavier than lithium, which involves looking back to the beginnings of the Universe.

Test yourself

1 Describe the relationship between power and energy.

2 Explain why the National Grid uses very high voltages to transmit energy.

3 Explain what is meant by renewable and non-renewable energy sources.

4 Describe why burning fossil fuels is thought to be increasing the temperature of the Earth's atmosphere.

5 Explain how red shift helps us to understand how far away stars are and how fast they are moving.

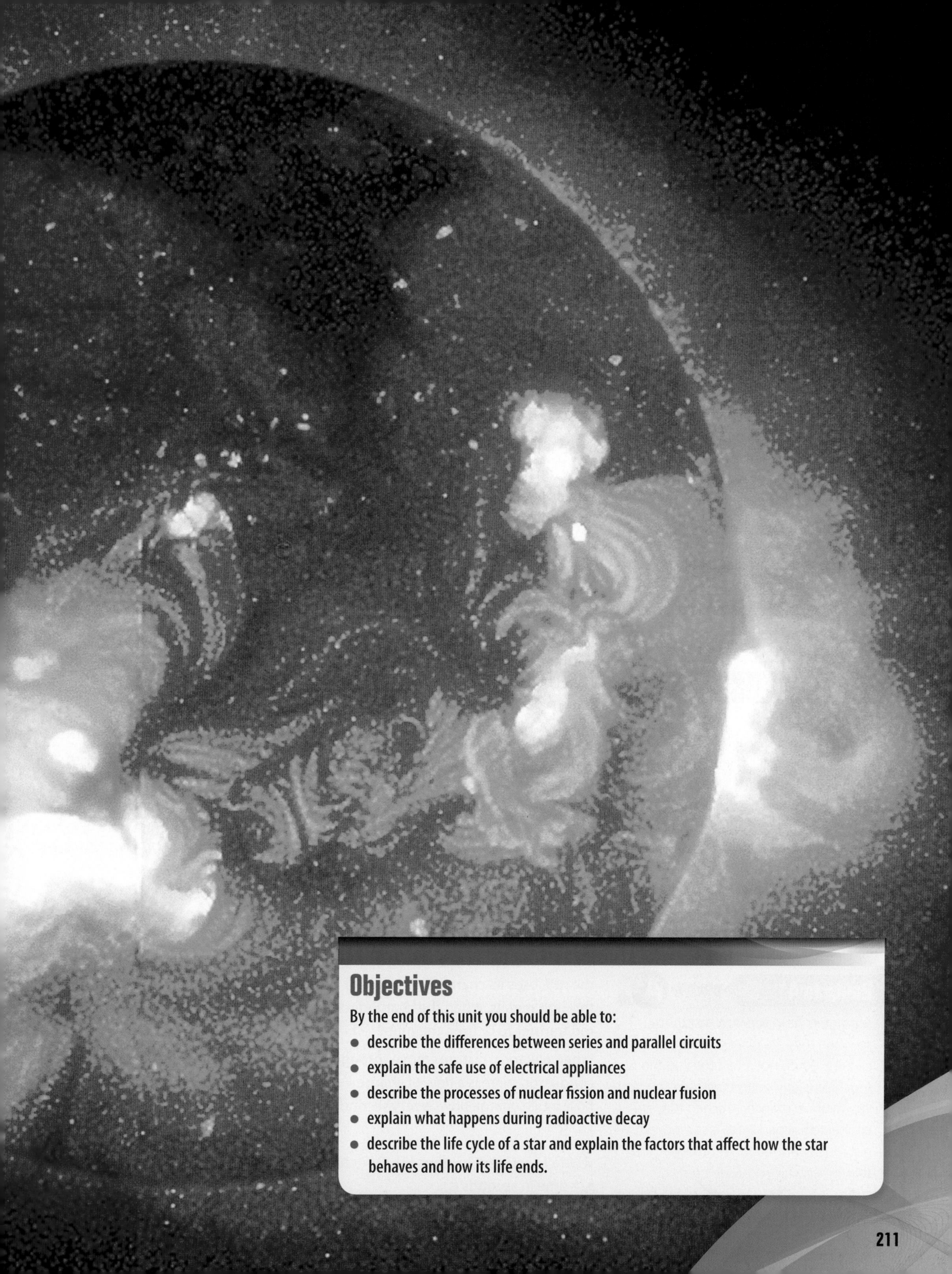

Objectives

By the end of this unit you should be able to:

- describe the differences between series and parallel circuits
- explain the safe use of electrical appliances
- describe the processes of nuclear fission and nuclear fusion
- explain what happens during radioactive decay
- describe the life cycle of a star and explain the factors that affect how the star behaves and how its life ends.

P2 3.1

Static electricity

Learning objectives

- explain what happens when certain insulating materials are rubbed together
- explain how materials can become charged with static electricity
- predict how statically charged materials will interact
- describe potential difference (voltage) in terms of energy transformed and flowing charge.

Charging up

Everyone has tried this at some time: if you rub a balloon on your jumper, you can stick it to a wall, or use it to pick up small pieces of paper. This is a simple demonstration of the effect of **static electricity**. When you rub the balloon, it becomes electrically charged and attracts other materials. You can charge up all kinds of **insulating** materials in a similar way.

An object becomes charged when negatively charged electrons are added to it or removed from it. Every atom has a positively charged nucleus surrounded by negatively charged electrons (see lesson P2 5.1). When an insulator is rubbed, electrons can either be rubbed off it or transferred onto it from the other material. Rubbing the balloon on your jumper moves electrons off your jumper and on to the balloon. The balloon becomes negatively charged and your jumper gets a positive charge.

If you bring together two electrically charged objects, there will be a force between them. If the two objects have the same charge, they will **repel** each other. Objects with opposite charges **attract** each other. So in the example above the negatively charged balloon clings to your positively charged jumper.

Taking it further

If you move a negatively charged object towards another object that is not charged, the electrons in the uncharged object are pushed away, leaving the surface positively charged. This means that a charged object will attract a neutral object.

Science skills

The apparatus in Figure 1 demonstrates how unlike static charges attract each other. The polythene rod gains electrons and becomes negatively charged when it is rubbed with a cloth. The acetate rod loses electrons when it is rubbed, and becomes positively charged. The rods actually move to be closer to each other.

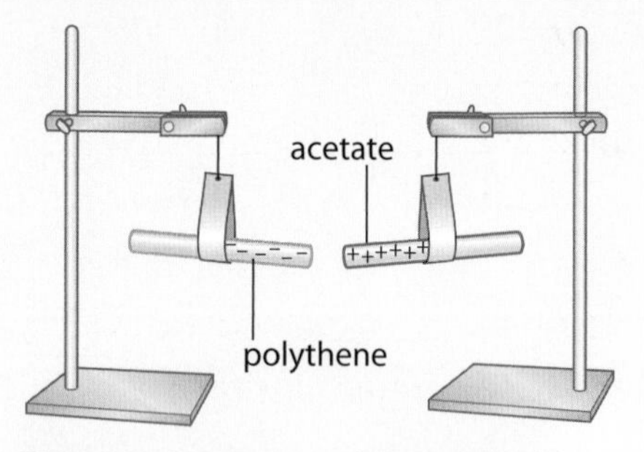

Figure 1 Apparatus to demonstrate that unlike charges attract.

a Describe what would happen if two polythene rods were used.

b The polythene rod is shown with a negative charge: this is because it has extra electrons. Explain how the acetate has become positively charged.

Taking it further

When you walk across a carpet and then go to touch a metal door handle, you might get a shock and there is sometimes a spark. This happens because the air in the small gap between your hand and the door handle suddenly starts to conduct electricity. The spark is a brief electric current travelling through the air.

Moving charges

If you touch a plastic rod charged with static electricity against a metal object, the rod loses its charge. This happens because electrical charges can move easily through metals. When the charged rod touches the metal, the electrons from the rod transfer to or from the metal.

When electrical charges move through a metal or some other electrical conductor, there is a flow of charge – an **electric current**. Current is a measure of how much charge flows every second. This can be expressed in an equation:

$$I = \frac{Q}{t}$$

where I = current in amps (A), Q = charge in coulombs (C) and t = time in seconds (s).

Route to A*

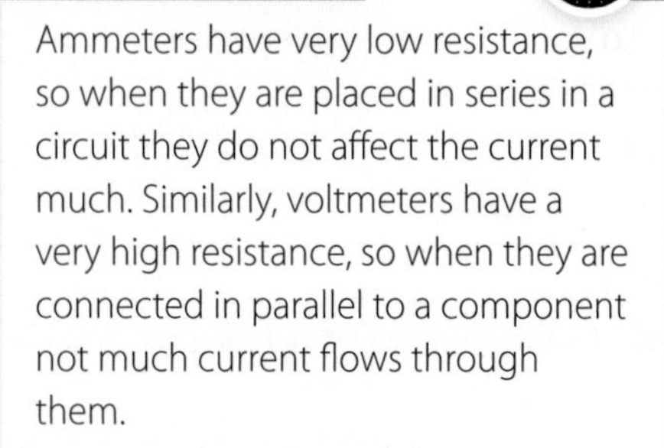
Ammeters have very low resistance, so when they are placed in series in a circuit they do not affect the current much. Similarly, voltmeters have a very high resistance, so when they are connected in parallel to a component not much current flows through them.

Electrical charge is measured in **coulombs** (C). A coulomb is the charge of just over 6 000 000 000 000 000 000 electrons, or 6×10^{18} electrons. Time is measured in seconds, so electrical current could have units of coulombs per second. In fact, the unit of current is the amp, A. A current of 1 amp = a flow of 1 coulomb of charge per second.

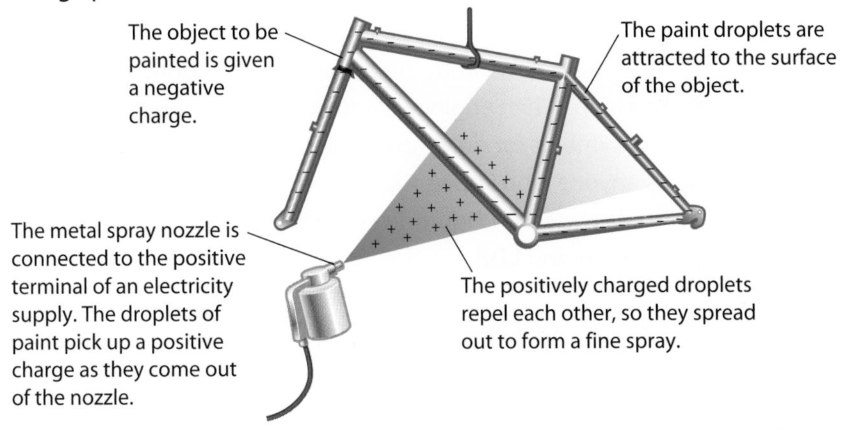

Figure 2 Electrostatic paint sprayers use electric charge to produce an even coat of paint.

Science in action

When plastics are recycled, they have to be separated into different types, and it can be hard to tell them apart. Researchers at the University of Southampton have developed a simple device called 'TriboPen' that measures the static electric charge produced when a plastic is rubbed. Different plastics can be identified by the size of the charge and whether it is positive or negative.

Charge, p.d. and energy

In an electrical circuit, a battery or other electricity supply produces a constant supply of electrons, which flow around the circuit. When the electrons flow through components in the circuit, such as a light bulb, the electrons transfer energy, or do work. The **potential difference** (p.d.), or voltage, between two points in a circuit is the work done, or energy transferred, per coulomb of charge passing through the circuit. This can be written:

$$V = \frac{W}{Q}$$

V is the p.d. in volts (V), W is the work done or energy transferred in joules (J), and Q is electrical charge in coulombs (C).

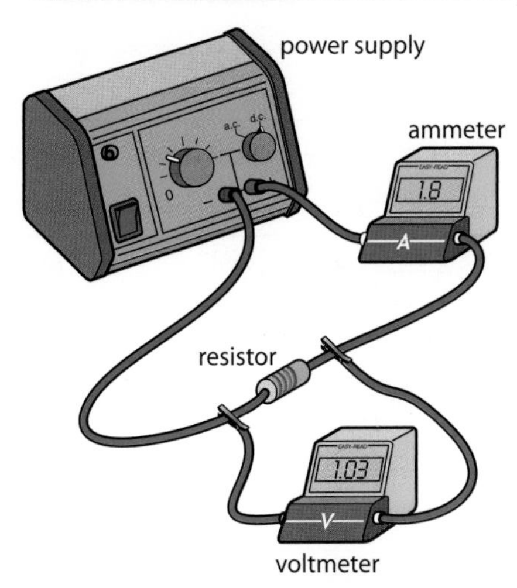

Figure 3 A voltmeter measures the change in electrical potential, so you connect it across an electrical component.

Questions

1. If 2 coulombs of charge transfer 6 joules of energy in a heater, what p.d. was applied?
2. Two pieces of sticky tape are pulled apart. Explain why it is possible that if they are held near each other again, they will attract each other.
3. Explain how an insulating material can become positively charged.
4. In an experiment, it was found that a plastic rod was attracted to a positively charged object. Use the information on these pages to explain what material the rod could have been. Explain your answer.
5. A charge of 0.2 A flows in a torch. Another torch has the same amount of current flowing through it. Explain how it is possible for the bulb to be brighter in one torch than the other. Use the relationship between voltage, charge and energy to help you explain your answer.
6. Car manufacturers use electrostatic spraying equipment to reduce paint waste and get a more even finish when spraying metal panels. Explain why the spray from the paint gun will spread over a wider area if the paint particles are all positively charged.
7. A voltage of 6 V is placed across a resistor. The current that flows is 600 mA. Calculate the amount of charge that flows in 8 seconds and the energy transformed into heat and light during that time.
8. Memory chips in computers are very sensitive to damage from static electric charges. People touching these components wear wrist bands that are earthed (allow current to flow to the ground). Explain where the electric charge comes from and how the wrist band removes the problem.

P2 3.2

Current, potential difference and resistance

Learning objectives

- interpret and draw circuit diagrams using common circuit symbols
- explain current–potential difference graphs
- calculate current, potential difference and resistance using the formula $V = IR$
- describe some uses of LDRs and thermistors.

Circuit symbols

Every electrical device, from a torch to a mobile phone, is based on an electric **circuit**. In any circuit the current flows from the power source, around the circuit and back in an unbroken loop. When engineers and scientists design circuits, they use an agreed set of symbols for drawing up their design. The symbols are a shorthand that allows anyone who knows them to 'read' the circuit diagram. Most of the circuit symbols you see in Figure 1 should be familiar but you may not have seen the **thermistor** and **LDR**.

Symbol	Function
switch (open) switch (closed)	Turn the current in a circuit off and on.
cell battery	Create a p.d. across a circuit. A **battery** is two or more **cells**.
diode	Allows current to flow one way but not the other.
resistor variable resistor	Resistors help regulate current flow in a circuit.
LED	Light-emitting diode.
lamp	Lights up when a current passes through it.
fuse	Breaks current flow if current gets too large.
voltmeter ammeter	Measure p.d. (voltmeter) and current (ammeter).
thermistor LDR	A thermistor is a temperature-dependent resistor. An LDR is a light-dependent resistor.

Figure 1 Common circuit symbols.

Potential difference, current and resistance

When you apply a **potential difference** (p.d. or voltage) to a circuit, electrical current flows around it. The amount of current that flows depends on the p.d. The larger the p.d. applied, the bigger the current through the circuit, if nothing else changes.

A second factor that can affect current flow through a circuit is the **resistance** of the **components** in the circuit. If the resistance is low, current flows easily. If a component has high resistance, it is hard for current to flow through it.

Ohm's Law

Current, potential difference and resistance are related to each other through Ohm's Law:

$V = IR$

where V = potential difference in volts (V), I = current in amps (A) and R = resistance in ohms (Ω).

The circuit in Figure 2 shows the set-up for measuring current flow and p.d. for a resistor. Current flow through the resistor is measured with an ammeter, and the p.d. across it with a voltmeter. If you measure the current flow at a range of p.d. values, you can plot a graph of potential difference against current.

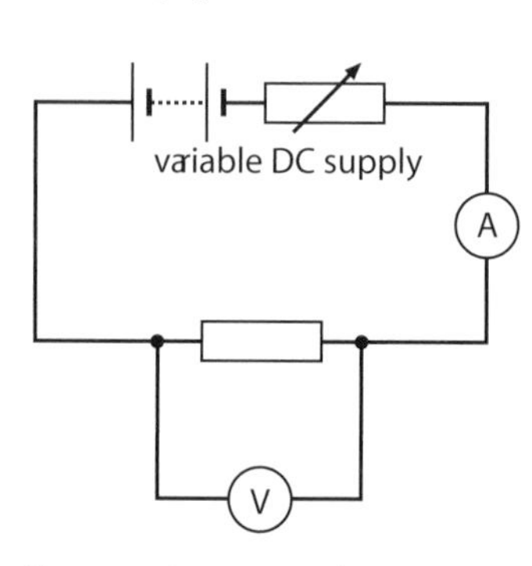

Figure 2 Measuring the current through a resistor at different p.d. values.

If a material obeys Ohm's Law then the current through it is directly proportional to the p.d. applied to it if the temperature of the material is held constant. A graph of p.d. and current for the circuit in Figure 2 is shown in Figure 3.

Finding resistance

We can find the resistance of any component in a circuit by measuring the current flow through it and the potential difference across it. By rearranging the Ohm's Law equation we get:

$$R = \frac{V}{I}$$

$$\text{resistance} = \frac{\text{p.d.}}{\text{current}}$$

We can use this equation to calculate the resistance.

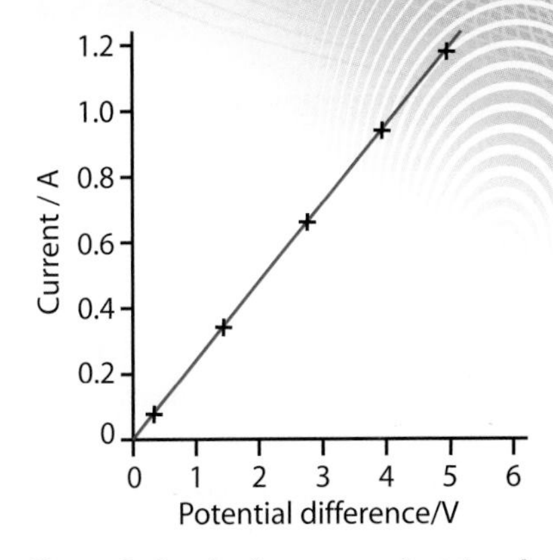

Figure 3 Graph of current against time for the circuit in Figure 2.

LDRs and thermistors

Electrical engineers use resistors to control the flow of current in an electrical circuit. Sometimes these are **fixed resistors** and sometimes they are **variable**, in which case the resistance of the component can change.

Two very useful types of variable resistor are the light-dependent resistor (LDR) and the thermistor.

A thermistor is a temperature-dependent resistor. Thermistors are often used in devices that control temperature. As it gets warmer the resistance of the thermistor gets lower. This electrical change can be used to activate a switch, for example to turn off the central heating.

The resistance of an LDR changes with light levels. The brighter the light, the lower the LDR's resistance. LDRs are often found in circuits that turn on automatically at night time, for example security lights. As it gets darker, the resistance of the LDR increases. This electrical change is used to switch on a lamp.

Examiner feedback

Notice how the graph in Figure 3 is a straight line through the origin. It shows that p.d. and current are directly proportional to each other.

Science skills

The circuit shown in Figure 4 has a p.d. of 6 V across a 24 Ω resistor.

The current that flows can be calculated by rearranging the equation

$V = IR$

This becomes $I = \frac{V}{R}$

The current is therefore $\frac{6}{24} = 0.25$ A.

a What would be the effect on the current of doubling the p.d.?

6 V
A
24 Ω
V
6 V

Figure 4 Circuit showing a 6 V battery and 24 Ω resistor.

Questions

1. Calculate the current through a 60 Ω resistor if a p.d. of 12 V is applied to it.
2. Use the data in Table 1 to draw a graph of p.d. and current for a resistor.

Table 1 Potential difference and current for a resistor.

p.d./V	0	1	2	3	4	5
Current/A	0	0.2	0.4	0.6	0.8	1.0

3. Describe the sequence of events in an electric oven controlled by a thermostat as it comes up to temperature and then maintains that temperature.
4. Draw a graph of p.d. against current for a resistor of 2000 Ω. Use p.d. values between 0 and 10 V.
5. Calculate the current through a resistor of 3700 Ω if a p.d. of 6 V is applied to it.
6. As it slowly gets dark at night, the resistance of an LDR changes. This change eventually triggers a light to come on. If the current though the LDR is 0.2 mA when the light turns on, and the p.d. across the LDR is 5 V, what is the resistance of the LDR?
7. A common type of thermistor is referred to as a negative temperature coefficient thermistor (NTC). As the temperature rises, the resistance of the thermistor falls. Why do you think it is said to have a negative temperature coefficient?
8. Describe an investigation you could do to find out how the resistance of a thermistor changes for temperatures between 0 and 100 °C. Include information about the equipment used, the method and safety issues.

A*

Non-ohmic conductors

Learning objectives

- describe the shape of a current–potential difference graph for a filament bulb and a diode
- explain how filament bulbs and diodes behave as the current through them changes
- explain why resistance changes in a filament bulb
- evaluate the use of different forms of lighting in terms of cost and energy efficiency.

Filament bulbs

In some electrical components, current and potential difference (p.d.) are not proportional to each other. Such components are called non-ohmic devices, because they do not follow Ohm's Law.

An example is a **filament bulb**, an ordinary light bulb. The **tungsten** filament in the bulb conducts electricity quite well, but its resistance changes as it warms up. The more current that flows through the filament, the hotter it gets. As it gets hotter the resistance of the filament increases, so it gets more difficult to push current through the circuit.

Figure 1 shows current flow through a filament bulb as the p.d. across the bulb changes. The current–p.d. graph is curved. As the p.d. across the bulb is increased, its resistance increases.

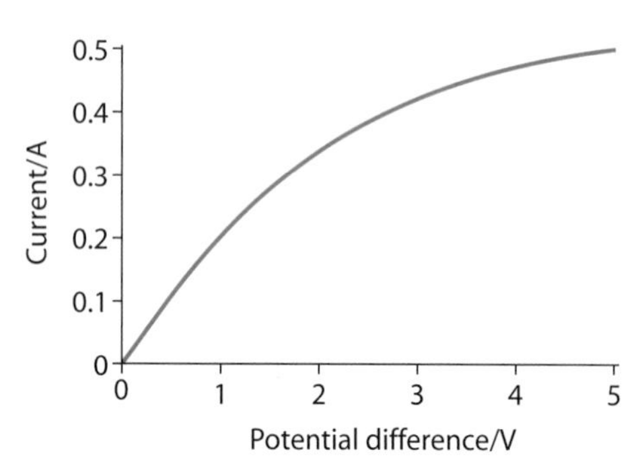

Figure 1 A filament bulb does not obey Ohm's Law.

Why the resistance changes

The current through the tungsten filament in a bulb is carried by electrons. As they pass through the filament the electrons *do work*. This means that some electrical energy is transferred through heating the filament. The atoms that make up the filament vibrate more, and the filament gets hotter. This increased vibration makes it harder for the electrons to get through, and the resistance of the filament increases.

Positive and negative current

Many components behave in exactly the same way no matter which way round they are put in a circuit. If you record the current through a component at several p.d.s, then change the power supply round so that the current flows the opposite way, the new current values are recorded as negative.

Science skills

Make a copy of Table 1 and use the graph in Figure 2 to fill in the blank values for the current flowing through a 5 Ω resistor.

Table 1

p.d. / V	−5	−4	−3	−2	−1	0	1	2	3	4	5
Current / A											

The values for the current are not affected by the direction of flow.

a Use the graph to work out what current would flow if a p.d. of 3.7 V was applied.

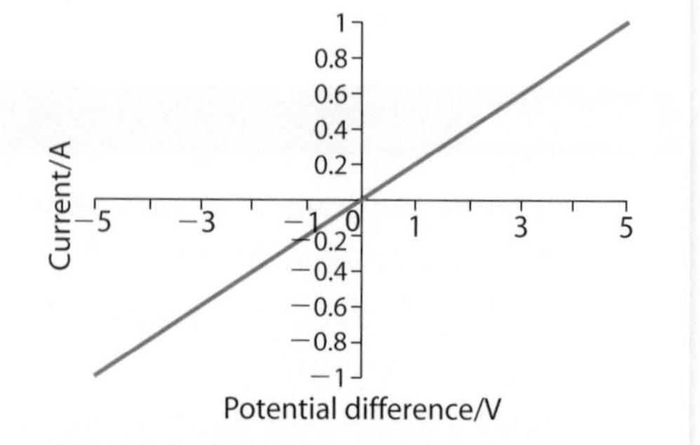

Figure 2 Resistors behave the same whether the p.d. is positive or negative.

Diodes and LEDs

Unlike other electrical components, a diode does not work the same both ways round. A diode is like a one-way valve for electricity. It will allow current to pass in one direction but not the other. In its forward direction a diode has a very low resistance. In the reverse direction it has a very high resistance. This makes the shape of the current–p.d. graph very different from that of a resistor or a filament bulb.

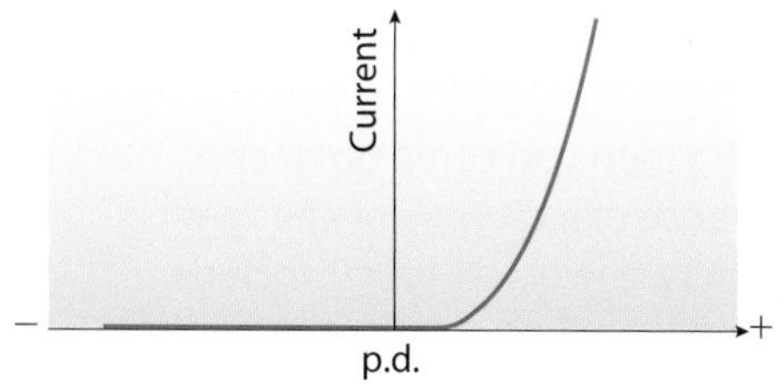

Figure 3 A diode only allows current to flow in one direction.

Like diodes, light-emitting diodes (LEDs) only conduct current in one direction. Figure 4 shows two LEDs in a circuit. One LED will allow current to pass, and it lights up. The other LED is connected the 'wrong way round', so it does not conduct current and does not light up.

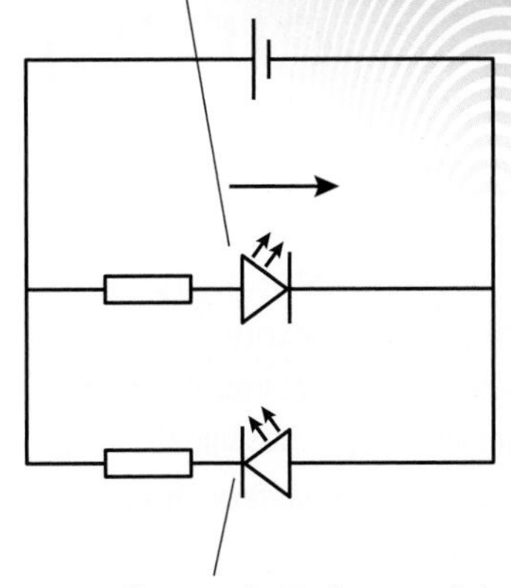

Figure 4 In this circuit, only one LED lights up.

Science skills

b Produce a graph using the data in Table 2.

c Using the graph you have plotted, find the current that flows for a p.d. of 1.8 V and a p.d. of −1.8 V.

Table 2 Positive and negative values for a circuit with a diode in it.

p.d./V	−2.5	−2.0	−1.5	−1.0	−0.5	0	0.5	1.0	1.5	2.0	2.5
Current/A	0.0	0.0	0.0	0.0	0.0	0.0	0.0	0.1	0.2	0.3	0.4

Better lighting

Filament bulbs are cheap to make but they are very inefficient. About 90% of the electrical energy they use goes into heating the filament rather than producing light. Compact fluorescent 'low energy' bulbs convert about 40% of their energy into light. This is four times better than a filament bulb. They are more expensive to make, but they use less electricity and can last 10 times as long.

LEDs are even more **efficient** than compact fluorescent bulbs, and they last hundreds of times longer than filament bulbs. They are used in torches and for some car lights. LED light bulbs are now available for the home, but they are very expensive at present.

The first LEDs were red like this. They now come in all colours and sizes.

Questions

1. Describe what happens to the temperature and the resistance of a filament bulb as the current flowing through it increases.
2. Use the graph in Figure 1 to produce a table that shows p.d., current and resistance for the filament bulb.
3. Describe why a diode could be used to stop damage being done to an electronic circuit by somebody putting the batteries in the wrong way round.
4. Describe the most striking difference between the current–p.d. graph for a diode compared with other components.
5. Explain why the batteries in a torch with an LED will last longer than in a torch with a filament bulb.
6. When lampshades are used, care needs to be taken to make sure that the bulb does not overheat the lampshade. Explain why this is more of a problem with a filament bulb than with compact fluorescent bulbs or LEDs.
7. Describe how these two statements can both be true:

   'If a larger p.d. is placed across a bulb, more current flows.'

   'The current flow is not proportional to the applied p.d. for a filament bulb.'
8. Use the information in Table 3 to describe the arguments for and against replacing an ordinary filament bulb with either a compact fluorescent bulb or an LED. The light output from the different bulbs is roughly the same.

Table 3 Information gathered by a student.

Type of device	Cost of purchase	Power consumption	Typical running costs per year	Lifetime of bulb
filament bulb	£1.00	75 W	£13.75	1000 hours
compact fluorescent bulb	£2.50	15 W	£2.75	15 000 hours
LED	£66	8 W	£1.50	50 000 hours

A*

Potential difference in series circuits

Learning objectives

- explain how to find the total potential difference in a series circuit
- explain why cells are usually connected together facing in the same direction
- describe how current flows in a series circuit
- calculate the potential difference (p.d.) across each component in a series circuit.

Series circuits

In a **series circuit**, there is only one route for electric current to flow along; there are no branches or splits. This is a bit like an athlete running round a track.

Cells and batteries

In science we call two or more cells connected in series a battery. A 12 V car battery, for example, has six cells of 2 V each, connected in series.

If you look at Figure 1(a), you can see that the cells are all connected facing the same way, with the positive pole on the right. When cells are connected like this, the total potential difference (p.d.) is all the individual cell p.d.s added together – in this case 4.5 V. In Figure 1(b) one of the cells has been put in facing the opposite way. Now the total p.d. is (3–1.5 V) = only 1.5 V.

(a) all cells facing the same way

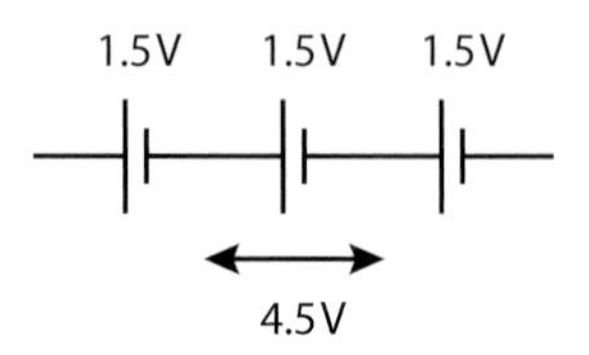

(b) one cell facing the opposite direction

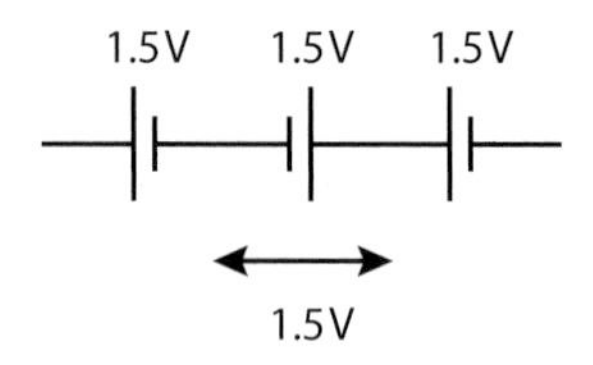

Figure 1 What happens when one cell is the wrong way round?

Current path

In a series circuit like the one in Figure 2, there is only one route the current can take to get from one pole of the cell to the other. This means that the current has to be the same everywhere in the circuit. It also means that if the circuit is broken anywhere, the current stops everywhere.

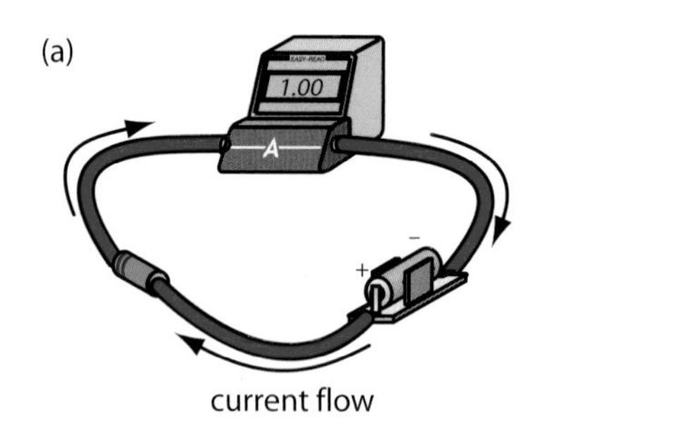

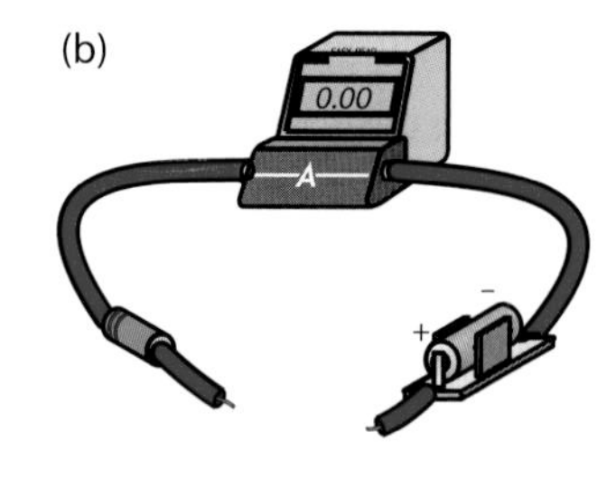

Figure 2 Current is the same everywhere in a series circuit (a). If there is a break in the circuit (b) no current flows.

Resistors in series

The larger the resistance in a circuit, the smaller the current flow will be. To find the total resistance in a series circuit you add together the values of the resistors:

$$R_t = R_1 + R_2 + \text{etc.}$$

where R_t is the total resistance in ohms (Ω), R_1 is resistance 1 and R_2 is resistance 2, etc.

If you look at the circuit in Figure 3 the total resistance is 30 Ω (10 + 20).

From the total resistance and the p.d. of the battery, you can calculate the current:

$$I = \frac{V}{R}$$

$$\text{so } I = \frac{6}{30} = 0.2\text{ A}$$

The current is the same throughout the circuit.

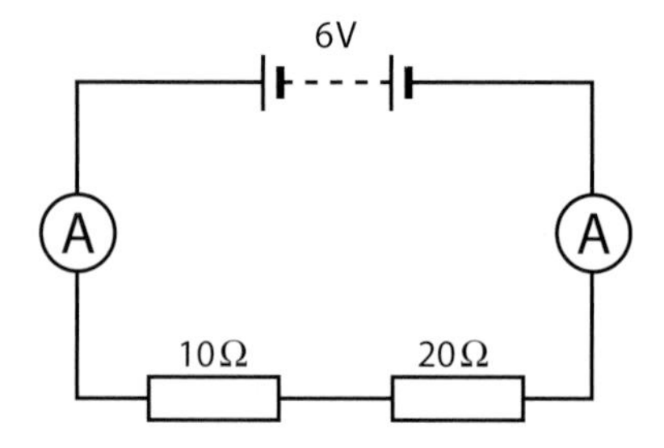

Figure 3 Two resistors in a circuit. The current at A and B is the same.

Calculating the potential difference across one of the resistors

In a series circuit, the total p.d. from the power supply is shared between the components.

The resistors in Figure 4 all have the same current through them because it is a series circuit. The total resistance of the circuit is 6 Ω.

Each component will take its share of the total p.d.:

potential difference across each resistor

$$= \frac{\text{total p.d.} \times \text{resistance of resistor}}{\text{total resistance}}$$

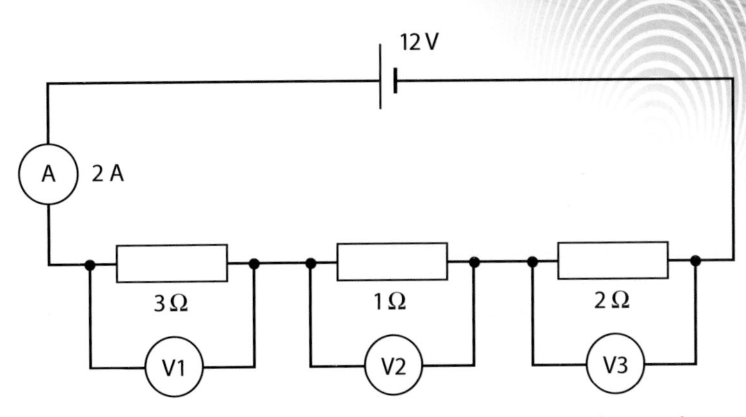

Figure 4 The p.d. across the resistors splits in the same ratio as the size of the resistors.

V1	V2	V3
p.d. $= \frac{12 \times 3}{6} = \mathbf{6\,V}$	p.d. $= \frac{12 \times 1}{6} = \mathbf{2\,V}$	p.d. $= \frac{12 \times 2}{6} = \mathbf{4\,V}$

If you add up these three p.d.s you can see that the total p.d. is 12 V, the p.d. of the cell.

Example 1

You can also work out missing p.d.s easily. In Figure 5 the missing p.d. is 3 V.

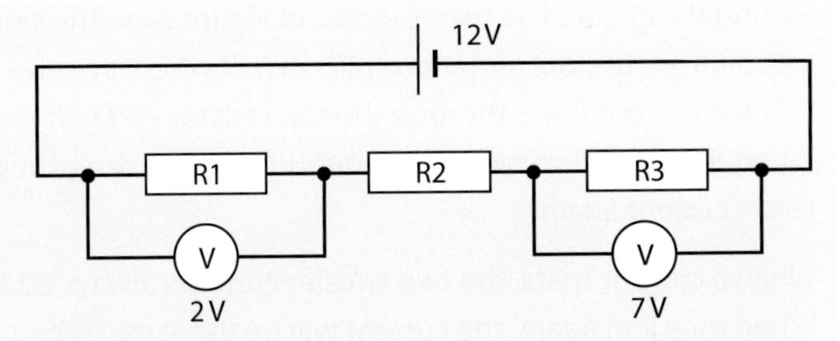

Figure 5 Three resistances, one missing p.d. All the p.d.s add up to the supply p.d. The other two p.d.s add up to 9 V so simply take away 9 from 12.

Taking it further

The actual speed of an electron flowing in a wire is usually quite slow. When you switch on a torch, the effect reaches the bulb almost instantly, but the actual electrons are moving surprisingly slowly. It could take minutes or even hours for an individual electron to travel from the switch to the bulb.

Questions

1. Draw a circuit with four identical cells in series that will give an output p.d. of 0 volts.
2. Draw a circuit with an 18 V battery made up of 3 V cells and two resistors in series. One resistor is 200 Ω and the other is 400 Ω. Calculate the p.d. across each resistor.
3. Redraw the circuit in Figure 3 with a third resistor in series of 5 Ω. Now calculate the p.d. across each resistor.
4. If two 1500 Ω resistors were in series in a circuit with a 3 V supply, what current would pass through them?
5. In a torch there are four cells, each 1.5 V. One of the cells is put in the wrong way round. Describe what the effect of this would be in terms of p.d. supply, current and how well the torch would work.
6. If a 200 Ω resistor is in a series circuit with a 10 V supply and another resistor, what resistance would the other resistor have to be if the current in the circuit is 0.02 A?
7. A particular type of copper wire has a resistance of 34 Ω per kilometre. Calculate the resistance of a 50 cm length.
8. Use drawings and written explanations to describe how resistances in a series circuit add together.

Parallel circuits

Learning objectives

- explain that p.d. is the same across each component of a parallel circuit
- explain how current splits in parallel circuits
- calculate the current in all parts of a parallel circuit.

What is a parallel circuit?

A **parallel circuit** contains at least one place where the current splits into two or more separate currents.

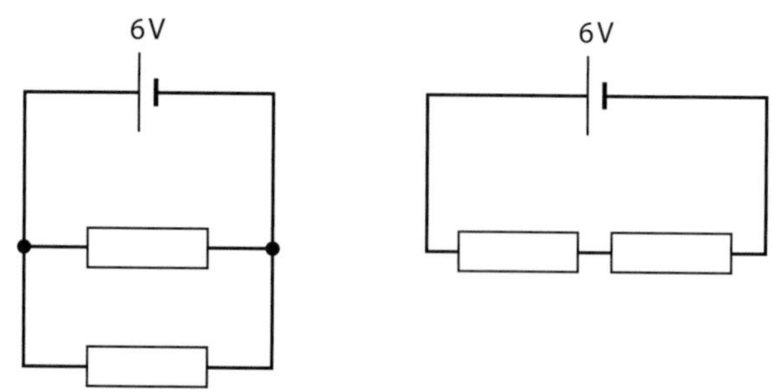

Figure 1 In the circuit on the left the components are connected in parallel. In the circuit on the right they are connected in series.

In series circuits the current is always the same but the p.d. is different across each component. In a parallel circuit, the p.d. is the same across each component, while the current is different in different parts of the circuit.

Science in action

The headlamps of a car are wired in parallel, so if one headlamp blows the other stays lit. The power output of modern headlamps is so high that they need special quartz glass because the bulbs get so hot they would break normal glass. They also contain a **halogen** gas, which makes the bulb last longer.

Current in a parallel circuit

For a parallel circuit it is useful to think about what the current does as it moves around the circuit. The two resistors in Figure 2 are the same size. The current sets off from the cell and then splits in half when it comes to a fork in the circuit. Half the current flows through the top resistor and half though the bottom one. When the currents have passed through the resistors, they join up, forming one larger current again.

When a current splits, the two smaller currents always add up to the larger one. When they join again, the current will be the same as before they split.

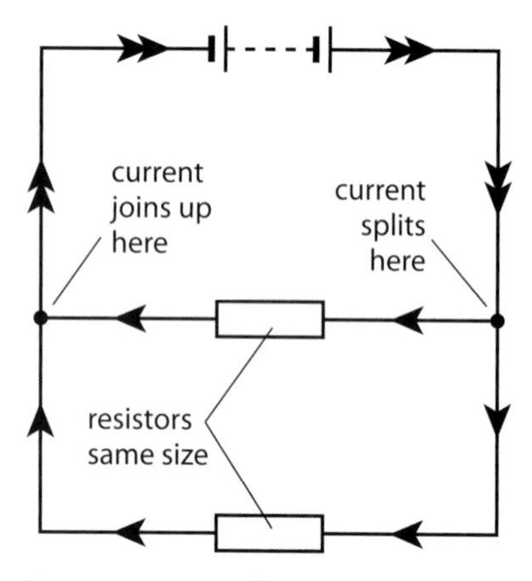

Figure 2 In a parallel circuit current splits and rejoins.

Calculating the current split

In parallel circuits, the resistance of each branch of the circuit determines how the current splits. The larger the resistance of the branch, the smaller the current that flows down it.

Example 1

In Figure 3, one branch of the circuit has a p.d. of 12 V across a 100 Ω resistor. By Ohm's Law $\left(I = \frac{V}{R}\right)$ this gives a current of $\frac{12}{100} = 0.12$ A.

The second branch of the circuit has a smaller resistance of 50 Ω. The p.d. is the same, so the current is $\frac{12}{50} = 0.24$ A. The total current flowing from the cell is therefore $0.12 + 0.24 = 0.36$ A.

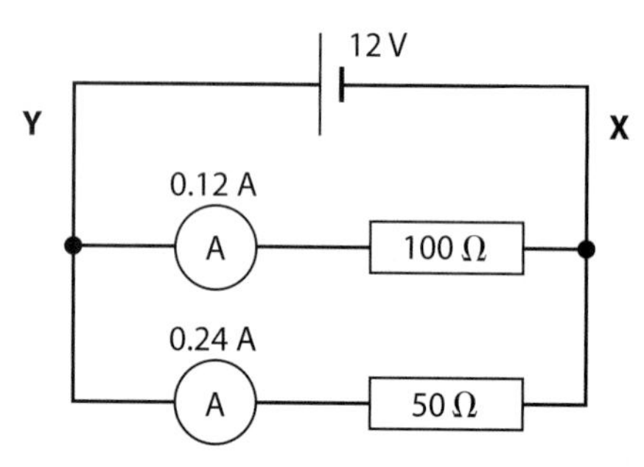

Figure 3 The current at points X and Y is 0.36 A.

Operating parts of a circuit independently

Imagine two small rivers that join together to form a larger one. If you dammed one of the rivers upstream of the join, what would be the effect? The second small river would still flow and would be unaffected by the dam. The flow of the main river would obviously be smaller.

Parallel circuits behave just like these rivers. Changing the flow in one branch of the circuit has no effect on the other branch, but it does affect the main current flow. In the circuit in Figure 4, each bulb has a switch that controls it. There is also a switch on the left that turns all the bulbs on or off. If each bulb has 0.1 A flowing through it when it is lit, the current flowing from the battery could be 0, 0.1 A, 0.2 A or 0.3 A, depending on how many bulbs are lit.

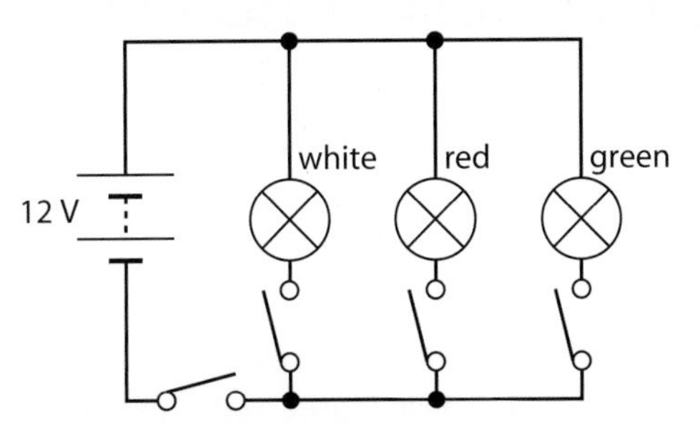

Figure 4 The current through each bulb changes depending on which switches are on.

3 A
3 A
X
2 A
current recombines here
current splits up here
1 A

Figure 5 All the bulbs are identical, but bulb X shines brighter than the other two.

What happens when bulbs are put in parallel?

If all the bulbs in Figure 4 are switched on, they will all be equally lit. This is because the p.d. across each one is the same and the bulbs are identical. In Figure 5, the bulbs are still all identical, but this time there are two of them on one branch and one on the other. The p.d. across the lower branch is the same as for the branch with a single bulb, but the resistance is greater because there are two bulbs. From Ohm's Law $\left(I = \frac{V}{R}\right)$, if the p.d. is the same but the resistance increases, the current is reduced.

Science skills

Look at Figure 5. Imagine what would happen if a switch was placed in the circuit next to bulb X. Describe the effect of opening the switch on the currents and the bulbs' brightness.

Questions

1 **(a)** Draw a circuit with a 6 V supply and two resistors in parallel, each of 100 Ω. **(b)** Calculate the current through each resistor and the total current out of the 6 V supply.

2 **(a)** A current of 4.4 A flows from a battery and then the circuit splits into two branches. If the current along one branch is 3.3 A, what will the current along the other branch be? **(b)** If the resistor in the branch with the larger current was 100 Ω, what would the resistance of the resistor in the other branch be?

3 For the parallel circuit in Figure 6, calculate the following: **(a)** the current passing through the single resistor in the top branch of the circuit **(b)** the total resistance of the lower branch **(c)** the current passing through the lower branch **(d)** the total current flowing out of the battery.

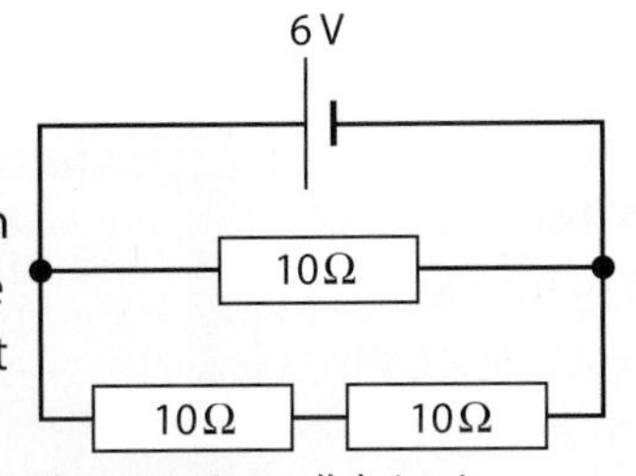

Figure 6 A parallel circuit.

4 A parallel circuit has one bulb in each of two branches. As current flows, one bulb glows brighter than the other. Suggest why this might be and describe the currents that flow in each part of the circuit.

5 Draw a parallel circuit with two branches, each containing a resistor. The p.d. of the battery is 10 V. The current flowing out of the battery is 0.125 A. One of the resistors must be 100 Ω. Calculate the resistance of the other resistor and draw the circuit.

6 Look at the circuit in Figure 5. In an experiment, one of the bottom bulbs blew and the other went out, but the top one remained lit. One of the students replaced the blown bulb with a piece of wire. Describe the effect of doing this and explain why this happens.

7 As a bulb gets older, the glass starts to darken. This is because the tungsten of the filament is starting to vaporise and is deposited on the glass. If the surface of the glowing filament is boiling away, the thickness of the filament must be changing. What effect do you think this will have on the current that flows and the performance of the bulb?

8 Look at Figure 5 again. In fact, the current through the single bulb will be slightly less than double the current through the branch of the circuit with two bulbs in. Explain why the current in one branch is not double the current in the other.

P2 4.1

Electricity at home

Learning objectives

- compare and contrast d.c. and a.c. electrical supplies
- describe the mains electricity supply in the United Kingdom
- calculate the period and frequency of an a.c. supply from a graph of potential difference against time.

Direct current

If you connect a cell or battery to other components to form a circuit, current will flow. The current flows in one direction only from positive to negative. This type of current is called **direct current** or **d.c.** If you were to look at a graph of the potential difference (p.d.) of a battery or cell plotted against time it would be a straight line as in Figure 1(a). The advantage of this type of electricity supply is that low-voltage electronic circuits can use it. Batteries produce low-voltage d.c. electricity, so many electronic devices can be portable.

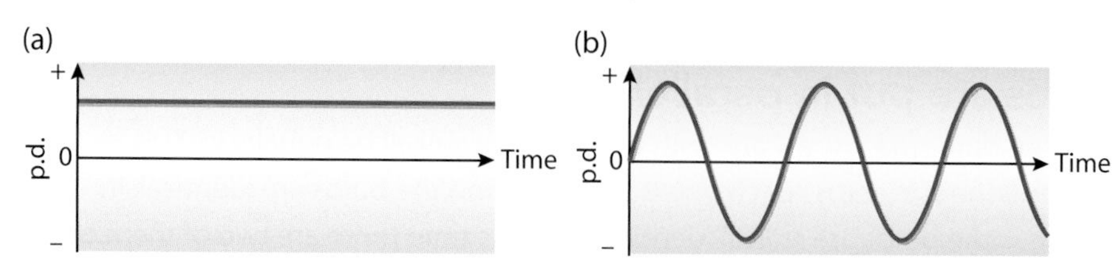

Figure 1 (a) In a d.c. supply the voltage is unchanging. (b) With a.c. it changes all the time.

Science in action

HVDC (high-voltage direct current) is a fairly new way of transmitting electricity that is better than a.c. in some situations. Direct current transmission is better for long undersea power lines. It is also better for transmitting electricity long distances in remote areas, and between two networks that have different a.c. systems.

Alternating current

With **alternating current** or **a.c.** electricity, the direction of the current changes many times per second. Electricity that is supplied to houses and factories is a.c. In the UK, the a.c. electrical supply has a **frequency** of 50 Hz. Frequency is a measure of how many times something happens in 1 second. It is measured in cycles per second, or hertz (Hz). In the case of the UK electricity supply, there are 50 complete waves in 1 second. The time taken for one complete wave is called the **period**. The UK supply has a period of 0.02 s.

One big advantage of a.c. electricity is that the p.d. or voltage can easily be changed using devices called transformers. Over long distances, low-voltage power lines lose a lot of energy. So for long-distance transmission, the p.d. is 'stepped up' to very high voltages, which greatly reduces the energy losses.

A disadvantage of a.c. electricity is that many electrical appliances run on d.c. If they are run on mains electricity, the a.c. supply has to be changed into d.c. This adds to the cost of manufacturing the appliances. The peaks in an a.c. voltage mean that you can get a more severe shock than with a d.c. voltage of the same value.

Science in action

The 'shape' of a human voice that can be seen on a CRO can be used for voice recognition.

Measuring p.d.

If you use a **cathode ray oscilloscope (CRO)** to measure the voltage of an electrical supply, there is a lot of information you can work out from the image.

The screens in Figures 2 and 3 show the p.d.s from two electrical supplies. Figure 2 shows a d.c. supply from a battery, and Figure 3 shows an a.c. supply from the mains.

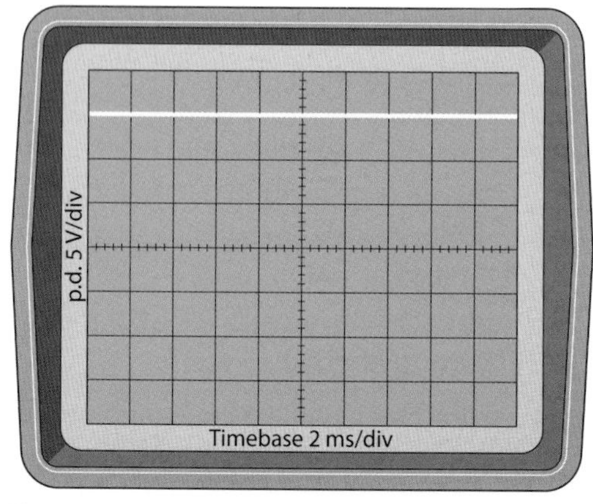

Figure 2 A CRO trace for a 15 V battery. The p.d. stays constant at 15 V (d.c.).

Taking it further

The cathode ray is produced by an electron gun that accelerates electrons from a glowing wire. The beam of electrons produced is moved around the screen by being deflected by electric fields.

You can see that the traces have different scales on the *y*-axis. In Figure 2 the d.c. scale is 5 V/division. This means that every square up the screen represents 5 V and so the p.d. of the battery is $3 \times 5 = 15$ V. The timebase setting is not important because the p.d. does not change.

In Figure 3 the *y*-axis scale for the a.c. supply is 100 V/division, so the peak voltage is 320 V. The timebase setting is 2 ms/division, so each square across the screen represents 2 milliseconds (0.002 seconds). The whole wave takes 10 divisions, so the time for one wave is $10 \times 0.002 = 0.02$ s. This is the period of the wave.

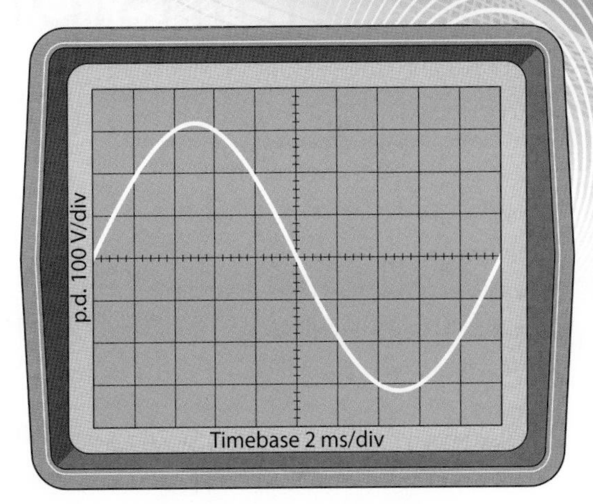

Figure 3 A CRO trace of alternating current.

Science skills

a Make a copy of Figure 3. Now add an a.c. voltage wave that has a frequency of 100 Hz and a peak voltage of 125 V.

Questions

1. Sketch a cathode ray oscilloscope trace for a 6 V battery.
2. Sketch a cathode ray oscilloscope trace for an a.c. supply with a peak voltage of 30 V and a frequency of 10 Hz.
3. A student makes the following statement to the teacher: 'If the current is negative for half of the time if you use mains electricity, surely an electric fire will not work, because half of the time the electricity is going backwards so that will cool the fire down again.' Explain why the student is wrong.
4. Describe the main advantages and disadvantages of mains electricity as a power source.
5. An a.c. mains supply drops to zero volts for an instant, 100 times each second. Use your understanding of the mains supply to explain how this can be true when the frequency of the supply is only 50 Hz.
6. If the p.d. of a mains supply drops to zero for an instant, 100 times a second, explain why a filament light bulb will not go out 100 times a second.
7. Describe how to calculate the frequency of a wave on a cathode ray oscilloscope.
8. The p.d. quoted for mains electricity is 230 V but the peak p.d. is 320 V. Use the cathode ray oscilloscope image shown in Figure 3 to help you explain why this might be.
9. High-voltage power lines are nearly always fatal if people come into contact with them. Every year some people die in accidents involving overhead power lines. Power cables could be buried underground to protect people from danger. This would also reduce numbers of unsightly pylons in the landscape. However, burying cables is much more expensive than overhead transmission. Write a letter either arguing the case for buried cables or defending the decision to use overhead power lines.

Examiner feedback

It is important that you understand what a complete wave on the oscilloscope screen looks like. Many students get confused and think that the line moves from zero volts up and back to zero volts in one wave. This would only be half a wave.

Route to A*

If a wave with a frequency of 250 Hz was displayed on the screen in Figure 3, how many complete waves would fit across the screen?

Route to A*

Why is it necessary for the cathode ray oscilloscope to have a range of settings for volts per division up and down the screen?

Science in action

The insulators that hold the cables on electricity pylons are shaped so that when it rains current does not flow through water collecting on the surface of the insulators.

Mains plugs

Learning objectives

- explain why most domestic electrical appliances are connected to the mains supply
- explain the design and the structure of electrical cables and plugs
- describe the way that a three-pin plug is wired.

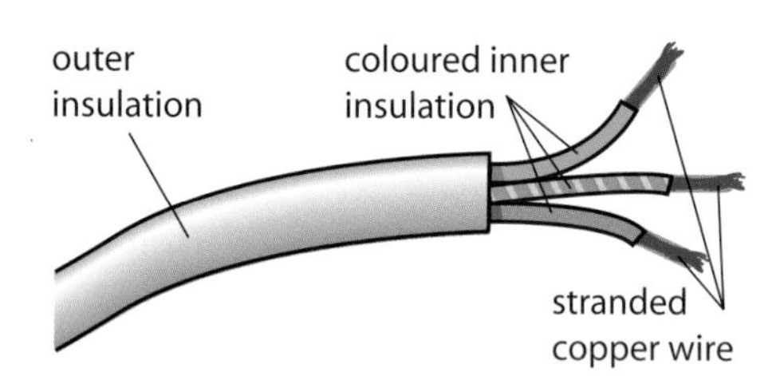

Figure 1 The structure of electrical cable.

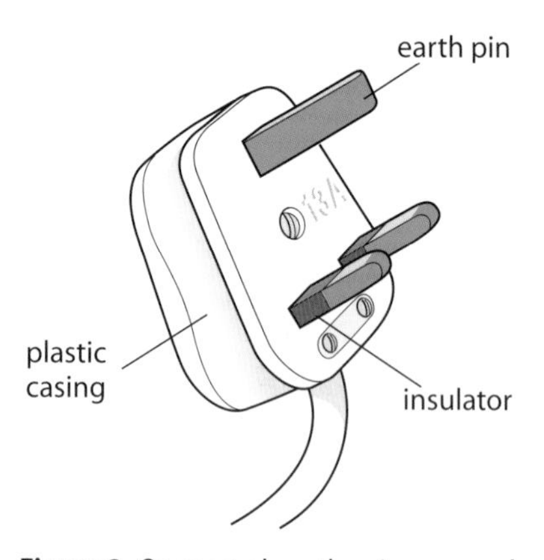

Figure 2 On most plugs the pins are made of brass, which does not rust.

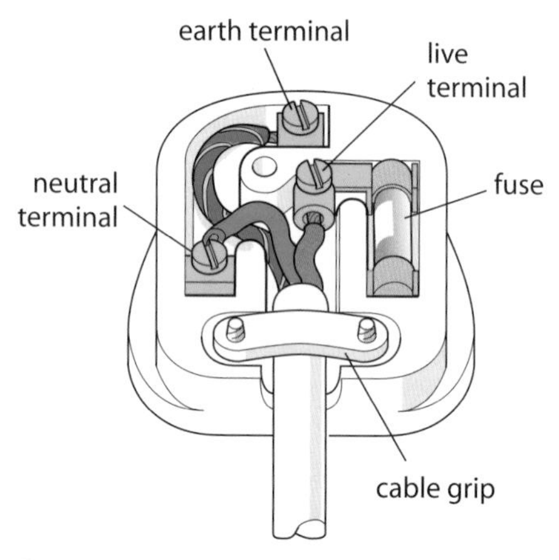

Figure 3 A correctly wired three-pin plug.

Why use the mains?

Mains electricity provides a continuous and powerful source of energy. Appliances such as cookers, washing machines and electric heaters always use mains electricity because a cell would not be powerful enough. Other appliances use the mains because the supply is continuous, that is, it does not run down. Appliances that use rechargeable batteries, such as mobile phones and laptops, also rely on mains electricity, as they need it to recharge their batteries.

The structure of electric cable

Electric cable used in the home has three main parts.

1. Stranded copper wire. This is the core of the cable. Copper is a very good electrical conductor. The bundle of thin copper strands can be bent and coiled without breaking.
2. Coloured inner insulation. The colour indicates which wire is which and insulates the copper from other wires within the cable.
3. Outer insulation. This is stiffer and harder than the inner insulation. It provides strength and extra protection.

Plugs

Plugs are used to connect electrical devices to the mains. The body of a plug is usually made from a tough insulating plastic to protect the user from the electrical contacts inside. Connecting pins made from a conducting metal are designed to fit into holes in an electrical socket, which connects to the mains circuit of the house.

Most modern plugs have three connecting pins: live and neutral pins, which connect to complete the electrical circuit, and an **earth** pin (see lesson P2 4.3). The live and neutral pins have an insulator around part of them to stop children touching the pins when they are partly removed from the socket.

Wiring a plug

When a plug is wired up, the colours of the wires in the cable must be connected in the correct way.

- Brown is the **live** wire and is connected to a **fuse** (see lesson P2 4.3), which in turn is connected to the live pin.
- Blue is the **neutral** wire and is connected to the neutral pin.
- Green and yellow is the earth wire and is connected to the earth pin.

There are some rules that must be followed when wiring a plug.

- Always check the correct colour is attached to the right pin.
- Make sure the wires are not stretched inside the plug.
- Make sure that the screws on the pins grip the copper wire firmly.
- No bare copper wire should be visible inside the plug.
- The cable grip should hold the outer insulation firmly.
- The correct fuse should be used for the appliance power rating (see lesson P2 4.3).

Which cable?

The type of cable an appliance needs depends on the power and its type of insulation. When large currents pass down a cable the copper wires heat up, so appliances that take a lot of current need thicker wires.

Science skills

Table 1 The cross-sectional area of copper wire needed for different current ratings.

Wire cross-section/mm^2	Current/A	Maximum power/W
1.0	10	up to 2400
1.25	13	up to 3120
1.5	15	up to 3600
2.5	20	up to 4800
4.0	25	up to 6000

a What thickness of cable would you need to use for a 3 kW kettle? Explain your answer.

b If a cooker needs to draw a current of 22 A, what thickness of cable is needed?

Science in action

If an extension lead on a reel is used for a portable device, it is a good idea to unwind all of the cable from the reel, even if the whole length is not needed. Otherwise, if a large current flows, it is possible that the cable will heat itself up and melt the insulating plastic.

Questions

1 In Figure 4, each plug has a fault. Explain what is wrong in each case.

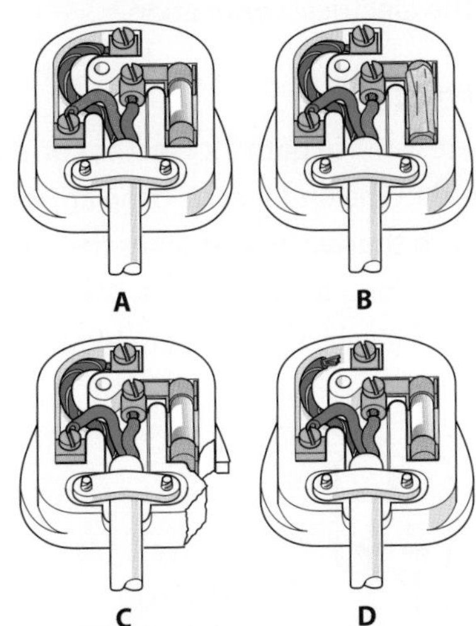

Figure 4 Faulty plug wiring.

2 Earth wires have green and yellow stripes. Explain why this might help people who have difficulty seeing differences between colours.

3 Explain why modern plugs have insulating plastic shrouds around part of the live and neutral pins.

4 Explain why the cable for a television needs to be stranded but a cable buried in the wall does not.

5 The inner insulation in a cable is soft and flexible. Explain why it would be dangerous to grip the inner insulation with the cable grip rather than the outer insulation.

6 As we learned in lesson P2 4.1, power cables carried by electricity pylons are made from aluminium with a steel core down the centre. Suggest why aluminium is used rather than copper and why there is a steel core to the wire.

7 From Table 1, plot a graph of maximum power against cross-sectional area. Use the graph to work out what would be the minimum cross-sectional area to use with a 4 kW heater.

8 Many electrical products are now supplied with moulded plugs that cannot be taken apart, although the fuse can still be replaced. Describe the advantages and disadvantages there might be for the consumer.

9 An electric iron has a special type of cable which often has a woven covering on it. What do you think this is for and why would ordinary electrical cable be unsafe to use?

P2 4.3

Fuses, earth and circuit breakers

Learning objectives

- explain how the earth wire and fuse work together to protect people
- describe the safe use of electrical appliances
- evaluate the use of fuses and circuit breakers.

Electrical safety

If an electrical appliance is faulty, it can become dangerous in one of two ways. The fault might reduce the resistance of the circuit, so a much larger current than normal might flow. If this happens the appliance overheats and may catch fire.

Alternatively a fault could cause a part of the electrical wire to touch a metal part of the appliance, such as the casing. If this happens somebody touching the appliance could get an electric shock, which could be fatal.

Fuses, earth wires and circuit breakers are safety devices that protect against these faults.

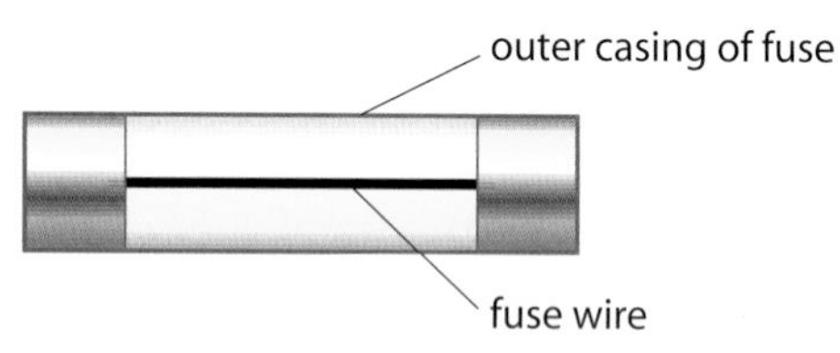

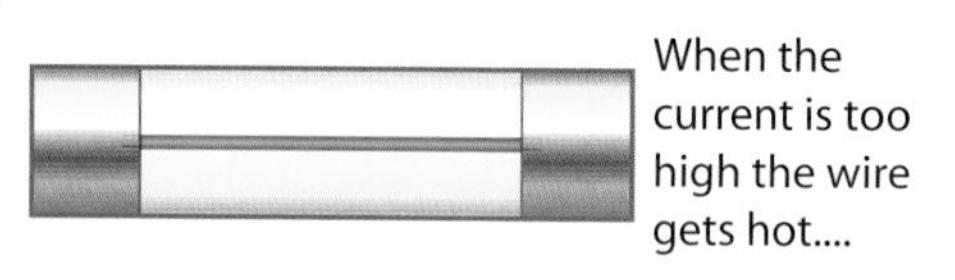

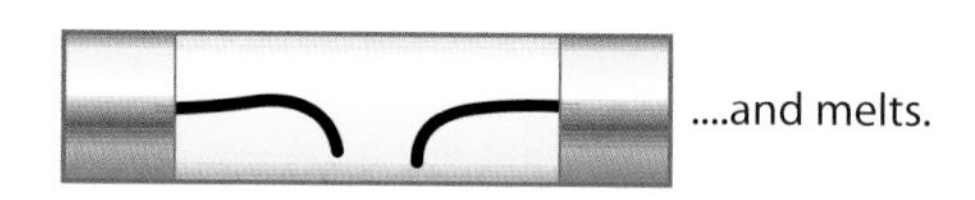

Figure 1 Fuses 'blow' when the current gets too high.

Science in action

There are many types of fuses, and it is not just the current rating that varies. Some fuses are made to blow quickly and some more slowly depending on the application. The size and shape of a fuse also varies depending on the current it needs to carry.

A large current fuse.

What does a fuse do?

A fuse is a small safety device that contains a length of wire that is designed to melt if the current in a circuit gets too high. Every fuse has a rating, which is the maximum current that it can carry. If the current gets above the fuse rating, the wire gets hot and melts, breaking the electrical connection.

What does an earth wire do?

Very simply, the earth wire is a low-resistance path for electric current to flow through. The earth wire is connected to the metal parts of an appliance that you might touch. If there is a fault in which the live wire touches the metal casing of a device, then a large current flows along the earth wire. Because the current through the device is briefly very high, it causes the fuse to blow and so makes it safe.

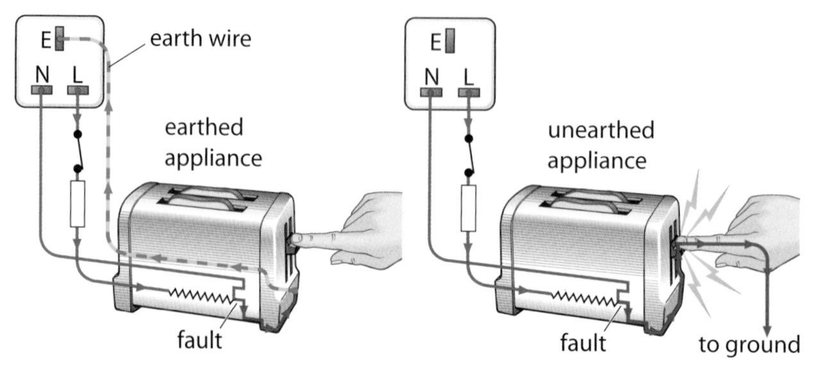

Figure 2 An unearthed toaster could give someone a nasty shock.

Some appliances have a plastic casing rather than a metal one. They do not need an earth wire because there are no metal parts that can become electrified. This kind of device is **double-insulated**. It uses a two-core cable because it does not need an earth wire.

Circuit breakers

Circuit breakers are similar to fuses, but they are more sensitive and operate much faster. Residual current circuit breakers (RCCBs) work by sensing any difference in current between the live and neutral wires. In a normal circuit the current in both wires is the same. If the RCCB senses a difference, it disconnects the electricity.

An RCCB can be plugged into an ordinary mains socket.

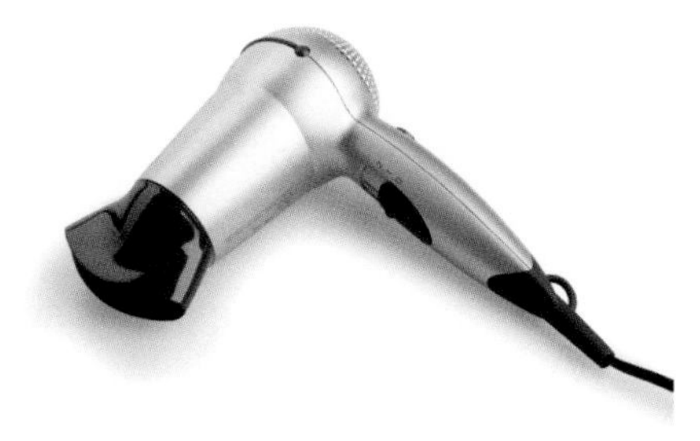

This hair dryer is double-insulated.

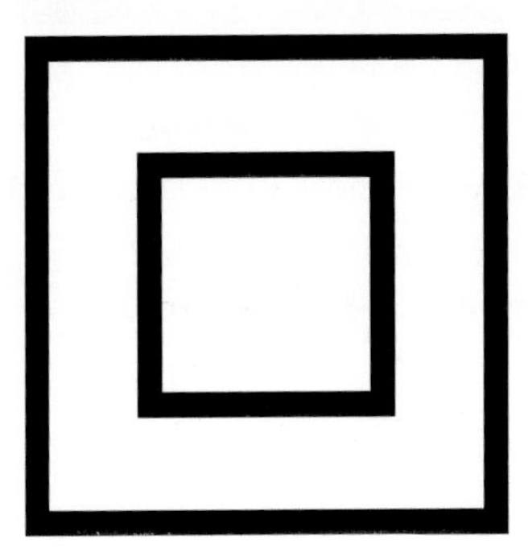

Figure 3 A double-insulated device has this symbol.

Imagine someone mowing the lawn with an electric lawnmower. They accidentally run over the electric cable and cut it. The bare ends of the cut cable are very dangerous. If they touch someone's skin the person could be killed by the shock. An RCCB would cut off the electricity and stop this happening.

Questions

1. Explain why it is more important to fit an RCCB to a hedge trimmer than to a lamp used in a bedroom.
2. Why is it that extension cables have the current ratings that they are designed for clearly marked on their packaging?
3. Even double-insulated appliances have a fuse fitted. How can a fuse work without an earth wire?
4. Cables for cookers are more expensive than those used for a kettle. Explain why this is.
5. When you wire a plug, it is important to use the correct fuse. Why do you think you should use a 3 amp fuse and not a 13 amp fuse for an appliance that usually takes only 0.5 amps?
6. An RCCB operates much more quickly than a normal fuse. Why might this be important in the case of an accident with a lawnmower?
7. Old houses used to have fuse boxes containing fuse carriers with fuse wires in them. If a fuse blew, the fuse wire had to be replaced. Most modern houses have circuit breakers that can be reset by flipping a switch. Why do you think circuit breakers have been introduced?
8. Evaluate the use of circuit breakers compared with fuses in domestic power supplies.
9. Replacement fuses can be bought in standard current ratings. The most common ones are 3 A and 13 A. Why do you think that there is not a 5.4 A fuse on sale for an appliance that uses 5.3 A of current?

A*

Examiner feedback

It is important to remember how RCCBs disconnect the current. An RCCB detects any difference in current between live and neutral wires.

Science in action

The test button on an RCCB simulates a fault in the appliance being used and the RCCB disconnects the power. The power can be reconnected using the reset switch.

Science in action

An RCCB makes use of an electromagnet. These types of devices depend on the magnetic field that is always generated when an electric current flows. The strength of the magnetic field can be increased by using a soft iron core, and by winding the wire in lots of coils.

P2 4.4 Current, charge and power

Learning objectives

- explain that current is the flow of charge
- describe the relationship between charge, current and time
- calculate power from current and p.d. values.

Examiner feedback

Make sure that you know the following quantities and units.

- Power (P) is measured in watts (W).
- Potential difference (p.d) or voltage (V) is measured in volts (V).
- Current (I) is measured in amps (A).
- Resistance (R) is measured in ohms (Ω).
- Time (t) is measured in seconds (s).
- Charge (Q) is measured in coulombs (C).
- Energy (E) is measured in joules (J).

Energy and power

In lesson P2 2.2 we saw that the power of a device is a measure of the rate at which it transfers energy. A 650 W microwave oven will cook a meal, but an 800 W oven will do it more quickly.

You can calculate the power of a device by measuring how much energy it transfers per second using the formula:

$$P = \frac{E}{t}$$

where P = power in watts (W), E = energy in joules (J) and t = time in seconds (s).

The power of an electrical device can be calculated another way, by measuring the current and the potential difference (p.d.):

$$P = I \times V$$

where P = power in watts (W), I = current in amps (A) and V = potential difference or p.d. in volts (V).

So, a car headlight that draws a current of 3 A and operates at 12 V has a power of 3 A × 12 V = 36 W.

Current and charge

We saw in lesson P2 3.1 that electric current is a flow of electric charge. The charge is carried by electrons. The amount of charge flowing in a given time can be found from the equation:

$$Q = I \times t$$

where Q = charge in coulombs (C), I = current in amps (A) and t = time in seconds (s).

This is a simple rearrangement of the equation $I = \frac{Q}{t}$ from lesson P2 3.1.

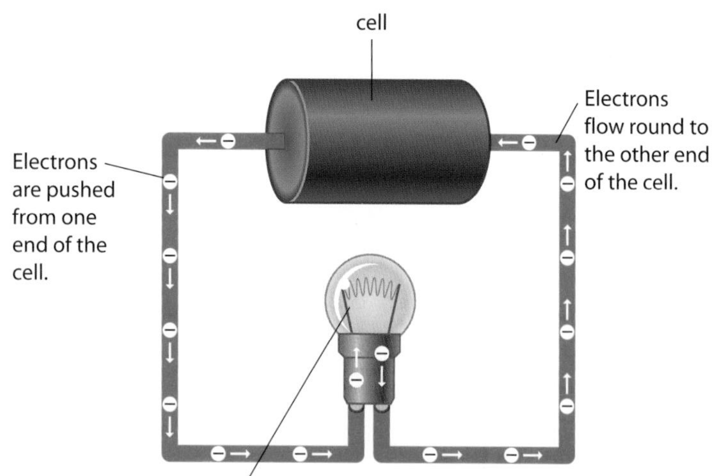

Figure 1 Flow of electrons around a circuit.

Electrons doing work

Electrons are 'pushed' through an electrical circuit by the battery or other electricity supply. Potential difference (voltage) is a measure of the electrical 'push'. As they move through a circuit, electrons transfer energy. The harder the electrons are pushed, the more energy is transferred. This is why an electrical resistor gets hot when current flows through it. The electrons transfer energy by heating as they move through the resistor.

The amount of energy transferred by one coulomb of charge (lots of electrons) depends on the p.d. that pushes it.

$$E = V \times Q$$

where E = energy transferred in joules (J), V = p.d. in volts (V) and Q = charge in coulombs (C).

Example 1

If the lamp shown in Figure 2 is on for 20 seconds, we can calculate not only the charge that flows, but also the power output and the total energy transferred.

$Q = I \times t = 0.4\,\text{A} \times 20\,\text{s} = 8\,\text{C}$

$P = V \times I = 12\,\text{V} \times 0.4\,\text{A} = 4.8\,\text{W}$

$E = V \times Q = 12\,\text{V} \times 8\,\text{C} = 96\,\text{J}$

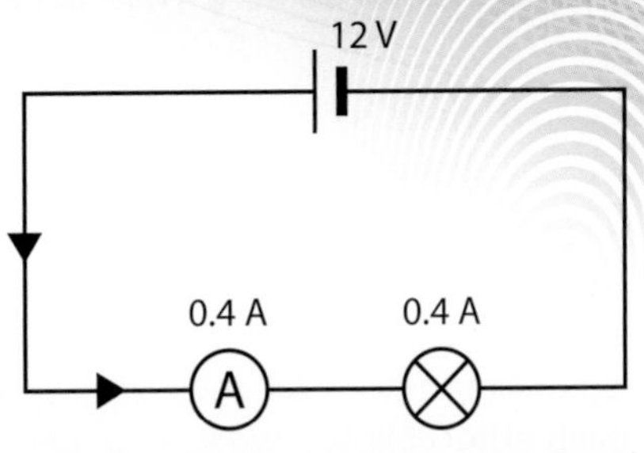

Figure 2 We can calculate the charge flowing, power and total energy output of this circuit.

Science skills

Table 1 Typical current, p.d. and power measurements for some electrical devices.

Device	p.d. / V	Current / A	Power / W
cycle lamp	3	0.6	1.8
child's night light	230	0.02	4.6
car headlamp	12	3	36
table lamp	230	0.26	60
kettle	230	13	3000

a Which device needs the thickest cable?

b Using the idea that energy output depends on charge and p.d., explain why a child's night light gives out more energy per second than a cycle lamp, even though the current is smaller in the night light.

Taking it further

The definition of an ampere is:

A constant current that produces a force equal to 2×10^{-7} newtons per metre of length between two conductors.

In this definition, the conductors are two straight parallel conductors in a vacuum, 1 m apart. They have infinite length and a negligible circular cross-section.

Questions

1. What is the equation that links energy, potential difference and charge?
2. How much energy will be transferred by a charge of 20 C pushed by a p.d. of 6 V?
3. A current of 250 mA flows through a 2 kΩ resistor. Calculate the charge that will flow through the resistor in 1 minute.
4. Calculate the energy output from a bulb that has a resistance of 3 Ω when a 12 V battery is connected to it for 1 minute.
5. How much energy would a 12 V car battery transfer if it provided a current of 40 A for 4 s?
6. As a torch battery runs down, the brightness of the bulb reduces. Describe what will be happening to the potential difference across, the current through and power output of the bulb. Link these three quantities by using any appropriate equations in your explanation.
7. Calculate how long it would take a bulb with a resistance of 0.1 Ω to give out 50 J of energy, if the p.d. across it was 0.4 V.
8. Draw circuit diagrams for two simple series circuits. Both have a 10 Ω resistor. One circuit has a 10 V cell and the other a 5 V cell. Calculate the current that flows, the charge that flows in 5 seconds and the power output of each circuit. Compare the energy output of the two circuits and explain why, although the voltage has doubled, the power output is not proportional to the voltage.

A*

Examiner feedback

It will be useful to know how to use the following equations:

- $P = \frac{E}{t}$
  where P = power in watts (W), E = energy in joules (J) and t = time in seconds (s).
- $P = V \times I$
  where P = power in watts (W), I = current in amps (A) and V = potential difference or p.d. in volts (V).
- $Q = I \times t$
  where Q = charge in coulombs (C), I = current in amps (A) and t = time in seconds (s).
- $E = V \times Q$
  where E = energy transferred in joules (J), V = p.d. in volts (V) and Q = charge in coulombs (C).

Assess yourself questions

1 Explain why it is only electrons that move when static electric charges are produced. *(1 mark)*

2 In an experiment, two pieces of the same type of plastic were rubbed separately by two different students using the same type of cloth. If the two pieces of plastic are brought together will they attract each other or repel each other? Explain why. *(2 marks)*

3 Lightning is an example of static electricity discharging between clouds and the ground. When lightning strikes, a large current flows for a very short time and this can damage buildings. Explain how a lightning conductor can protect a building. *(2 marks)*

4 Damp air conducts electricity much more easily than dry air. Would it be easier to do static electricity experiments on a dry or damp day? Explain your answer. *(2 marks)*

5 Draw a series circuit containing a cell, a fuse, a bulb and an open switch. *(4 marks)*

6 **(a)** Draw a parallel circuit containing a cell and two 100 Ω resistors in parallel. *(2 marks)*

(b) If the cell has a p.d. of 4 V, what is the current through each resistor? *(2 marks)*

7 If a current of 3 A passes through a light bulb that has 5 V across it, calculate the following:

(a) the power of the light bulb *(1 mark)*

(b) the resistance of the light bulb *(1 mark)*

(c) the energy output by the light bulb in 4 minutes. *(1 mark)*

8 If only 2.5 V was put across the light bulb in question 3, what do you think the current would be? Explain your answer. *(2 marks)*

9 Two resistors are in series and the p.d. across them both together is 9 volts. One resistor has a resistance of 800 Ω; the other has a resistance of 400 Ω. Calculate the following:

(a) the total resistance of the resistors *(1 mark)*

(b) the current flowing through the resistors *(1 mark)*

(c) the p.d. across each resistor. *(1 mark)*

10 Refer to Figure 1.

(a) Describe what will happen to the resistance of the LDR as it gets darker. *(1 mark)*

(b) What will happen to the LDR's share of the voltage as it gets darker? Explain why. *(2 marks)*

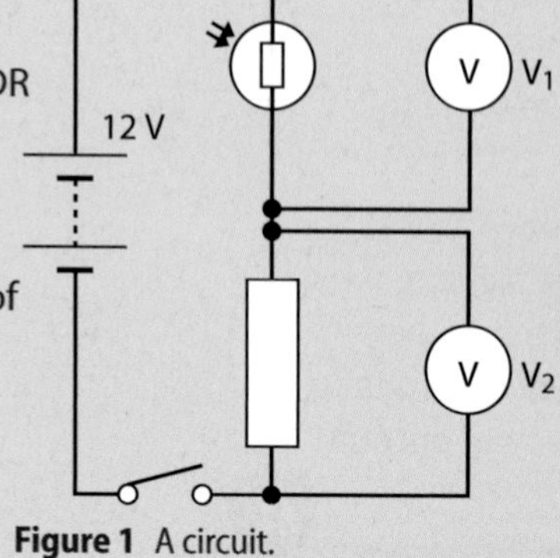

Figure 1 A circuit.

(c) If the LDR had a resistance five times that of the resistor, what would V_1 and V_2 be? *(2 marks)*

11 Calculate the current in the circuit in Figure 2. What is the value of the p.d. labelled V_{out}? *(3 marks)*

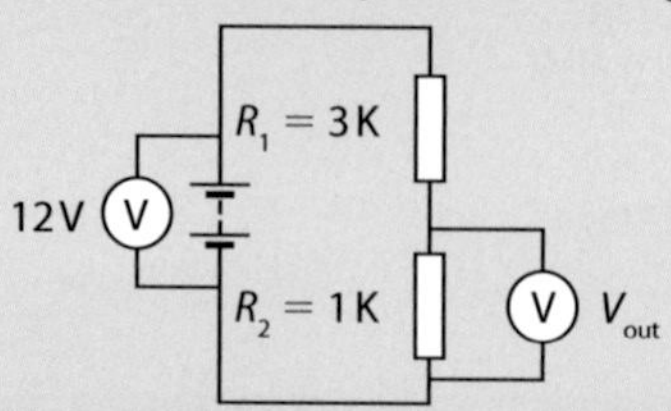

Figure 4 A circuit.

12 An a.c. current is applied to a diode and a bulb in series.

(a) Describe the way that the bulb will light up. *(2 marks)*

(b) Sketch a graph of current against time for the current in the bulb. *(1 mark)*

13 Explain why a cell is a better option for a small torch than using mains electricity. *(1 mark)*

14 Describe the function of the circuit in Figure 3. Explain what happens to the current in the circuit and the brightness of the bulb as the temperature changes. *(2 marks)*

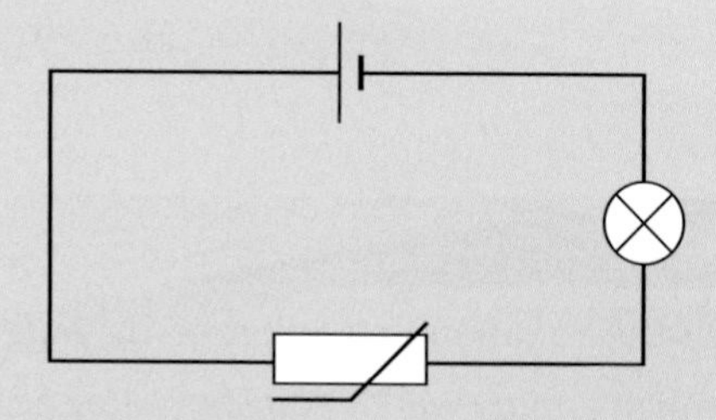

Figure 3 A circuit.

15 A student built a series circuit with two light-emitting diodes in it. She made sure that they pointed in opposite directions in the circuit. She said 'If I have one diode pointing in each direction, then at least one of them will light up.' Explain why this reasoning is wrong. *(2 marks)*

16 An LDR is to be placed in series with a fixed-value resistor of value 500 Ω. If the LDR and resistor have a p.d. of 9 V across them, what will be the voltage across the LDR at:

(a) 10 am?

(b) 4 pm?

(c) 8 pm?

(3 marks)

Table 1 LDR's resistance changes during the day.

Time of day	Resistance of LDR/Ω
8.00 am	100
10.00 am	110
12.00 noon	95
2.00 pm	130
4.00 pm	90
6.00 pm	120
8.00 pm	540

17 Looking at the values for the resistance of the LDR in question 16, can you suggest a reason why the resistance is higher at 2 pm than at 4 pm? *(1 mark)*

18 Figure 4 shows a circuit with two resistors in series.

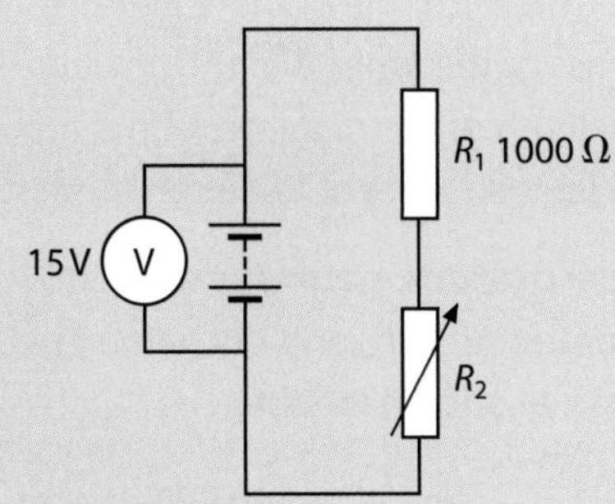

Figure 4 A circuit.

(a) Describe what will happen to the current that flows in this circuit as the variable resistor (R_2) changes from 0 Ω to its maximum value of 5000 Ω. Include maximum and minimum values for the current. Show your working. *(3 marks)*

(b) What will be the maximum and minimum values for the p.d. across the 1000 Ω resistor as R_2 is changed? *(2 marks)*

(c) The variable resistor is adjusted so that the current through the circuit is 0.02 mA. What is the new value of the resistance R_2? *(1 mark)*

(d) The variable resistor is adjusted until the p.d. across the fixed resistor is 3.6 V. What is the value of R_2 now? *(1 mark)*

19 Look at Figure 5. Describe the p.d.s represented in as much detail as possible. *(3 marks)*

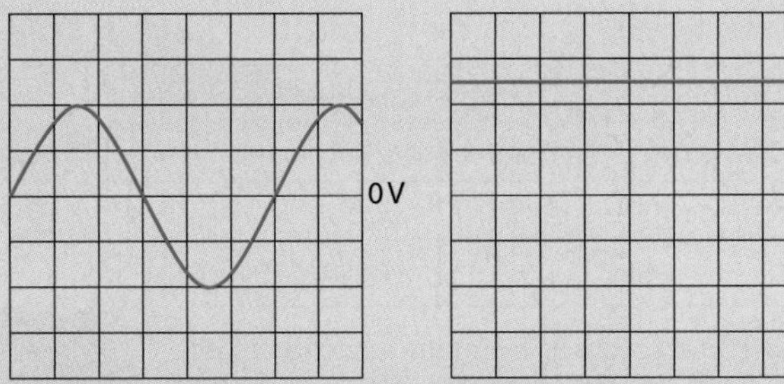

Figure 5 The time base is 5 ms/division and the p.d. sensitivity is 5 V/division.

20 A student is investigating the resistance of a series of wires made from different materials. Describe what actions he should take to ensure that his experiment is a fair test. *(2 marks)*

21 A piece of resistance wire was used to make a component with a resistance of 1.75 Ω. The label on the reel of wire noted that the wire had a resistance of 4 Ω per metre. What length of wire would be needed? *(2 marks)*

22 A student placed a voltmeter in series with a light bulb, rather than in parallel with it. Explain why the voltmeter would not work like this. *(2 marks)*

23 A digital ammeter reads '4.23'. If the meter is set to measure up to 10 mA, what is this current in amps? *(1 mark)*

24 Look at the two images of three-pin plugs. For each one describe the fault in the wiring. *(4 marks)*

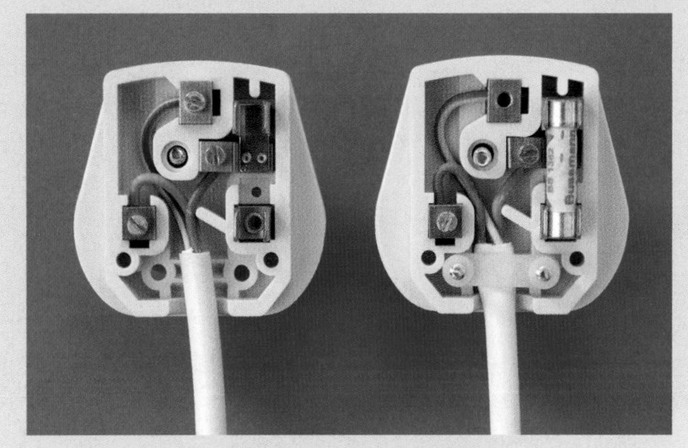

25 A 300 W grinder is powered by UK mains electricity. What size fuse would it need: 1 A, 3 A or 13 A? Explain your answer. *(2 marks)*

26 Describe how an earth wire and a fuse work together to protect the user from an electrical fault in a coffee grinder and explain why using too high a fuse rating in this device could be dangerous. *(4 marks)*

27 A student was investigating the flow of current in very thin fuse wires. Describe what happened to the current, and the wire, as he slowly increased the voltage across the wire. *(2 marks)*

28 Safety advice recommends that people should not use too many appliances from the same socket. Explain why this is dangerous. *(1 mark)*

29 Describe how an RCCB circuit breaker can help prevent injury in the garden. *(1 mark)*

30 A hand dryer has two electrical functions. It has a heating element and a motor that drives a fan to move air. The heating element is a coil of wire that warms up when electric current flows through it.

(a) If the hand dryer has a power of 2300 W, what current will flow through the heating element if the voltage is 230 V? *(1 mark)*

(b) Some modern hand dryers use very low-power heaters, but much more powerful motors, so that the air speed is much higher and knocks the water off your hands. Explain why will this be a cheaper option to run than hand dryers that use slower, hotter air? *(2 marks)*

What is an atom like?

Learning objectives

- describe a model of the atom
- describe the particles that make up an atom
- explain why the overall charge of an atom is zero
- describe the formation of ions and the electrostatic forces between them
- describe the relationship between atomic number, atomic mass, isotopes and the particles that make up an atom.

Atomic structure

Atoms are the building blocks of all matter. The basic details of atomic structure were worked out in the early twentieth century. All atoms have a very small, positively charged nucleus at the centre, with one or more negatively charged **electrons** orbiting it. Most of the mass of the atom is in the nucleus. The nucleus is very small: it is only about one 10 000th of the diameter of the atom. Electrons are much smaller even than the nucleus, so most of the atom is empty space.

Atoms contain three types of particle: **protons**, **neutrons** and electrons. The nucleus is made up of positively charged protons and uncharged neutrons. Both particles have roughly the same mass, as shown in Table 1.

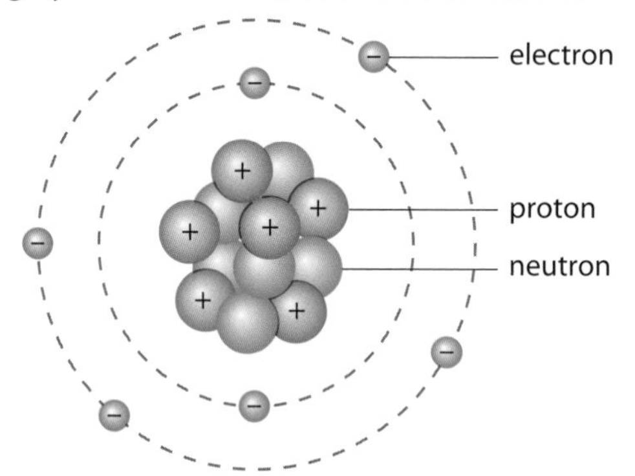

Figure 1 Structure of an atom. The nucleus is not to scale; if it was, the electrons would be 2.5 kilometres away.

Electrons are much lighter than protons and neutrons. The negative charge of an electron is equal and opposite to the positive charge of a proton.

Atoms are neutral: they have no overall charge, so the number of electrons is always the same as the number of protons.

Table 1 Properties of particles in an atom.

Name of particle	Mass	Charge	Location
proton	1 atomic mass unit	+1	nucleus
neutron*	1 atomic mass unit	0	nucleus
electron	$\frac{1}{1800}$ atomic mass unit	−1	orbiting the nucleus

*Neutrons are very slightly heavier than protons, by about 1 electron mass.

Examiner feedback

Be really sure you have understood the properties of the three particles that make up an atom. You can work out almost everything else about atomic structure from this.

Science skills

a Use Table 1 to describe what you would get if you combined one proton and one electron. How does this compare with a neutron?

Atomic number and mass number

If you look at Figure 1 again, you will see that the atom shown has six protons in the nucleus. Each element has a different number of protons in its nucleus. The number of protons in the nucleus of an atom is known as its **atomic number**.

Carbon is the element with atomic number 6, so the atom in Figure 1 is a carbon atom. Any atom with six protons must be an atom of carbon.

The atom will automatically have six electrons around it to make it neutral.

The **mass number** is the total number of particles in the nucleus. For the carbon atom shown in Figure 1 the mass number is 12, as there are six protons and six neutrons.

Examiner feedback

Make sure that you have thoroughly understood the difference between atomic number (number of protons) and mass number (total number of protons plus neutrons).

Isotopes

All carbon atoms have six protons. If they didn't they would not be carbon atoms. However, they do not all have the same number of neutrons. Most carbon atoms contain six neutrons. They are called carbon-12 atoms, because they have a mass number of 12 (there are 12 particles in the nucleus). However, some carbon atoms contain six protons and eight neutrons. These are carbon-14 atoms: they have a mass number of 14.

Where atoms have the same number of protons but different mass numbers (different numbers of neutrons) they are called **isotopes**.

Forming ions

For a variety of reasons, some atoms either gain or lose electrons. This means that they are no longer atoms, but have become **ions**. If they have gained one or more electrons then they have an overall negative charge. If they have lost one or more electrons then they have an overall positive charge. For example, a lithium atom can lose an electron and become a lithium ion. Overall the ion has a single positive charge because it has two negatively charged electrons but three positively charged protons.

Ions with opposite charges attract each other, while those with like charges repel one another. So positive ions attract negative ions and repel positive ions, while negative ions attract positive ions and repel negative ions.

Questions

1. Use the periodic table to describe the number of protons, neutrons and electrons in an atom of helium.
2. Hydrogen has three isotopes: hydrogen, sometimes called protium, with a mass number of 1, deuterium with a mass number of 2 and tritium with a mass number of 3. Sketch the nuclei of each of these isotopes, and note the numbers of each type of particle.
3. Describe how an atom of chlorine changes if it gains an electron.
4. An atom of uranium-238 has 92 protons. How many electrons and neutrons does it have? Explain how to find these two numbers.
5. An atom of uranium-235 also has 92 protons. Describe how it is different to an atom of uranium-238.
6. If a particle had four atomic mass units, two of which were neutrons, but had no electrons, what would the particle be?
7. A proton is about 1.5×10^{-15} metres across. A hydrogen atom is about 1×10^{-10} metres across. Some people compare the structure of the Solar System to the way an electron orbits a nucleus. The Sun has a diameter of about 1.3 million km. Neptune, the outermost planet, is about 4.5 billion km away. Comment on how the sizes of the bodies and their separation compare.

A*

Examiner feedback

Make sure that you can explain what an isotope is in an examination.

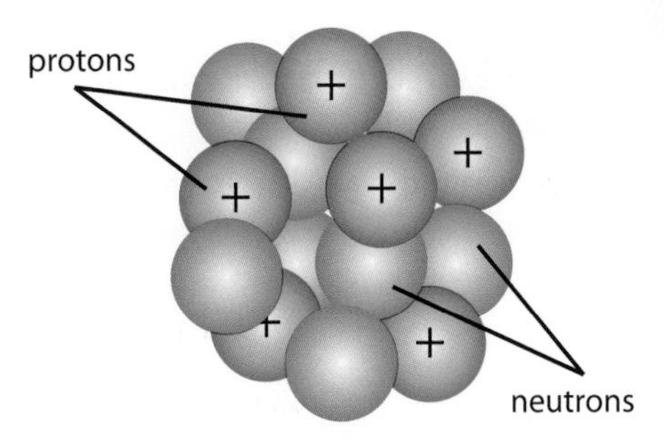

Figure 2 The nucleus of a carbon-14 atom.

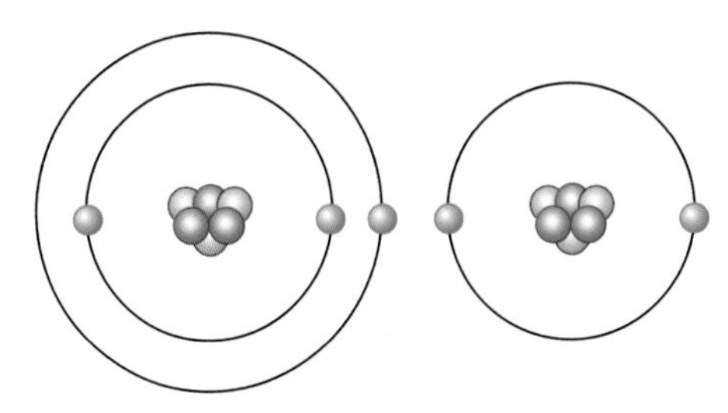

Figure 3 If a lithium atom loses an electron it becomes a positive lithium ion.

Taking it further

Some periodic tables give two numbers for each atom. One is the number of protons; the other number is the mass. The mass number is not a whole number for most elements. This is because the mass number is the average of the different isotopes of the element, taking into account their relative abundance on Earth. For example, carbon has a mass number of 12.011 because the majority of the carbon on the Earth is carbon-12. There is a small percentage of other carbon isotopes.

Nuclear radiation

Learning objectives

- explain what nuclear radiation is
- describe the different types of nuclear radiation
- describe the most common sources of background radiation
- describe the changes that occur in the nucleus during alpha and beta emission.

The discovery of radioactivity

In 1896 Henri Becquerel accidentally discovered that uranium salts could blacken a photographic plate, even when kept in the dark. His research assistant Marie Curie discovered that the uranium salts constantly gave out some kind of radiation. She called this kind of radiation **radioactivity**.

Radioactive materials **emit** radiation all the time, no matter what is done to them. Heating, cooling, chemical reactions, high and low pressure, etc. have no effect on radioactive emissions.

What is radioactivity?

The radiation emitted from a radioactive substance comes from the atomic nucleus. In most atoms, the combination of protons and neutrons in the nucleus is stable: it does not change. Radioactive elements have an **unstable nucleus**. The atoms give off radiation in a process known as radioactive **decay**. This changes the type of element the atom is. Eventually the decay will lead to a stable element.

Many very heavy elements, such as uranium and radium, are radioactive because their large nuclei are unstable. Some lighter isotopes (see lesson P2 5.1) are also radioactive.

Practical

Scientists often use a Geiger counter to detect how radioactive a sample is.

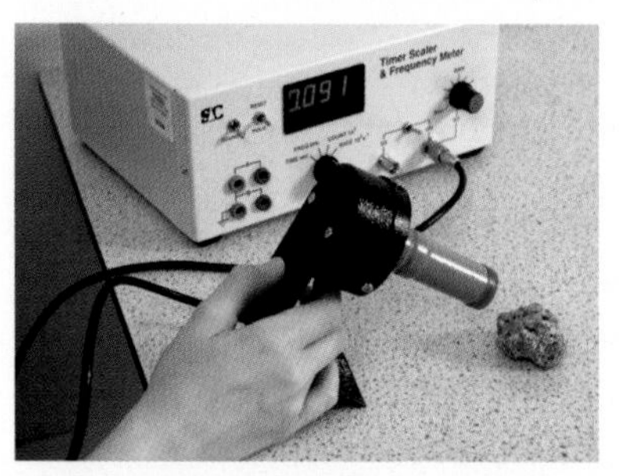

Testing a possible radiation source with a Geiger counter.

When you take measurements of radiation from a source, the background radiation will also be measured. The way to overcome this problem is to take the background count first and subtract this from the reading taken from the source. This will give a measure of the source activity itself.

a A student uses a Geiger counter to measure the radioactivity of a source and records the count as 45 counts per minute. She also performs a background radiation count and finds it is 26 counts per minute. What is the activity of the source?

Background radiation

There is a constant, low level of radiation around us at all times, known as **background radiation**. This comes from several sources. The Earth contains many elements that are radioactive and these constantly give out small amounts of radiation. There is also some low-level radiation from space, known as cosmic radiation. A small amount of radiation comes from manmade sources such as atomic bomb tests, or accidents in which radioactive materials escape. By far the largest source of background radiation is naturally occurring radon gas.

Alpha radiation

Not all radioactive substances produce the same kind of radiation. Radium is a radioactive metal that was first isolated by the scientists Marie and Pierre Curie. When radium decays, it produces alpha radiation. The radium nucleus spits out an alpha particle, made up of two protons and two neutrons.

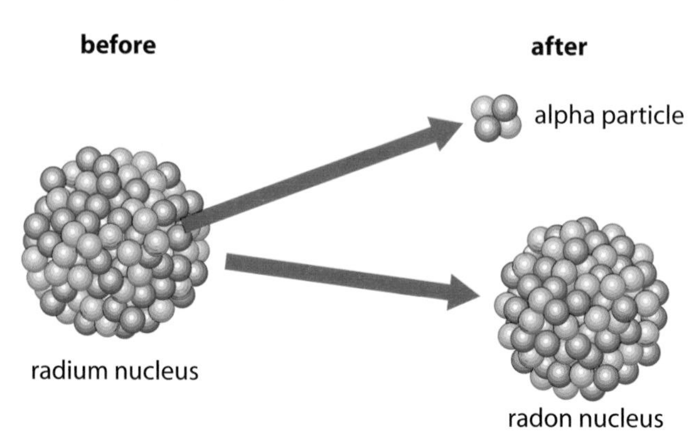

Figure 1 The alpha decay of radium.

In Figure 1 a radium nucleus gives out an alpha particle. This means that the nucleus of the atom has two fewer protons and therefore the nucleus is that of a different atom – in this case, the gas radon. The nuclear equation is shown below.

$$^{224}_{88}\mathrm{Ra} \longrightarrow {}^{220}_{86}\mathrm{Rn} + {}^{4}_{2}\alpha$$

The top number next to each symbol is the mass number; this drops by four. The bottom number is the atomic number, which drops by two.

An alpha particle is effectively a helium nucleus, so an alternative equation for this decay is:

$$^{224}_{88}\mathrm{Ra} \longrightarrow {}^{220}_{86}\mathrm{Rn} + {}^{4}_{2}\mathrm{He}$$

Beta decay

A beta particle is a fast-moving electron. You may remember that there are no electrons in the nucleus of an atom, so where does the electron come from? The answer is that it comes from the decay of a neutron into a proton and an electron. The nuclear equation for the reaction is:

$$^{14}_{6}\mathrm{C} \longrightarrow {}^{14}_{7}\mathrm{N} + {}^{0}_{-1}\beta$$

Notice that the mass number has not changed. The number of particles in the nucleus is still 14. However one neutron has turned into a proton, so now there is one more proton and one less neutron. The atom changes from carbon (six protons) to nitrogen (seven protons).

Questions

1. What is the difference between an alpha particle and a helium atom?
2. Radon causes most of our background radiation. Where does radon come from?
3. Use the periodic table to work out what uranium-234 becomes if it gives out an alpha particle.
4. What type of particle does a beta particle have the same mass and charge as?
5. Some students describe alpha emission as making the atoms 'move down the periodic table' and beta emission as making the atoms 'go back up'. Explain what they mean.
6. A Geiger counter can be used to measure how radioactive a source is. Why is this not the same thing as measuring the total amount of radiation that a source is emitting?
7. Uranium-238 decays into thorium. Thorium decays into protactinium, which decays back to uranium again. Use the periodic table to find out what particles are being emitted and what mass numbers are involved for these elements.
8. Write nuclear equations for each step of the process described in question 7.

A*

Examiner feedback

Remember, it is the atomic number that determines which element an atom is.

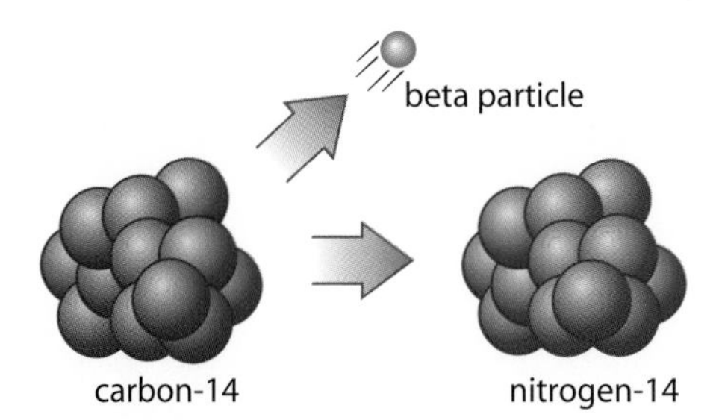

Figure 2 Carbon-14 decays to nitrogen by beta emission.

Science in action

It is possible to date the rocks of the Earth using uranium-238. As it decays, uranium-238 transforms eventually into lead-206 which is stable. By comparing the relative amounts of these two elements, it is possible to work out the age of the rock. This only works on very old rocks because the half-life of uranium-238 is 4.5 billion years.

Taking it further

Chlorine has many isotopes, but there are two that are more common. About 25% is chlorine-37 and 75% is chlorine-35. On the periodic table, the element chlorine has a mass number of about 35.5, which takes into account the abundance and masses of the isotopes.

Ionising power

Learning objectives

- describe the different types of radiation
- explain why they have different penetrating powers
- describe how they behave in electric and magnetic fields.

Gamma radiation

We discussed alpha and beta decay in lesson P2 5.2. A third type of radioactive decay produces gamma radiation. Uranium-238, for example, releases gamma radiation. Rather than producing atomic particles, this type of decay gives off high-energy electromagnetic waves, more energetic than X-rays.

Table 1 The properties of the three different types of nuclear radiation.

Type of radiation	Alpha	Beta	Gamma
Symbol	α	β	γ
Image			
What it is	2 neutrons and 2 protons	an electron	electromagnetic waves, similar to X-rays
Mass	4 atomic masses	$\frac{1}{1800}$ of an atomic mass	no mass
Charge	+2	−1	no charge
Range in air	a few centimetres	a few metres	a very long way
What stops it	thin paper	thin aluminium	some passes through thick lead

Examiner feedback

Make sure that you can recall all of the information in Table 1, as this is the basis of many examination questions.

Ionising radiation

Alpha, beta and gamma radiation are all types of ionising radiation. This means that they ionise atoms ('knock off' electrons to form charged ions). An alpha particle can ionise thousands of air molecules as it travels, slowing down as it does so. A beta particle will ionise only a few atoms before it loses its energy, and gamma rays only ionise one atom or molecule each.

Taking it further

It is very difficult to be specific about the size of the nucleus of an atom. This will vary depending on the actual element that is being measured. Because it is so difficult to measure the nucleus of an atom, different scientists and laboratories quote slightly different values.

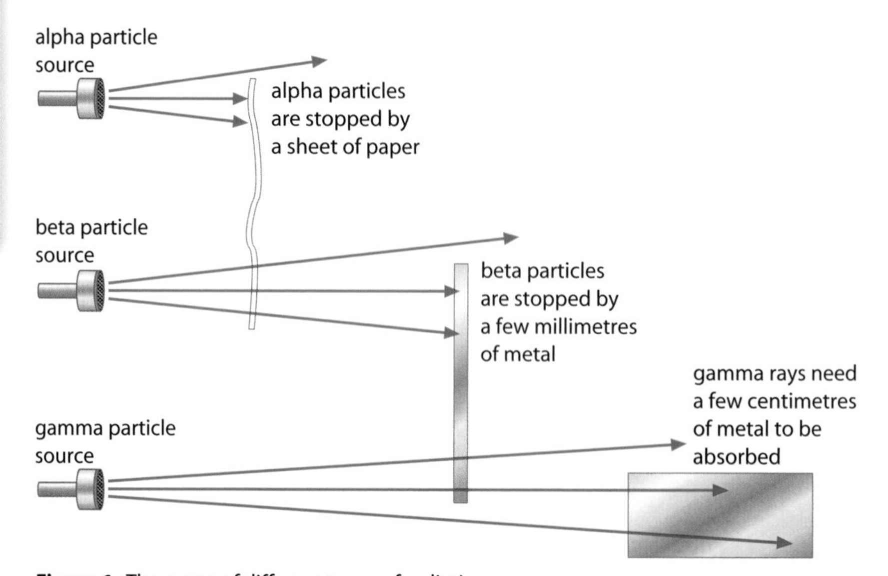

Figure 1 The range of different types of radiation.

Radioactive dangers

Radiation from radioactive substances can be dangerous to humans. The main danger is from ionisation of atoms. Ionisation within living cells can damage or kill the cell.

From Table 1, gamma radiation seems more dangerous than other types because it has a long range and can penetrate even thick lead. However, over a short distance alpha radiation can cause more damage than either gamma or beta radiation. For example, swallowing an alpha source is very dangerous because it ionises the cells very close to the source, whereas gamma radiation inside the body causes less local damage because it tends to travel further before ionising cells, or even escapes the body altogether without ionising any cells.

Bending radiation

Alpha and beta radiation are both made up of charged particles. This means they are affected by electric and magnetic fields. Gamma rays have no charge so they are unaffected by such fields.

Alpha particles are positively charged so they are attracted to negatively charged metal plates. Beta particles are negatively charged so they are attracted to positively charged plates.

In a magnetic field, alpha and beta particles are deflected in opposite directions because they have opposite charges. They are also deflected by different amounts (see Figure 4). Why does this happen?

An alpha particle has twice the charge of a beta particle, but over 7000 times as much mass. This means that it is very difficult to deflect, compared with a beta particle.

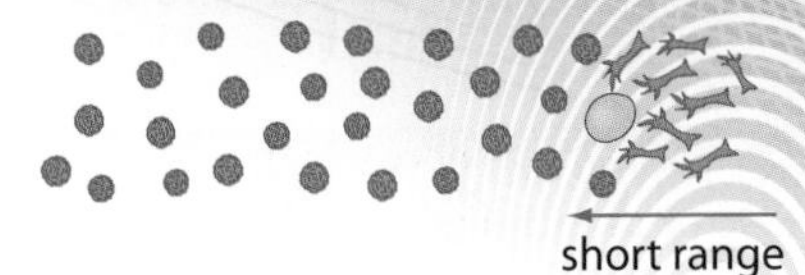

An alpha particle is like a boulder smashing through the trees.

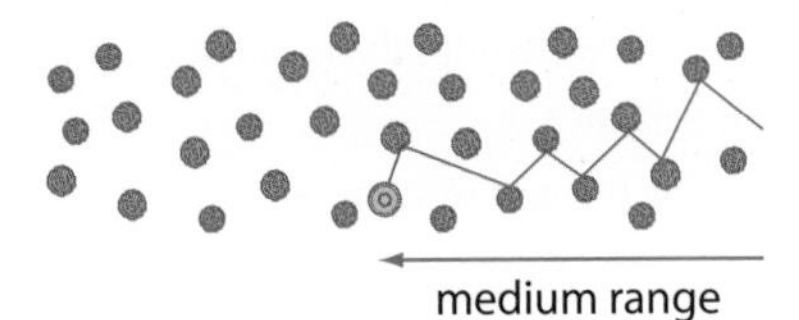

A beta particle is like a discus thrown through the trees.

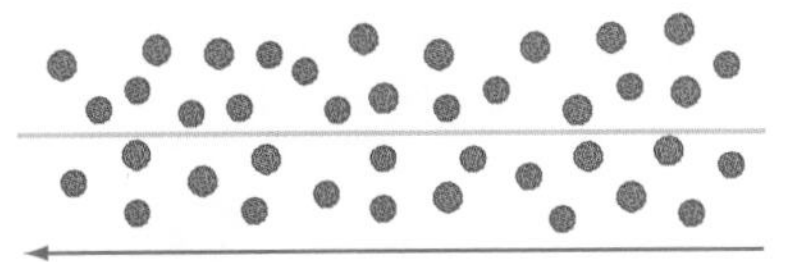

A gamma particle is like a laser beam shining through the trees.

Figure 2 Ionising power. If we imagine an alpha particle as a boulder crashing through the forest, beta radiation is like a discus bouncing off the trees, and gamma radiation is a high-powered laser beam.

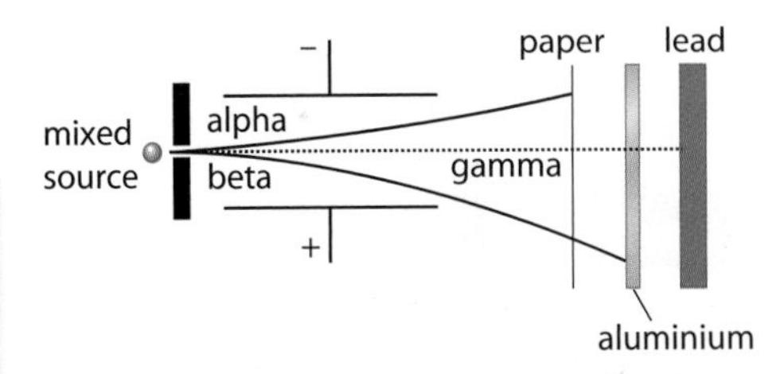

Figure 3 Alpha, beta and gamma radiation in an electric field.

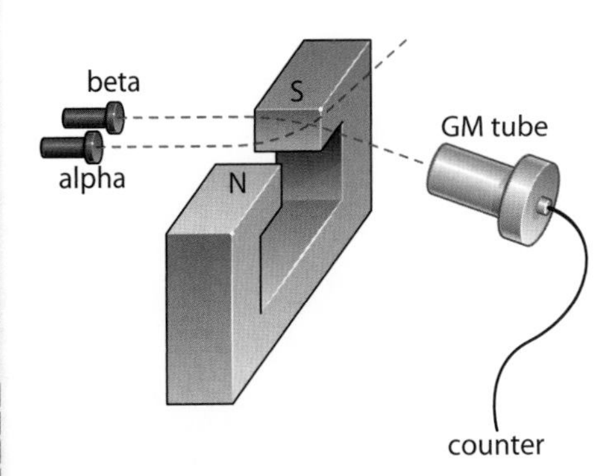

Figure 4 Effect of a magnetic field on alpha and beta radiation.

Questions

1. A student asks you to describe the size of each type of ionising radiation. Describe the 'size' of each type of radiation to the student.
2. Some people think that an alpha particle is a helium atom. Explain why they are mistaken.
3. Explain why beta particles have a much larger range in air than alpha particles.
4. Explain why being far away is the best protection against gamma radiation, while a paper towel will protect against alpha particles.
5. Gamma radiation is pure energy, just like light. Gamma radiation comes in packets called photons. Each photon's energy must all be transferred at once when it interacts with an atom. How many atoms do you think a gamma ray will ionise? Explain your answer.
6. Describe what you think will happen to an alpha particle that travels close to the nucleus of an atom but does not ionise the atom.
7. Explain why ionising radiations tend to travel less far in dense materials like lead compared with less dense materials like air.
8. The analogy of the radiations travelling through the forest in Figure 2 has several shortcomings. Explain one of these shortcomings.

A*

P2 5.4

Alpha particles

Learning objectives

- explain how alpha-scattering experiments give us evidence for the structure of the atom
- explain how theories change because of new evidence
- describe the uses of alpha particle sources.

Useful but dangerous

The ionising properties of alpha radiation have found many uses. Experiments involving alpha particles have also helped to change the way that scientists view atoms. But there are hazards associated with alpha radiation.

Alpha particles in use

One of the most common uses of alpha particles is in smoke detectors, as shown in Figure 1. An alpha particle source in the detector ionises the air around it. The levels of charged particles in the air are measured by an electronic circuit. If smoke gets into the detector, it neutralises the charged particles and so the number of charged particles in the air is reduced. The detector picks up this change, and the smoke alarm goes off.

Alpha particles ionise the air and these charged particles move across the gap forming a current.

Siren will sound when the detector current falls.

Smoke enters smoke detector.

Am-241 alpha source

Americium-241 source gives off a constant stream of alpha particles.

detector

battery

Smoke in the machine will absorb ions so current falls.

A detector senses the amount of current of ionised particles.

Figure 1 Smoke detectors rely on a source of alpha radiation.

When radioactive materials decay, they produce heat as well as radiation. This property is used to make long-lasting 'batteries' known as **thermoelectric generators**. These generators use a source of alpha radiation, usually plutonium-238, to heat up a set of thermocouples, which are devices that produce electricity when heated. Thermoelectric generators have been widely used in unmanned spacecraft.

The *Galileo* space probe travelled to Jupiter. Its instruments were powered by a thermoelectric generator.

Alpha particle sources are also used in the treatment of some cancers and to get rid of static electricity.

Examiner feedback

It is worth remembering at least one positive use of alpha particles to give as an example in an examination answer.

Alpha-particle scattering

In 1904 the British scientist J. J. Thomson proposed the 'plum pudding' model of the atom. He suggested that an atom was a blob of positively charged matter, with negative particles scattered through it like raisins in a plum pudding. But in 1910 an experiment with alpha particles changed our ideas about the atom.

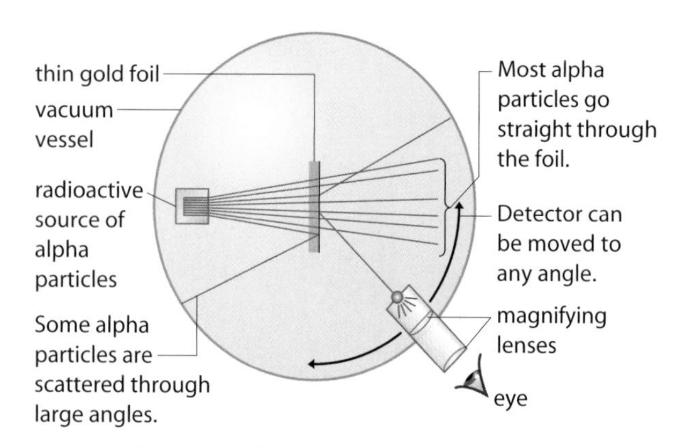

Figure 2 The design of Rutherford's alpha particle experiment.

The experiment was designed by New Zealand physicist Ernest Rutherford. A stream of alpha radiation was fired at a thin metal foil target. Rutherford expected that the 'plum pudding' atoms would deflect (bend) the alpha particles a small amount. What actually happened was that most particles went straight through the foil, a few were deflected, and some even 'bounced back' off the foil.

Scientists consider the evidence from experiments and put forward theories to explain their findings. Sometimes the evidence suggests that a theory needs changing. This is what happened with Rutherford's experiment. The plum pudding model could not explain the results. Rutherford therefore formulated a new model in which most of the atom is empty space, and nearly all the mass is in a tiny, positively charged lump at the centre.

Dangerous gas

The majority of the background radiation we experience comes from natural sources, and the biggest of these is alpha radiation emitted by the gas radon. Alpha particle radiation is the most dangerous to get inside your body, so breathing in radon is very dangerous.

Radon is produced by the radioactive decay of uranium and radium. These elements are found in rocks. As they decay into radon, the gas seeps up from the ground. Normally this is not too much of a problem, but in some areas where the rocks contain uranium, radon gas can collect inside houses and buildings. Special measures need to be taken to prevent this, as shown in Figure 3.

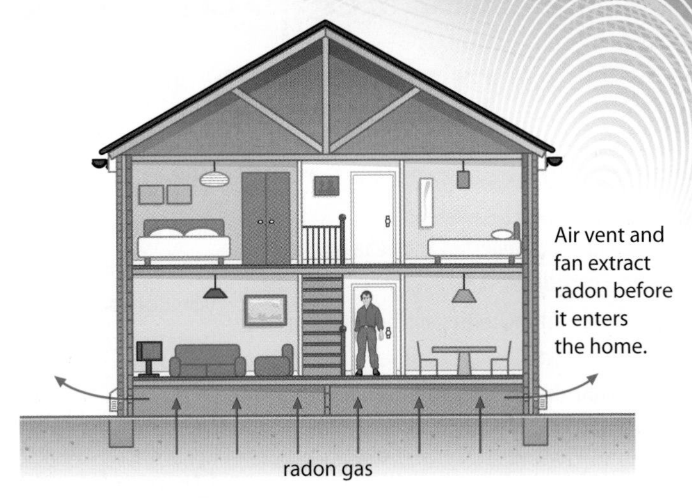

Figure 3 Coping with radon gas.

Science skills

Figure 4 shows a map of how radon is distributed around the country.

a How do you think local authorities should respond to the information shown?

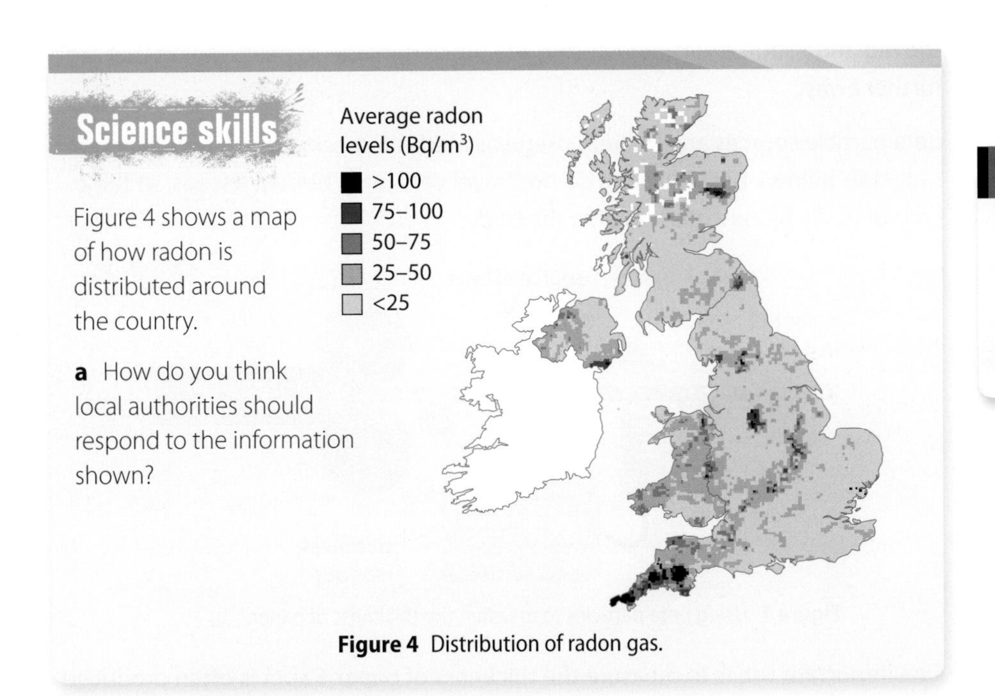

Figure 4 Distribution of radon gas.

Examiner feedback

When answering questions about background radiation, it is worth remembering that radon gas is the major contributor to background radiation in this country.

Questions

1. Why would beta particles not give the same sort of results as alpha particles in a scattering experiment?
2. Why is it OK to have a radioactive alpha source in a smoke detector in your house?
3. Explain why the alpha particle experiment was conducted in a vacuum.
4. Why would a nuclear model for an atom predict that the vast majority of alpha particles would pass through a thin gold foil undeflected?
5. Alpha particle sources can be very damaging if they get inside you. Explain why this is and why they might be deliberately injected into a cancer patient.
6. What was it about the alpha-scattering experiment that made Rutherford revise the 'plum pudding' model of the atom?
7. Plutonium-238 is an almost pure alpha emitter. Explain why this is useful for the power source of an atomic battery that is to be used in a spacecraft.
8. Radon gas comes from radioactive material in the rocks below the Earth. Why is the radioactive rock itself not a danger?

P2 5.5

Beta particles and gamma rays

Learning objectives

- describe the properties and behaviour of beta particles
- describe the properties and behaviour of gamma rays
- describe the uses of beta and gamma sources.

Why are beta radiation sources useful?

Beta particles have the advantage of being fairly easy to detect while at the same time **penetrating** a reasonable distance into material that is not too dense. If you compare them with alpha particles, they are much more penetrating, partly because they are smaller. Compared with gamma rays they are much easier to shield, so they are very useful in applications where radiation must pass through thin material before being detected.

Uses of beta radiation sources

Beta sources (such as phosphorus-32 or iodine-131) can be very useful when treating patients with cancer. The beta particles damage the cancer cells close to the radiation source without causing too much damage to healthy cells that are further away.

Beta particle sources are not very useful as medical tracers (see 'Uses of gamma radiation' below). Beta particles do not travel very far in human tissues, so they cannot easily be detected outside the body.

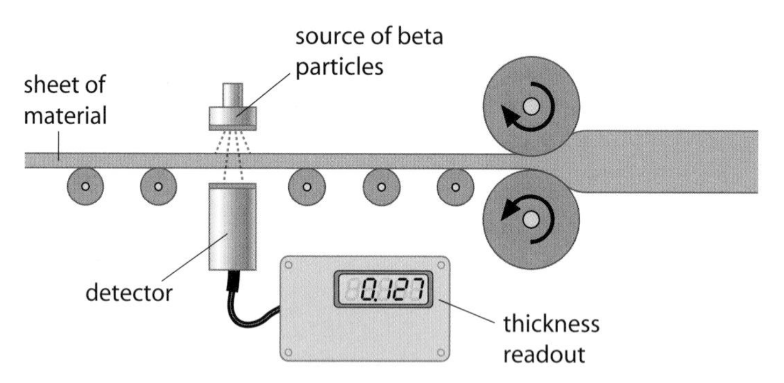

Figure 1 Using beta particles to measure the thickness of paper.

One important use is to measure the thickness of paper. Paper is often produced in a continuous roll, and manufacturers want to keep the thickness consistent. A device that physically touches the paper when measuring thickness risks tearing. However, if a beta particle source is placed just below the paper and a detector above, the thickness of the paper can be accurately measured by detecting how much beta radiation is absorbed by the paper. If the paper is too thick then the number of beta particles reaching the detector will drop; if it is too thin, the beta particle count will rise.

Scientists also use beta isotopes to study the flow of water in rivers. A small amount of a beta-particle-emitting isotope is placed in a water source and later the water in rivers is sampled to trace the path from the source. The amount of radiation is so small it does no damage to the environment.

Route to A*

As part of quality assurance, the engineers maintaining the equipment in a paper mill will calibrate the equipment that measures the paper thickness. Describe a method of calibration that could be used with the beta particle source.

Examiner feedback

Make sure you remember an example of an application of beta particles to use in answering examination questions.

Taking it further

Beta particles from different radioactive nuclei are emitted with different energies. This can mean that their range in air varies from just a few centimetres to several metres.

Uses of gamma radiation

Of the three types of ionising radiation, gamma rays are the most penetrating. This makes gamma sources very useful as medical **tracers**. A tracer is a substance that is injected into the body, and is detected from outside the body using a scanner of some type. Doctors can follow the progress of the tracer through the body, which can be very helpful in diagnosing some medical conditions.

Gamma rays can also be used to detect cracks in pipes. The source is placed in the centre of the pipe and the level of gamma radiation detected outside the pipe is recorded. If there is a crack, it will show up as a higher radiation count because there is less material to stop the gamma rays.

Figure 2 Gamma rays can be used to check for faults in pipelines.

Questions

1. Why are gamma ray sources more useful as tracers in humans than other types of ionising radiation?
2. Explain why alpha and beta particle sources would be useless for finding cracks in pipes.
3. Explain why getting an alpha or a beta particle source inside you is more dangerous than inhaling a gamma source.
4. Iodine-131 reduces in radioactivity after just a few weeks. Why does this make it useful for treating human cancer cells?
5. A student asked a question in class: 'If beta particles aren't stopped by paper, how can they be used to measure the thickness of paper?' Write a description of how beta particles are used to measure paper thickness that answers this question.
6. An ionising radiation detector contains a special gas. There is a window at the end of the detector that lets the radiation in so it can be detected. Why must the window be as thin as possible?
7. Cancer cells tend to accumulate phosphorus more than ordinary cells. Why is this useful in identifying cancers?
8. The thyroid gland concentrates iodine from the bloodstream. If radioactive iodine-131 is swallowed then it can kill unwanted cells in the thyroid gland. Why doesn't the radiation harm the cells in the blood?

Examiner feedback

Gamma radiation is quite hard to detect because it can go straight through the detector. Alpha radiation tends to be shielded by almost any matter so again it is hard to detect. Beta radiation is more penetrating than alpha so it can be easily detected when in a material such as water.

Science in action

Gamma rays are used to irradiate food. This kills all of the bacteria in the food so that it remains fresh in its packaging.

Route to A*

Gamma rays can be used in surgery. Beams of gamma rays are directed from different angles and meet at the place where the doctors want to kill cells. Because the gamma rays come from several different directions, each individual beam does not cause much damage to the healthy tissues but where they meet, the increased gamma ray dose kills the cells.

Taking it further

The reason that gamma rays only ever ionise one atom is that gamma rays are photons of energy, and this energy can only be transferred all at once.

Half-life

Learning objectives

- explain the term half-life
- calculate the activity of a radioactive source after given amounts of time
- explain why different half-lives make different sources suitable for different applications.

Examiner feedback

You need to understand that the half-life is the time taken for half of the atomic nuclei to decay, but as this is a random process no one can tell *which* nuclei will decay.

Radioactive decay

When the nucleus of an atom or a radioactive isotope gives out a radioactive particle it is said to have decayed. Radioactive decay is a random process. We cannot predict whether one particular nucleus will give out radiation in the next 5 minutes or in 5000 years. However, if we look at many millions of atoms, it is possible to work out an average rate of decay for a particular element or isotope. Some radioactive atoms decay quickly, while others decay much more slowly. One way of comparing the rate of decay of different radioactive materials is through their **half-lives**.

Half-life

The half-life of a radioactive isotope is the average time it takes for the number of nuclei in a sample to halve, or the time it takes for the count rate from a sample containing the isotope to fall to half its initial level. Imagine you have a box of 64 chocolates. If you eat 32 chocolates over the first 2 days, 16 chocolates over the next 2 days, eight chocolates over the next 2 days and so on, you could say that the half-life of the chocolates was 2 days, because every 2 days the number of chocolates left in the box is halved.

Science skills

Figure 1 shows the decay of a radioactive material with a short half-life.

a What would the radiation level be after 40 hours?

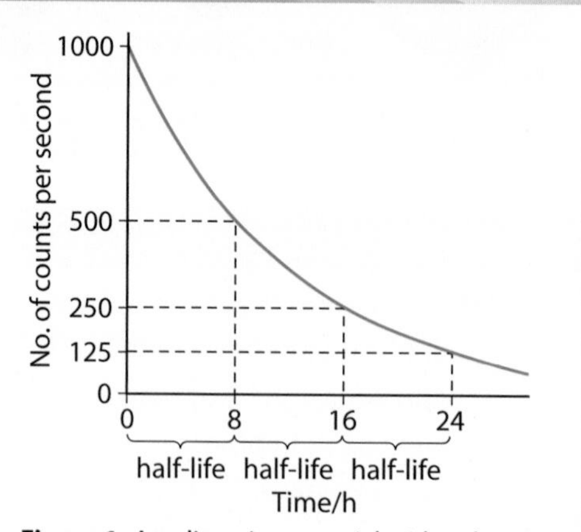

Figure 1 A radioactive material with a short half-life.

For a radioactive material, the half-life is the time it takes for half of the atoms in a given sample to decay, or for the activity of a source to halve. For example, sodium-24 has a half-life of 15 hours. If a sample of sodium-24 gave out 100 particles a minute when you started using it, after 15 hours this would have dropped to 50 particles a minute. After a further 15 hours the activity would be 25 particles per minute, and so on. After each 15-hour period, the activity halves.

Uranium-238 has a half-life of 4 468 000 000 years, quite a lot longer than sodium-24. It takes nearly 4.5 billion years for the activity of uranium-238 to be halved. This is such a long time that for most purposes the activity of uranium-238 can be considered to be constant.

Table 1 The half-lives of some radioactive substances.

Isotope	uranium-238	thorium-234	protactinium-234	uranium-234	thorium-226	radium-226	radon-222
Half-life	4.5 billion years	24.5 days	1.14 minutes	233 thousand years	83 thousand years	1590 years	3.825 days
Particle emitted	alpha	beta	beta	alpha	alpha	alpha	alpha

Using half-lives

The rate at which radioactive elements decay can be used as a way of working out the ages of things. Isotopes with very long half-lives are used to work out the age of the oldest rocks. Scientists can compare the amount of original isotope with the amount of the elements that it decays into, to work out how many half-lives have passed. The Earth is dated at around 4.6 billion years old, a little more than the half-life of uranium-238.

Carbon-14 is a useful element for calculating the age of things that were once alive. Carbon-14 has a half-life of about 5730 years. Scientists have found that the carbon dioxide in the atmosphere contains a fairly constant percentage of carbon-14. Plants get their carbon from carbon dioxide in the air, while other living things get their carbon directly or indirectly from plants, so living things all contain a similar percentage of carbon-14. Of course, when something dies, the carbon-14 it contains will slowly reduce because it is not being refreshed from the air. Scientists can use this decay to calculate the time when any living thing died, up to an age of about 60 000 years.

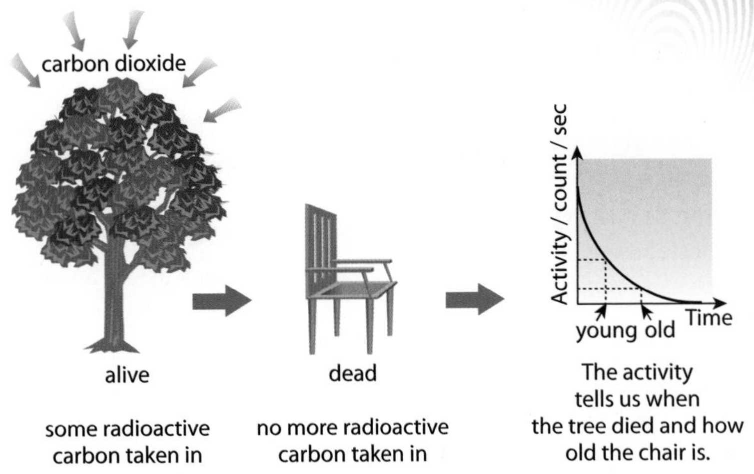

Figure 2 Dating once-living material using carbon-14 dating.

Questions

1. If there were 5 000 000 atoms of a radioactive element in a sample, how many would there be left after one, two and three half-lives?
2. If the half-life of a sample was 20 minutes, and the activity was 8000 counts per minute, what would the activity be after 2 hours?
3. Draw a graph of the activity of a source that has a half-life of 20 minutes. Start the activity at 640 counts per second.
4. Explain why scientists have to take background radiation into account when finding out the age of a piece of cotton used to wrap a mummy.
5. Explain why the production of radon gas from rocks is fairly constant.

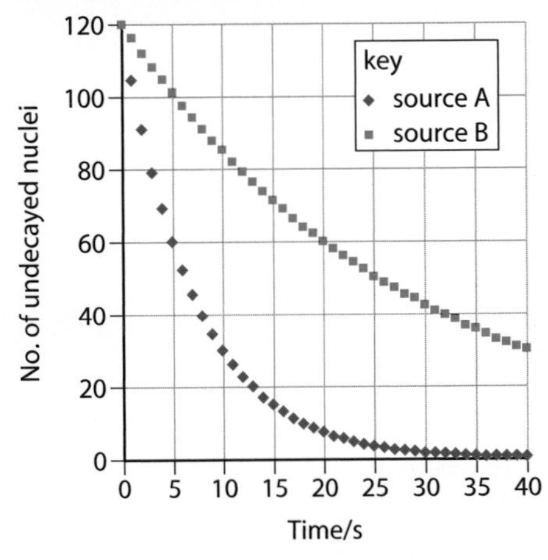

Figure 3 Radioactive decay of two substances.

6. From Figure 3, work out the half-lives of samples A and B.
7. Explain why radiocarbon dating using carbon-14 would not be very useful for finding out how long ago a tree died if it was known to be alive 5 years ago.
8. Describe how scientists must change the way they take measurements when trying to calculate the half-life of a very small sample of radioactive material.
9. Use the data from Table 2 to plot a graph and find the half-life of the sample under investigation. You can assume the background count is 25 per minute.

Table 2 Data from a radioactive source.

Time/days	No. of counts in 10 minutes
0	768
1	647
2	517
3	420
4	368
5	335
6	310
7	297
8	285
9	276
10	268

Hazards and safety in the radioactive environment

Learning objectives

- describe the hazards associated with working in an environment where there is ionising radiation
- explain what steps are taken to reduce these hazards.

Radiation is all around us

We are all exposed to background radiation from a variety of sources (see lesson P2 5.2), but it is advisable to keep our exposure to ionising radiation to a minimum.

Nuclear waste

Nuclear waste is an inevitable result of using nuclear fission to generate electricity. High-level nuclear waste is highly radioactive, and some of it has a long half-life. Governments are still working out the best way to store nuclear waste. All the options have their drawbacks: for example, if it is buried deep under the Earth, nuclear waste might eventually contaminate the water supply.

Some radioactive isotopes have very long half-lives and eventually decay into a gas. This means that storage of high-level radioactive waste has to be gas-tight for thousands of years. Security is also a problem: waste could be stolen and then used to try and blackmail governments.

Put it into deep boreholes in solid rock.

Dispose of it at sea in special containers.

Bury it under the seabed.

Continue to store it above ground.

Launch it into space.

Dispose of it at the edges of tectonic plates in the Earth so it is dragged into the Earth's mantle.

Figure 1 How should we dispose of nuclear waste?

Reactor accidents

In 1986 the nuclear reactor at Chernobyl in the Ukraine was destroyed during a devastating accident. Radioactive materials from the reactor were spread around the surrounding area. Some materials were carried thousands of miles by the wind. It is always difficult to be certain about the effect that radioactivity has on people's health, but many scientists think that some people suffered severe health problems because they were exposed to increased levels of ionising radiation. One of the effects of too much radiation is an increase in the number of people suffering from cancers. Some estimates suggest that the Chernobyl disaster caused cancers in many millions of people.

Decontaminating buildings after the Chernobyl disaster.

Science skills

Some areas of Europe are still affected by radioactive materials from the Chernobyl disaster. In parts of Wales, for example, caesium produces high levels of radioactivity in the grass. One effect seen in the area is the increased risk of thyroid cancer in children.

Figure 2 shows the rate of radioactive decay for caesium.

a From Figure 2, work out the half-life of caesium.

b If caesium activity at time zero is 16 times higher than is safe for human health, how long will it take for caesium activity to reach a safe level?

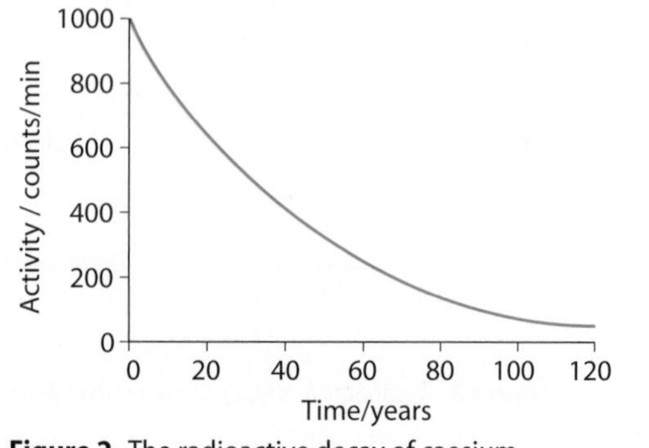

Figure 2 The radioactive decay of caesium.

How do we protect ourselves?

Wearing protective clothing can stop radioactive sources from getting onto or into your body. This does not prevent exposure to gamma radiation if radioactive sources are in the air or land on your clothing, but it does mean that radioactive materials can be washed off. The very best form of defence against ionising radiation is to keep as far away as possible from radioactive materials.

Working with radiation

People who work in nuclear power plants and other places where there are radioactive materials need to take special precautions.

Part of the danger is that it is very difficult for us to sense when ionising radiation is around us. It has no smell, we do not feel anything, and the effects are often not immediate. One solution is to wear radiation badges. Some of these contain photographic film that can be processed to check how much radiation a person has been exposed to. Clothing is also important. Workers may wear overalls that are checked frequently for contamination.

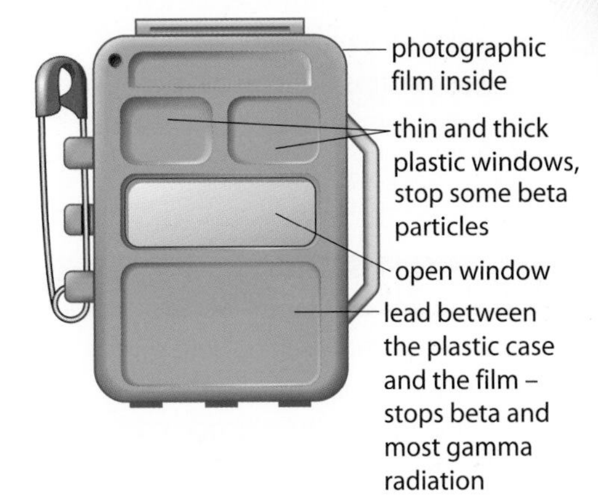

Figure 3 A radiation badge.

Healthcare professionals often work with ionising radiation. A patient having an X-ray or other treatment is only exposed to a small radiation dose. However, a technician, doctor or dentist who deals with this kind of treatment every day could easily get a high dose over just a few weeks. The higher the dose of radiation, the more risk there is of health problems. They therefore take special care to keep their exposure to a minimum. People taking X-rays, for example, operate the X-ray machine at a distance, from behind a wall or screen.

Examiner feedback

The exposure to ionising radiation from artificial sources is a small percentage of the total background radiation that people are exposed to. The fraction made up by fallout from nuclear weapons testing and accidental releases into the environment is also very small.

Dangers in perspective

It is important to understand that the dangers of radiation from medicine, industry and power generation are small compared with many other aspects of our modern lives, although many people are concerned that an event like Chernobyl could happen again.

Questions

1. Explain why distance is the best protection against ionising radiation.
2. X-rays are damaging to human tissues. Why can a patient be exposed to X-rays but a dentist leaves the room when they give you a dental X-ray?
3. Why would workers in a nuclear power plant wear radiation-sensing badges while at work even though they never directly enter the reactor?
4. Why is it impossible to be totally safe from ionising radiation?
5. Why do radiation dosage badges have a series of different screens covering the sensitive photographic plate?
6. Some radioactive waste will need to be stored for thousands of years. Explain why it has to be stored for so long.
7. A student made the comment: 'All radiation badges do is to tell you if you have had too much radiation. Surely by then it is too late, you have been exposed to the radiation and the damage is done.' Answer this comment, explaining why radiation badges are used.
8. Explain why it is difficult to be certain that a cancer diagnosed in someone from the Chernobyl area is a result of their exposure to ionising radiation from the accident.

P2 6.1 Nuclear fission

Learning objectives

- describe the process of nuclear fission
- list the most commonly used nuclear fuels
- describe how a nuclear reactor works
- evaluate the advantages and disadvantages of nuclear energy.

What is fission?

'Fission' means splitting. **Nuclear fission** is splitting the nucleus of an atom. The process is used to produce energy in nuclear power plants.

Only a few materials release energy when their nuclei are split. The material most commonly used in nuclear power plants is uranium-235. Figure 1 shows how fission occurs. The process starts when the nucleus of a uranium-235 atom absorbs a neutron. This makes the nucleus unstable. It splits into two parts and releases a number of neutrons. A large amount of energy is also released.

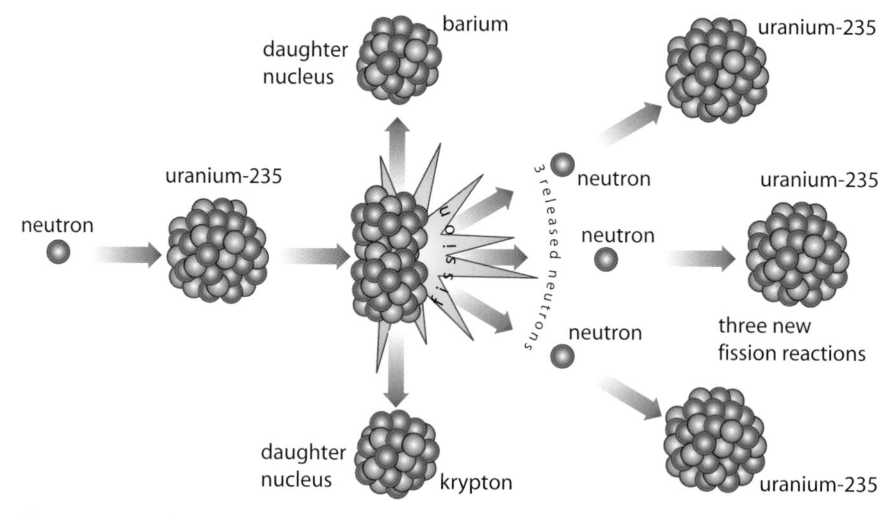

Figure 1 In the fission process uranium-235 splits to form barium and krypton. The neutrons released can go on to be absorbed by other uranium nuclei that then also split.

Taking it further

The energy that each nucleus gives out when it splits is measured in electron volts (eV). 1 electron volt is the same as 1.6×10^{-19} J.

Examiner feedback

It is worth making a special point of being able to draw a sketch that shows the fission process and the chain reaction.

Once fission has started, the process keeps itself going. Some of the neutrons that are produced by splitting nuclei go on to be absorbed by other uranium nuclei. The nuclei of these atoms also split, and give out more neutrons, which go on to split more nuclei. This process is called a **chain reaction**.

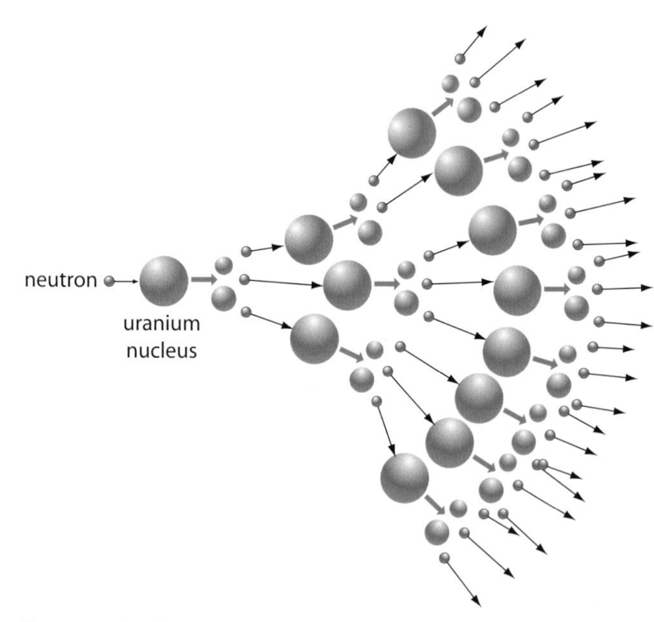

Figure 2 The chain reaction.

Nuclear fuels

Naturally occurring uranium contains less than 1 per cent uranium-235. The rest is almost all uranium-238. Uranium-238 is not suitable as a **nuclear fuel** because it does not undergo fission very easily. To make a useful nuclear fuel, engineers have to enrich the uranium by increasing the percentage of uranium-235. The enriched fuel contains 3–5 per cent uranium-235.

Plutonium-239 is also used as a nuclear fuel. Although it does not occur naturally, it is formed in nuclear reactors. Plutonium-239 offers an alternative to uranium, and some people see using plutonium as a way to get rid of some radioactive waste (see lesson C2 5.7). However, many scientists think it is a very dangerous fuel to use. Plutonium was used in the atomic bomb that was dropped on Nagasaki in 1945 and killed over 50 000 people.

Nuclear reactors

At the centre of a nuclear reactor is the core. It contains rods of enriched uranium fuel surrounded by a **moderator**. A moderator is a substance such as **heavy water** or graphite that slows down the neutrons given out by uranium-235. If the neutrons are not slowed, they are not captured by other uranium nuclei and the chain reaction breaks down.

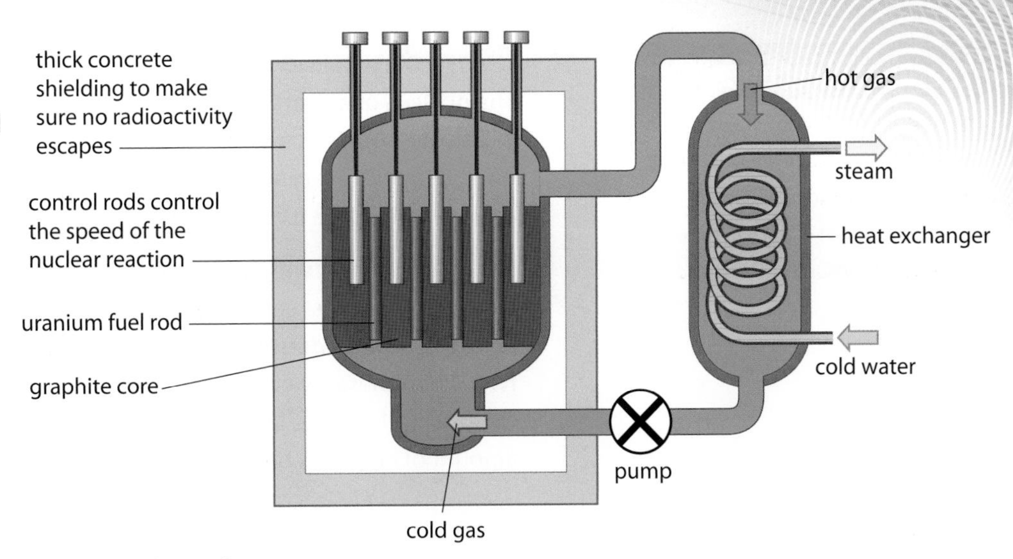

Figure 3 Inside a nuclear reactor.

The reactor core also contains **control rods** that control the speed of the chain reaction. The control rods are made of a material that absorbs neutrons. If they are pushed into the core, they slow the fission reaction. As they are pulled out, the rate of the chain reaction increases.

The energy that is released from the reactor is transferred to water in a heat exchanger. The water becomes steam, which can be used to drive a steam turbine and generate electricity.

Route to A*

One of the costs that is sometimes not taken into account when considering the cost of electricity production using nuclear fuels is the **decommissioning** costs. These can be very high indeed.

Pros and cons

Most electricity in the UK is generated by burning **fossil fuels**. There are only finite reserves of these fuels, and burning them produces carbon dioxide. Nuclear fission produces no carbon dioxide and known fuel reserves will last many years. However, nuclear power plants are expensive to build, and fission produces radioactive waste that must be stored safely for many years. Renewable energy sources produce little or no carbon dioxide, but they have other drawbacks, for example solar power can only be produced during the day.

Questions

1. What are the two fissionable materials used in nuclear reactors?
2. Why must naturally occurring uranium be 'enriched' before it can be used as a fuel?
3. Use a block diagram to describe the fission process for uranium-235.
4. Why is uranium-238 not suitable for use as a nuclear fuel?
5. Describe what is meant by the term 'chain reaction'.
6. Describe the main advantages and disadvantages of nuclear fission as an energy source compared with using fossil fuels or renewable energy sources.
7. A thousand billion uranium nuclei splitting would give about 32 J of energy. This is nowhere near enough energy to make a cup of tea. Explain why it is that 1 kg of uranium fuel can produce the same energy as thousands of tonnes of coal.
8. Explain why the uranium-235 inside a fuel rod undergoes fission by absorbing neutrons emitted from other fuel rods rather than by absorbing neutrons emitted within the same fuel rod. Describe how engineers have made use of this to be able to control the rate of the chain reaction in the reactor.

Nuclear fusion

Learning objectives

- describe the process of nuclear fusion
- explain how stars produce energy
- evaluate the advantages and disadvantages of nuclear fusion as an energy source.

Everyday fusion

The nucleus of any atom is positively charged. This means that any two nuclei will always repel each other. It takes a lot of energy to push two nuclei together, but it happens every second of every day in the heart of our Sun and the process has been going on for several billion years. This process is nuclear **fusion**.

Fission only works with heavy elements, but fusion usually involves much lighter elements. The most common fuel in fusion is hydrogen. Hydrogen has two heavier isotopes: **deuterium**, with one neutron in the nucleus, and **tritium**, with two. When these two isotopes are pushed together hard enough, the nuclei fuse together to form the heavier element helium. During this nuclear reaction, a vast amount of energy is released. There is also one neutron left over.

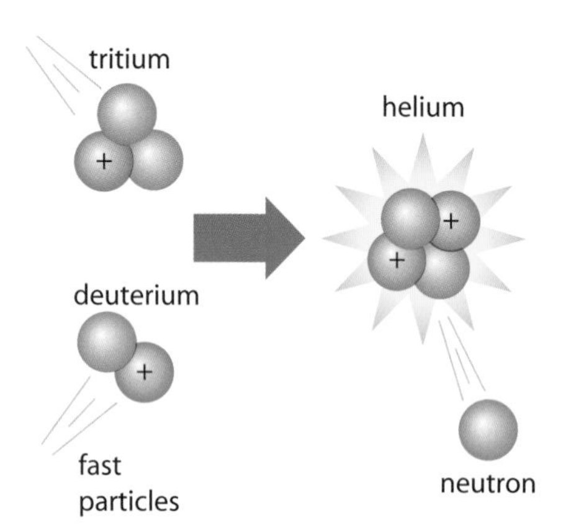

Figure 1 The fusion process.

Conditions in stars

As has already been mentioned, forcing nuclei together takes a huge amount of energy and would not normally happen. However, the conditions in the centre of the Sun and all other stars are ideal for fusion. Very high temperatures mean that the atoms are moving at extremely high speeds and colliding with high energy.

Gravitation is the force that starts fusion happening in a star, and contains it so that the reaction keeps going rather than flying apart in a vast explosion. The great mass of a star means that it has a very strong gravitational field. The gravitational force squashes the high-energy hydrogen atoms together. The combination of high temperature and great pressure creates conditions in which fusion can occur. The heat released during fusion keeps the reaction going.

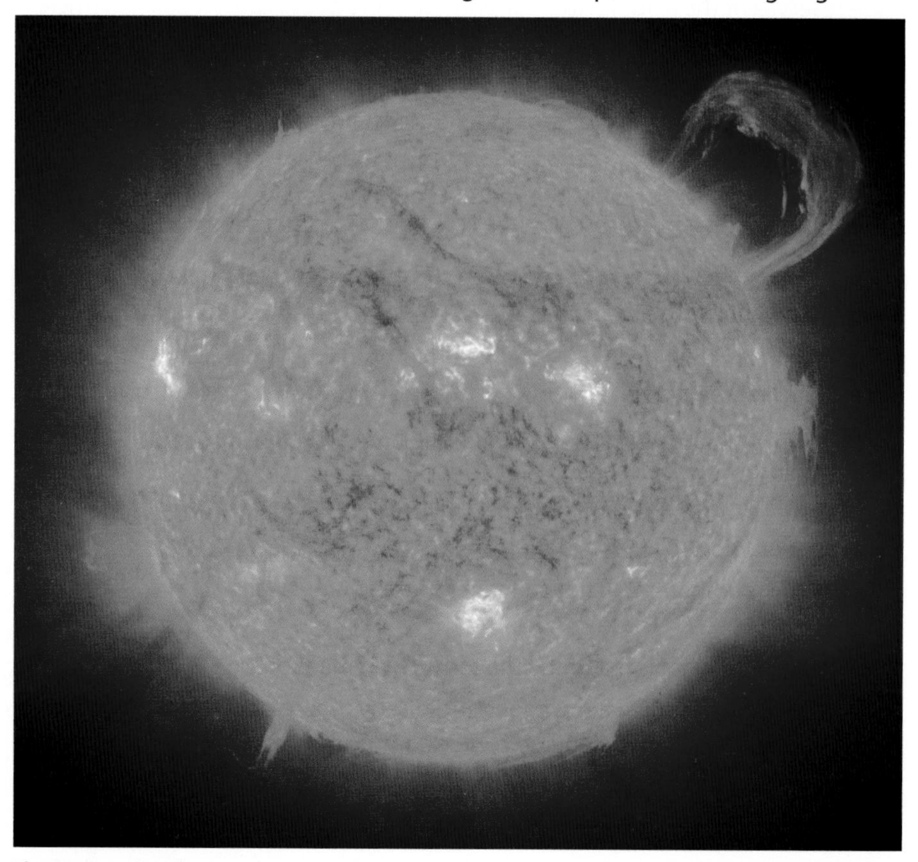

The Sun is a giant fusion reactor.

Examiner feedback

Many students get confused about the difference between fusion and fission. Make sure you are able to describe each process.

Taking it further

Scientists can use the famous equation $E = mc^2$ to calculate the amount of mass that is converted to energy during nuclear fission and fusion processes. 1 mg of matter is converted into nearly 10^{11} J of energy.

More about our star

The Sun has a mass of about 2×10^{30} kg. Most of this is hydrogen. During its life, the Sun will convert about 0.7 per cent of its mass into energy, at a rate of 4 million tonnes each second. The fusion of hydrogen in the core of the Sun generates huge amounts of energy. This energy heats the centre of the Sun and is transferred outwards. It can take 100 000 years or more for this energy to reach the surface. Because space is mostly a vacuum, the only way for energy to leave the surface of the Sun is through electromagnetic radiation, including visible light. Light from the Sun takes a little over 8 minutes to get to the Earth.

Route to A*

When deuterium and tritium fuse together, an alpha particle and a neutron, both of high energy, are produced. It is this energy that can be used to heat water and so produce steam for electricity generation.

Fusion reactors

If we could build fusion reactors here on Earth, they would supply huge amounts of energy from tiny amounts of fuel. The hydrogen fuel could be made from seawater, and the waste product is helium, which is harmless.

Scientists have been working on fusion reactors for over 50 years. To get fusion to work, the hydrogen must be heated to roughly 100 million °C. This is nearly 10 times hotter than the heart of the Sun. This has been achieved in experimental reactors, but for many years these reactors produced less energy than was needed to get them going. Larger-scale test reactors are now being built, but the costs are enormous.

Taking it further

In the Sun's core, helium is formed in a complicated process. To form helium, two protons fuse together to form deuterium – hydrogen with one neutron – and give out other particles. The deuterium atom and a proton then fuse to form a helium-3 atom as well as a gamma ray. If two helium-3 atoms fuse, they form a helium-4 atom and two protons.

Science in action

The high temperatures needed to make nuclear fusion happen would melt the walls of any container. Two promising solutions to the problem are being developed. One is to use strong magnetic fields to hold the super-heated hydrogen in place inside a doughnut-shaped container called a tokamak. The other is to fire an array of lasers at a tiny fuel pellet. The intense laser beam energy is used to make the fuel pellet implode. The energy released from the fusion of the fuel pellet is much larger than the energy needed to produce the laser beams.

Questions

1. What force keeps the fusion reaction in place inside the Sun?
2. Describe the difference between nuclear fusion and nuclear fission.
3. Why is it so hard to generate electricity on Earth using nuclear fusion?
4. What are the main advantages and disadvantages of fusion compared with other energy sources?
5. Why is it so difficult to get nuclei to be close to each other?
6. Describe the conditions in the centre of the Sun or a star that are needed to keep a nuclear fusion reaction going and explain why each of them is needed.
7. High-speed neutrons can turn materials that are non-radioactive into radioactive isotopes. Why might this be a drawback for fusion power?
8. Describe the arguments that an engineer working on future fusion power might use to persuade a government that they should put large amounts of money into fusion power research.

A*

Science in action

There are two types of atomic weapon. In an atom bomb the explosion is produced by suddenly putting enough plutonium in one place to begin a fission chain reaction. The hydrogen bomb is a fusion device, but an atom bomb has first to be exploded to get the extremely high temperatures needed for fusion to start.

The lives of stars

Learning objectives

- describe the life cycle of a star
- explain what causes some stars to become white dwarves and others to go supernova.

An area of star formation in the Eagle Nebula.

A star is born

Stars form inside huge clouds of gas and dust known as nebulae. A concentration of gas and dust in one area slowly grows as more material is pulled in by gravitational forces. As the gas and dust gathers, it begins to compress and warm up. The gathering material forms a spinning disc of hot gas, called a **protostar**. Eventually it becomes so hot at the centre of this disc that nuclear fusion begins, and a true star is formed. Once the star is burning brightly, it blows off the outer layers of gas and dust.

Science in action

The Spitzer telescope is one of four major NASA space telescopes. Its specially cooled detectors are designed to detect **infrared** radiation. The Spitzer telescope can 'see' objects buried deep in dust clouds that cannot be seen using visible light, for instance young stars starting life in 'stellar nurseries'.

Stars are stable for most of their lives

Once a star has formed, it gradually burns up its hydrogen fuel, turning it into helium. The fusion at the star's core produces enormous amounts of energy that radiate outwards, trying to tear the star apart. The star's massive gravity field works in the opposite direction, trying to collapse the star in on itself. The inward force of gravity and the outward pressure caused by the fusion reaction balance each other. This balance of forces keeps the star stable for millions of years.

A star in this phase of its life is called a main sequence star. The Sun is about halfway through its **main sequence** phase, which should last another 5 billion years or so.

The fuel runs out

Once a star has used up its hydrogen fuel, it is close to the end of its life. Fusion continues for a time using helium and other elements (see lesson P2 6.4). The star swells up to many times its former size, becoming a **red giant** or a **supergiant**. This stage can last tens of thousands of years.

Eventually all the star's fuel is used up, and fusion stops. Without the outward pressure from fusion, the core of the star collapses. What happens next depends on how massive the star is.

Taking it further

A light year is a measure of distance, and is the distance travelled by light in a year. Astronomers also use the astronomical unit (AU): the distance from the Earth to the Sun.

If light travels at about 300 000 000 m/s, approximately how far is one light year?

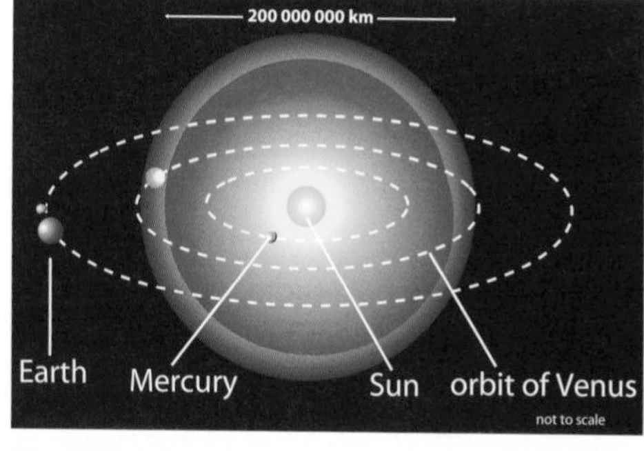

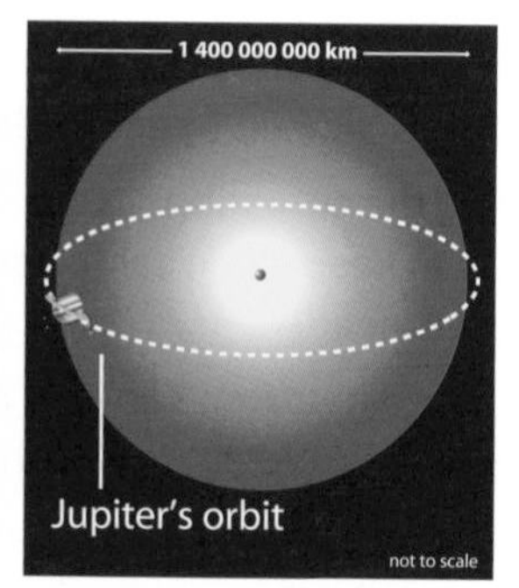

Figure 1 A red giant could reach almost to Venus's orbit. A red supergiant would reach almost to Jupiter.

Different endings

In a star of about the size of our Sun or smaller, the core collapses to form a small, dim star called a white dwarf. The atoms in the white dwarf are very densely packed. Its gravity is about 100 000 times greater than the Earth's. Over billions of years a **white dwarf** cools to a dull cinder – a **black dwarf**.

The ending of a large star, more than 1.5 times the mass of the Sun, is more dramatic. The core collapses even further, then rebounds in a massive explosion called a **supernova**. A supernova explosion can release as much energy in a few years as our Sun will release in its entire 10-billion-year life.

Although many stars have gone supernova, it is not something we see every day. Back in 1054 Chinese astronomers observed a very bright light that was visible during the daytime and lasted for well over a year. This supernova has now become the crab nebula.

Taking it further

The Crab Nebula contains the remnants of a supernova that left behind a neutron star. Neutron stars can be involved in dragging material from other stars and producing immensely powerful X-rays in the process.

Neutron stars and black holes

If a star is up to three times more massive than the Sun, it collapses to form a ball of matter called a **neutron star**. A neutron star is incredibly dense. It is about the size of the Earth, but its mass is greater than that of the Sun.

When a star more than three times bigger than the Sun explodes, it leaves behind a **black hole**. This is a small ball of matter even more dense than a neutron star. Its gravity is so strong that even light cannot escape from it.

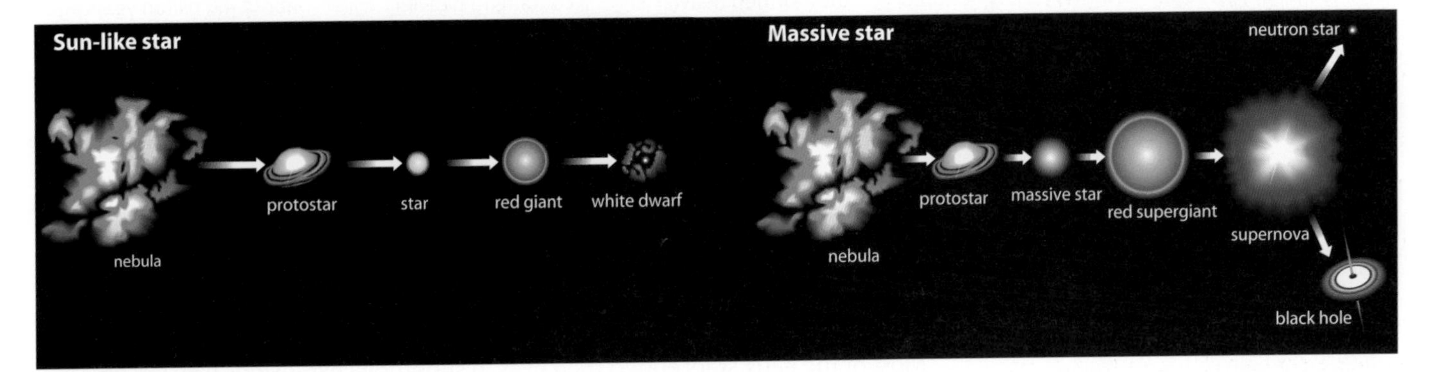

Figure 2 The lives of stars.

Questions

1. How big is an average-sized red supergiant star?
2. Why is it impossible to see black holes?
3. Why don't stars collapse under their own weight?
4. Describe what controls whether a star will end up as a neutron star or a white dwarf.
5. Explain how a white dwarf star can contain more mass than our Sun, but be much smaller.
6. What causes a star to start nuclear fusion?
7. Why do you think that the heavier a neutron star is, the smaller it is?
8. Explain how the two following comments can both be true.

   'A recent supernova was observed about 400 years ago.'

   'This supernova occurred over 20 000 years ago.'

P2 6.4 Where stars come from

Learning objectives

- describe how scientists think stars are formed
- explain where the heavier elements come from
- describe the process of planetary formation.

Once there was only hydrogen

The universe began about 14 billion years ago with a huge explosion called the **Big Bang**. Within seconds the universe appeared and for a short time it was too hot for any sort of matter to exist; the universe was pure radiation. After a few hundred thousand years, things had cooled enough for atoms to form. But only atoms of the simplest elements, mainly hydrogen, formed at this time.

The spread of hydrogen across the universe was fairly even at first. However, over millions of years some of it gathered into clouds under the pull of gravity. As they got bigger and denser, the gas clouds began to warm up. Eventually it became so hot at the centre of these clouds that nuclear fusion began, and the first stars were born.

Heavier elements

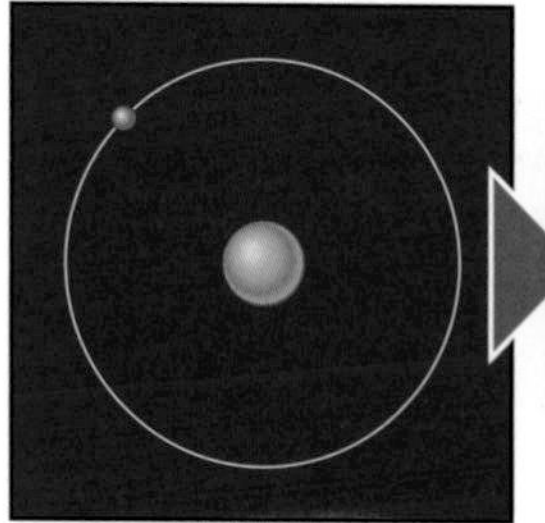

Hydrogen was formed over 13 billion years ago, about 400 000 years after the Big Bang.

The first stars were formed a few hundred million years after the Big Bang, purely from hydrogen.

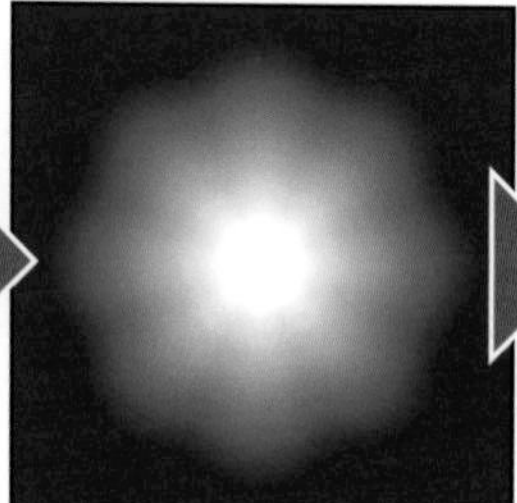

As the first stars got old they formed heavier elements and then exploded, scattering matter across space.

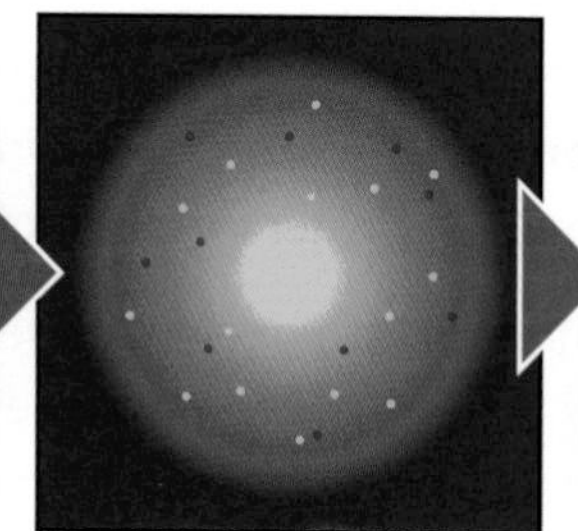

Newer stars formed containing heavier elements from older exploded stars.

Our Solar System formed about 4.6 billion years ago. The planets, asteroids and living things formed from the dust of exploded stars.

Figure 1 How the heavier elements in the universe formed.

All the elements heavier than lithium have been formed in stars. The first stars were mostly large and hot. They burned through their hydrogen fuel in a few million years. When it ran out, they started to fuse helium. Helium fusion can form carbon, oxygen and other heavier elements up to iron.

Most of the early stars ended their lives in supernova explosions (see lesson P2 6.3). In the immense heat and pressure of these explosions, the heaviest elements were formed. The force of the explosions spread the newly formed elements across space as clouds of gas and dust.

The process of forming heavier elements, which began with the first stars, continues today. All stars form the lighter elements by fusion. The heaviest element that stars form towards the end of their lives is iron. Heavier elements are produced in supernovas.

The Crab Nebula in the constellation of Taurus is the remnant of a supernova recorded by Chinese astronomers in 1054.

Science in action

The formation of the Moon cannot be observed because it happened billions of years ago, and it is obviously not possible to test out a theory by trying to form the Moon again. Scientists have used computer modelling to test out the theory that the Earth was hit by another planet-sized object, and the simulation results support the hypothesis.

Examiner feedback

You will need to have a good understanding of how heavier elements were formed, in particular that they were not formed during the Big Bang.

Forming planets

Some of the gas and dust that gathers to form a protostar may not become part of the star itself. It forms a disc of material around the young star. Over time, pieces of the matter clump together and eventually form planets, moons, asteroids and comets. The Earth was formed in this way around 4.5 billion years ago.

Artist's impression of the beginnings of our Solar System, with planets and asteroids forming around the young Sun.

Questions

1. What influence caused the first stars to form?
2. Why were early stars almost purely hydrogen?
3. Where did the carbon that we depend on for life come from?
4. Describe what will happen inside the Sun towards the end of its life that will create elements heavier than hydrogen.
5. Describe what must be happening to the percentage of hydrogen in the universe. Explain why you think this.
6. **(a)** Use the periodic table to list the elements that are formed by normal stars. **(b)** What is the name of the lightest element that cannot be formed by fusion in a normal star?
7. Explain why some scientists think that many of the early stars may have been more massive than our Sun.
8. The early stars formed from clouds of hydrogen. Explain why these early stars could not have had planets, and where the different atoms that make up our Solar System came from. Include examples of the formation of elements, including iron and gold.

A*

Science in action

Recent discoveries have changed our view about the formation of the Moon and the history of the Earth. Some scientists now think that the Earth was hit by another planet-sized object. A lot of rock was hurled into orbit around the Earth. It slowly clumped together to form the Moon.

Taking it further

It is no accident that the inner planets are small and rocky, while the outer ones are mostly gas. Close to the Sun, it was too hot for light gases, such as hydrogen and helium, to condense. Further away from the Sun, the newly-forming gas planets were able to 'sweep up' left over gas as they orbited the Sun.

ISA practice: speed and braking distance

A tyre manufacturer has produced a new type of tyre. Their scientists want to find out how far a car will travel from the brakes being applied to the car stopping, at different speeds, using the new tyre.

Section 1

1 Write a hypothesis about how speed affects braking distance. Use information from your knowledge of forces and braking to explain why you made this hypothesis. *(3 marks)*

2 Describe how you could carry out an investigation into this factor.

You should include:

- the equipment that you could use
- how you would use the equipment
- the measurements that you would make
- a risk assessment
- how you would make it a fair test.

You may include a labelled diagram to help you to explain the method.

In this question you will be assessed on using good English, organising information clearly and using specialist terms where appropriate. *(9 marks)*

3 Design a table that you could use to record all the data you would obtain during the planned investigation. *(2 marks)*

Total for Section 1: 14 marks

Section 2

Two students, Study Group 1, carried out an investigation into the hypothesis. Figure 1 shows their results.

Speed 32 km/h: took 5 m to stop
Speed 48 km/h: took 11 m to stop
Speed 64 km/h: took 21 m to stop
Speed 96 km/h: took 51 m to stop
Speed 112 km/h: took 71 m to stop

Figure 1 Study Group 1's results.

4 **(a)** Plot a graph of these results. *(4 marks)*

(b) What conclusion can you make from the investigation about a link between the braking distance and the speed of the car? You should use any pattern that you can see in the results to support your conclusion. *(3 marks)*

(c) Look at your hypothesis, the answer to question 1. Do Study Group 1's results support the hypothesis?

Explain your answer. You should quote some figures from the data in your explanation. *(3 marks)*

Below are the results from three more study groups.

Figure 2 shows the results from Study Group 2, two other students investigating the hypothesis.

Speed 32 km/h: took 3 m to stop
Speed 48 km/h: took 7 m to stop
Speed 64 km/h: took 17 m to stop
Speed 96 km/h: took 35 m to stop
Speed 112 km/h: took 55 m to stop

Figure 2 Study Group 2's results.

Study Group 3 carried out independent research into average stopping and braking distances for the ten most popular tyres by sales volume. Table 1 shows their results.

Table 1 Average stopping and braking distances for the top ten tyres at five different speeds.

Speed / km/h	Average stopping distance/m	Average braking distance/m
32	12	6
48	23	14
64	36	24
96	73	55
112	96	75

Figure 3, produced by Study Group 3, compares the braking distance of the new tyre with the average braking distance for the ten most popular tyres by sales volume.

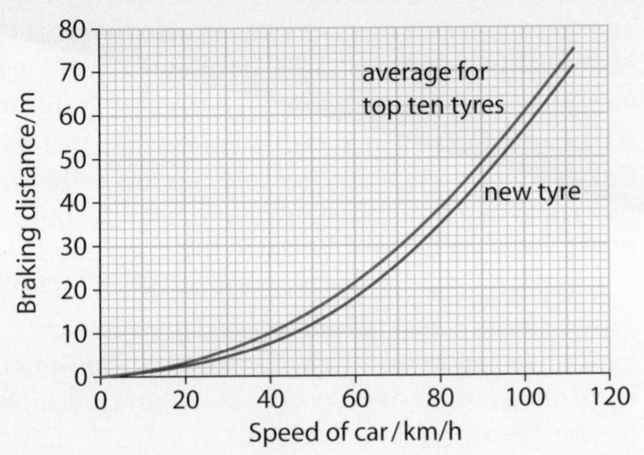

Figure 3 Graph of braking distance versus speed for the top ten tyres and for the new tyre.

Table 2 gives the results of Study Group 4, another group of students who investigated the hypothesis.

Table 2 Results of Study Group 4.

Speed / km/h	Braking distance/m			
	Test 1	Test 2	Test 3	Mean of tests
32	6	8	6	7
48	15	12	15	14
64	23	28	24	25
96	61	56	57	58
112	82	75	80	79

5 **(a)** Draw a sketch graph of the results from Study Group 2. *(3 marks)*

(b) Look at the results from Study Groups 2 and 3. Does the data support the conclusion you reached about the investigation in question 5(b)? Give reasons for your answer. *(3 marks)*

(c) The data contain only a limited amount of information. What other information or data would you need in order to be more certain whether the hypothesis is correct or not?
Explain the reason for your answer. *(3 marks)*

(d) Look at the results from Study Group 4. Compare the data from Study Group 1 with Study Group 4's data. Explain how far the data shown supports or does not support your answer to question 5(b).

You should use examples from Study Group 4's results and from Study Group 1. *(3 marks)*

6 **(a)** Compare the results of Study Group 1 with Study Group 2. Do you think that the results for Study Group 1 are *reproducible*?
Explain the reason for your answer. *(3 marks)*

(b) Explain how Study Group 1 could use results from other groups in the class to obtain a more *accurate* answer. *(3 marks)*

7 Applying the results of the investigation to a context.

The new tyre manufacturer wants to claim that their tyres are safer than the leading competitor's.
Suggest how ideas from the original investigation and the other studies could be used by the manufacturer support this claim. *(3 marks)*

Total for Section 2: 31 marks

Total for the ISA: 45 marks

Assess yourself questions

1 Describe the difference between fusion and fission. *(2 marks)*

2 Explain why alpha particles do not travel very far through air. *(3 marks)*

3 Explain the difference between the nuclei of carbon-14, carbon-12 and nitrogen-14. *(3 marks)*

4 What was it about the alpha-scattering experiment that made Rutherford come up with a new model for the structure of the atom? How was it different to the old model? *(3 marks)*

5 In an experiment, a radioactive source was directed towards a Geiger counter. The source was about 1 cm away from the counter. The following measurements were taken:

- nothing in the way – 500 counts per minute
- paper in the way – 510 counts per minute
- aluminium foil in the way – 11 counts per minute
- thick lead in the way – 9 counts per minute.

What sort of ionising radiation was this? Explain your reasoning. Why did the lead not reduce the counts to zero? *(4 marks)*

6 If the activity of a radioactive source is 5000 counts per minute and it has a half-life of 20 minutes, sketch a graph that shows its activity over the next 2 hours. *(5 marks)*

7 Uranium-238 emits alpha particles; this forms element X. Element X emits beta particles; this forms element Y. Element Y emits beta particles; this forms element Z. Element Z emits alpha particles. Use the periodic table to work out what elements X, Y and Z are and what Z turns into. Justify your answers. *(4 marks)*

8 Uranium-235 emits alpha radiation, and the isotope this forms emits beta particles. What elements are formed in these two decay activities? *(2 marks)*

9

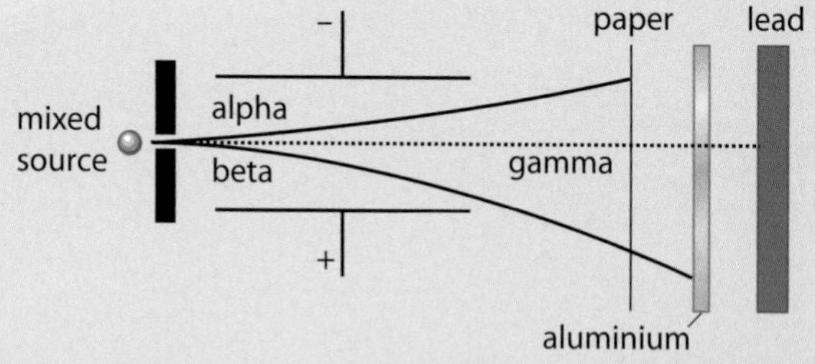

Figure 1 How different types of radiation are deflected by an electric field.

Look at Figure 1. Describe why the different particles are deflected in different directions and gamma is not deflected at all.

In this question you will be assessed on using good English, organising information clearly and using specialist terms where appropriate. *(6 marks)*

10 Explain why, although beta particles have only half as much charge as alpha particles, they are deflected more in a magnetic field. *(2 marks)*

11 An experiment in a nuclear laboratory found a particle that had the following properties:

- The particle was deflected by a magnetic field.
- The particle could not penetrate a sheet of paper.

Use your knowledge of the properties of radioactive particles to identify it. *(1 mark)*

12 An alpha particle is two neutrons and two protons, moving at very high speed. It has a lot of kinetic energy. As the alpha particle ionises atoms, it loses energy. What two things will happen to an alpha particle when it loses all of its kinetic energy and captures two electrons? *(2 marks)*

13 When some radioactive elements emit beta particles these particles have much more energy than beta particles from other types of elements. As a beta particle passes through air, it ionises molecules, losing kinetic energy each time it does so. Why do you think that the high-energy beta particles will have a greater range in air than low-energy ones? *(2 marks)*

14 A trainee technician in a science lab was told that the Geiger counter had a very delicate window at the end that could easily be damaged. To protect it, he put a plastic cup over the end. Why was this not a good idea? *(2 marks)*

15 Many science fiction films have used the idea that ionising radiation can cause animals to change into huge monsters. What properties of ionising radiation are the writers basing these ideas on? *(2 marks)*

16 Radium used to be used to make the hands of a watch glow in the dark, so that you could tell the time at night. These days, radium is banned for use in this sort of application. A replacement, tritium, is used instead. Tritium emits low-energy beta particles. Why would low-energy beta radiation be considered safe when used in instruments? *(3 marks)*

17 It is often stated that no changes to physical conditions can affect the half-life of a radioisotope. Why is this? *(2 marks)*

18 When measuring the half-life of very long-lived radioactive elements, it is difficult to make an accurate statement about the actual half-life of the material. Why is this? *(2 marks)*

19 Why do workers in nuclear power plants wear radiation badges? *(1 mark)*

20 Uranium is usually enriched using a centrifuge. The process partly separates the two isotopes by using the difference in density between them. Explain why it is impossible to separate the two isotopes chemically. *(2 marks)*

21 Why is uranium-235, which is only a small proportion of naturally occurring uranium, a suitable fuel for an atomic fission reactor, whilst the more common uranium-238 is not? *(1 mark)*

22 If you look at Figure 3 on page 155 you will see that a gas is circulated over the core of a nuclear reactor to transfer the heat to a heat exchanger. The heat exchanger then heats water to create steam that is sent to a turbine. Why have the designers not simply pumped the water straight over the core to create steam? *(2 marks)*

23 Control rods are used in nuclear reactors to control the rate of energy production.

(a) What do the control rods do as neutrons pass through them? *(1 mark)*

(b) How does this change the energy output of the radioactive core? *(1 mark)*

24 Describe the fusion process that scientists are trying to use on Earth to produce fusion power. *(4 marks)*

25 Nuclear fusion could solve all of our energy problems as well as reducing global warming. What science is there to support this statement? *(3 marks)*

26 Fusion power uses raw materials from the sea to produce electricity. This should surely mean that the energy it produces would have almost no cost. Why is this idea false? *(2 marks)*

27 A fusion reactor on Earth would need to run at a higher temperature than in the centre of the Sun. Why might this be? *(2 marks)*

28 What sort of stars become neutron stars? *(1 mark)*

29 In what stage of a star's life are very heavy elements formed? *(1 mark)*

30 Explain where the elements that make up our bodies came from. *(1 mark)*

31 At what point will a star start to turn into a red giant? *(1 mark)*

32 A white dwarf may have the same mass as our Sun but is much smaller because it has collapsed under its own gravitational pull. Why is the Sun larger than a white dwarf? *(2 marks)*

33 Some students researched the properties of stars and produced Table 1. Use their data to put the stars in order of their brightness as seen from the Earth. Note that how bright a star looks depends on its actual luminosity (brightness) and the square of its distance from the observer.

$$\text{apparent brightness} = \frac{\text{luminosity}}{\text{distance squared}}$$

Table 1 The students' data.

Name	Distance/light years	Luminosity
Alpha centauri A	4.4	1.5
Arcturus	37	215
Procyon	11	7
Rigel	770	57 000
Sirius	8.6	22

(5 marks)

34 A student researching the Sun found that the process of fusion converts mass into energy. The Sun is converting about 4000 million tonnes of mass into energy each second. They made the following comment to a science teacher: 'If the Sun is losing 4000 million tonnes each second, at that rate surely the Sun cannot last for long?' The science teacher disagreed. Why? *(2 marks)*

35 The Sun has a lifespan of about 10 billion years.

Look at Table 2 and describe what it shows about the lifespan of stars compared to their mass.

Table 2 Data on stars.

Mass of star/solar masses	Life expectancy/years
60	3 million
30	11 million
10	30 million
3	400 million
1	10 billion

(2 marks)

36 Betelgeuse is a red supergiant star with a mass many times that of the Sun. Although it is reaching the end of its life, it may well carry on for thousands of years yet.

(a) Explain why it is not fusing hydrogen to produce energy. *(2 marks)*

(b) What is likely to happen to Betelgeuse in the future? *(2 marks)*

(c) Why is it unlikely that we personally will be able to confirm this prediction? *(2 marks)*

37 A supernova is incredibly bright. Why is it very short-lived ? *(2 marks)*

38 If the Sun has a mass of 2×10^{30} kg, how much mass must a star have to end its days as a black hole? *(2 marks)*

39 A science teacher set a student the task of checking out the maths of the Sun's fusion processes. She gave her pupil the following data.

Mass of sun: 2×10^{30} kg

Mass to energy conversion rate: 4000 million tonnes/s

Percentage of Sun's mass that can be converted into energy: about 0.5%

At this conversion rate, how long will it be before the Sun runs out of fuel? *(6 marks)*

40 Many people have a great fear of the radioactive materials that we have put into the environment. Describe the science that can be used to explain why this fear might be justified and also why some of these fears might be misplaced. *(4 marks)*

GradeStudio Route to A*

Here are three students' answers to the following question:

As part of the safety monitoring in a laboratory that tests samples of radioactive ores, the scientists are required to check the equipment used for the tests for contamination. Describe how they might do this and explain why, although equipment can be exposed to high levels of radiation, it will not normally become radioactive itself.

In this question you will be assessed on using good English, organising information clearly and using specialist terms where appropriate. *(6 marks)*

Read the answers together with the examiner comments. Then check what you have learned and try putting it into practice in any further questions you answer.

B Grade answer

Student 1

You could check the equipment by using a radiation counter to see if it counted quickly. Radiation does not make equipment radioactive. Need to put the counter close to the equipment to count all the radiation given off.

It is better to use technical vocabulary such as 'Geiger counter' if possible.

The student fails to go on to explain what does contaminate the equipment. Contamination could occur if some of the radioactive material itself was to rub off onto the equipment.

It is possible the student knows why the counter should be held close to the equipment, but they do not mention that alpha particles have a short range in air.

Examiner comment

The student has a reasonable understanding of ionising radiation but has failed to take the opportunities to explain what contamination is and how it could happen. The comment about holding the counter close to the equipment may imply knowledge of the range of alpha particles in air but the student needs to be explicit. They have not demonstrated higher levels of understanding by linking statements and justifying them. They have failed entirely to mention background radiation and the part this plays in taking measurements.

- Read the question carefully.
- Put the key points into a logical sequence.
- Make sure that you use scientific understanding to support your comments. For example, make sure you demonstrate an understanding of background radiation, the difference between alpha, beta and gamma radiation and what contamination means.
- Make sure that links between parts of the answer are clearly made.
- Ensure you justify the statements you make.
- Be full in your explanation and extend explanations by using words and phrases like 'because' or 'so that'.
- Use connectives such as 'and' or 'this means' to make the sentences flow.

Grade answer

Student 2

Good use of correct terminology.

The different types of radiation that could be detected are mentioned but the student fails to explain what effect this has on the use of the counter, i.e. the counter would need to be very close to the equipment to detect alpha radiation, as it has a very short range in air.

The scientists would need to take background radiation counts so that the radiation count is accurate. The source could have fallen over and touched the equipment. The scientists could use a Geiger counter to check for unwanted levels of radiation. They should make sure to check for alpha as well as beta and gamma radiation. A piece of radioactive ore may have stuck to the surface. Background radiation is a bit random in the way it behaves so the scientists would need to be careful here.

The student describes a possible route of contamination but does not explain that this is about contamination.

Examiner comment

The student obviously has covered the main parts of the question but the answer is clumsy and disjointed. There is some use of correct vocabulary. There is evidence of understanding of how ionising radiation behaves and of the effects of background counts. However, the student fails to explain why the counter should be held close to the equipment: so that alpha particles that have a short range in air can be detected.

Grade answer

Student 3

Good application of knowledge about the short range of alpha particles in air.

To test for contamination of the equipment, scientists would need to use an ionising radiation detector such as a Geiger counter to check if any ionising radiation was being given off from the equipment. Background radiation levels would need to be taken into account. The counter would have to be held very close to the equipment to pick up alpha-particle emissions, because alpha particles have a very small range in air. Contamination could be caused if the source material itself came into contact with the equipment and some of it was left on the surface. The equipment would not become contaminated simply through exposure to ionising radiation as it is the radioactive isotope that emits radiation.

Examiner comment

This candidate has covered all the main points.

Contamination is when the radioactive source gets onto something. The use of equipment is clear. The student's understanding of alpha particle range is clearly linked to the way the equipment must be used. Background radiation is effectively referred to.

Examination-style questions

1 The figure below shows a car travelling along a level road. It also shows the two horizontal forces, P and Q, acting on the car. The total mass of the car and its driver is 2000 kg.

(a) At one point in the journey, force P is 3000 N and force Q is 2000 N. Calculate the acceleration, in m/s^2, of the car at this point in the journey. Write down the equation you use. Show clearly how you work out your answer. *(2 marks)*

(b) The car will eventually reach a constant speed. Explain why.
In this question you will be assessed on using good English, organising information clearly and using specialist terms where appropriate. *(6 marks)*

2 The figure below shows two cars, X and Y, on a roller coaster.

Car Y has a total mass of 300 kg and is stationary.
Car X has a total mass of 500 kg. It moves from rest down the track and collides with car Y.
The vertical height of the track is 5 m.

(a) **(i)** What energy transfer takes place as car X travels down the track? *(1 mark)*

(ii) Calculate the speed of car X, in m/s, just before it hits car Y. Assume that the track is frictionless. Write down the equation you use. Show clearly how you work out your answer. The gravitational field strength is 10 N/kg. *(4 marks)*

(b) Calculate the momentum of car X, in kg m/s, just before it hits car Y. Write down the equation you use. Show clearly how you work out your answer. *(2 marks)*

(c) The cars collide, stick together and move forward along the track. Calculate the speed of the cars along the track.
Write down the equation you use. Show clearly how you work out your answer. *(2 marks)*

3 The photos show an electric kettle and an electric hairdryer.

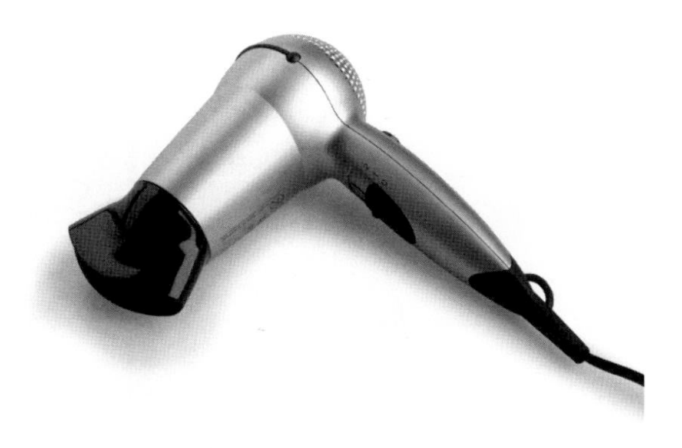

(a) Why does the kettle need to be earthed but not the hairdryer? *(1 mark)*

(b) Explain how earthing the kettle protects the user. *(4 marks)*

4 The graphs show show how the current through three different components varies with the potential difference across them.

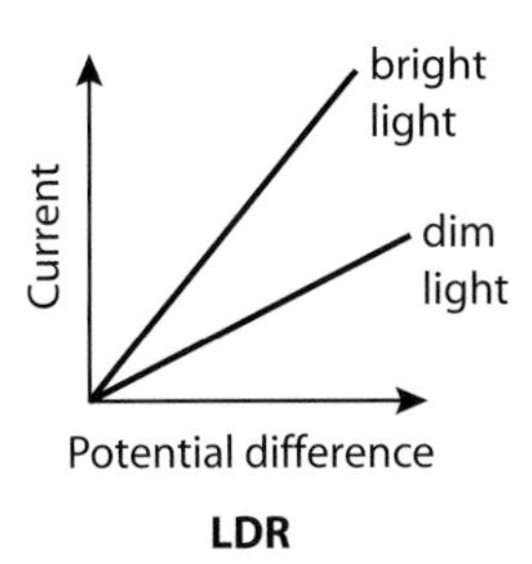

LDR

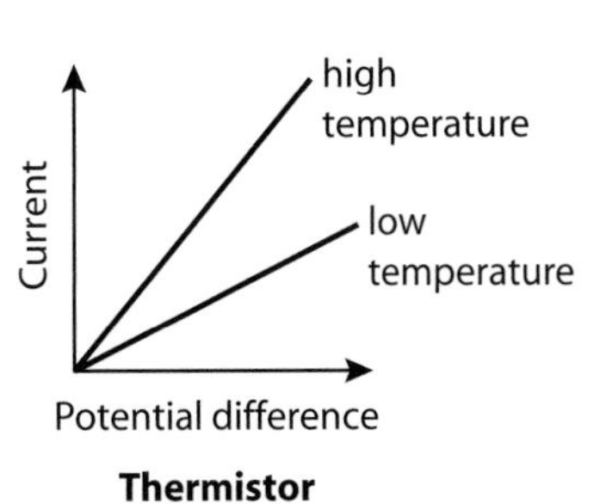

Thermistor

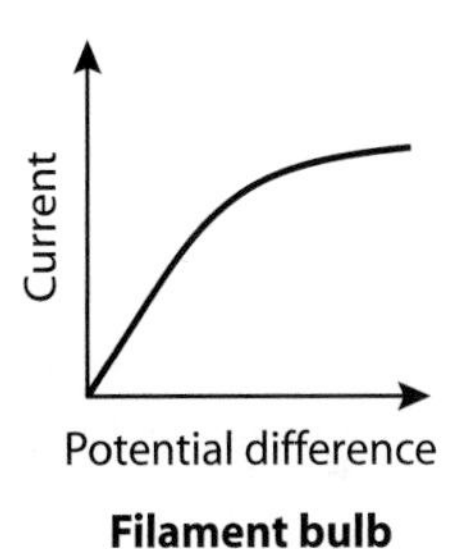

Filament bulb

(a) (i) Why have line graphs been drawn rather than bar charts? *(1 mark)*

(ii) Use the graphs to explain why an LDR can be used as a light sensor. *(2 marks)*

(b) Thermistors are used in digital thermometers. One digital thermometer can measure temperatures between −50.0°C and 200.0°C .

(i) What is the range of the thermometer? *(1 mark)*

(ii) What is the resolution of the thermometer? *(1 mark)*

(c) (i) How does the graph for the filament bulb show that the current is *not* directly proportional to the potential difference across it? *(1 mark)*

(ii) Explain why the resistance of the filament bulb increases with increasing temperature. *(2 marks)*

5 The pie chart shows the percentage of background radiation received from various sources.

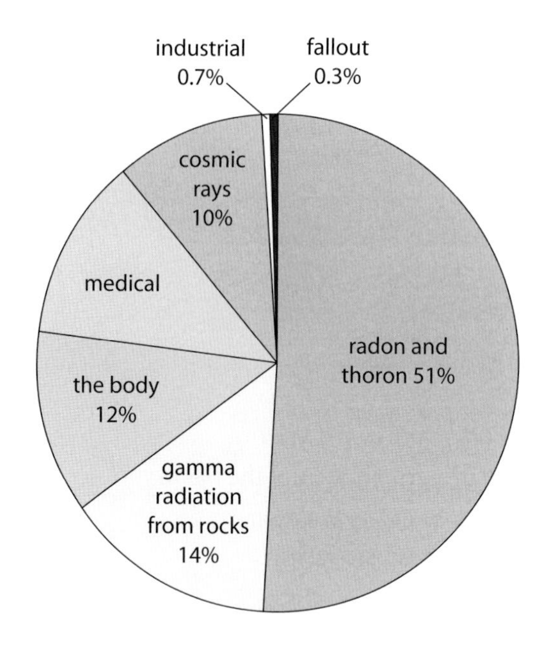

(a) **(i)** What percentage of the background radiation comes from medical sources? *(1 mark)*

(ii) How is radiation harmful to the human body? *(1 mark)*

(iii) Despite the harmful effects of radiation, radioactive isotopes are often used in hospitals. Give *two* uses of radioactive isotopes in hospitals. *(2 marks)*

(b) It is estimated that 2500 people in the UK die each year from radon-induced lung cancer. Radon is a radioactive gas.

Here are some facts about radon:

- Radon seeps from rocks and soils in the Earth.
- A common isotope of radon is radon-222.
- Radon-222 decays by emitting alpha particles.
- Radon-222 has a half-life of 3.8 days.

(i) Use the information above to explain why radon is so dangerous. *(3 marks)*

(ii) The equation for the decay of radon-222 is:

$^{222}_{86}\text{Rn} \rightarrow \,^{?}_{?}\text{Po} + \,^{4}_{2}\text{He}$

What are the atomic number and the mass number of the polonium isotope produced? *(2 marks)*

(iii) A sample of gas contains 8μg of radon-222. How much radon-222 will decay in 15.2 days? *(3 marks)*

6 The diagram below shows a source that emits three types of nuclear radiation. The radiation passes through an electric field between two charged plates.

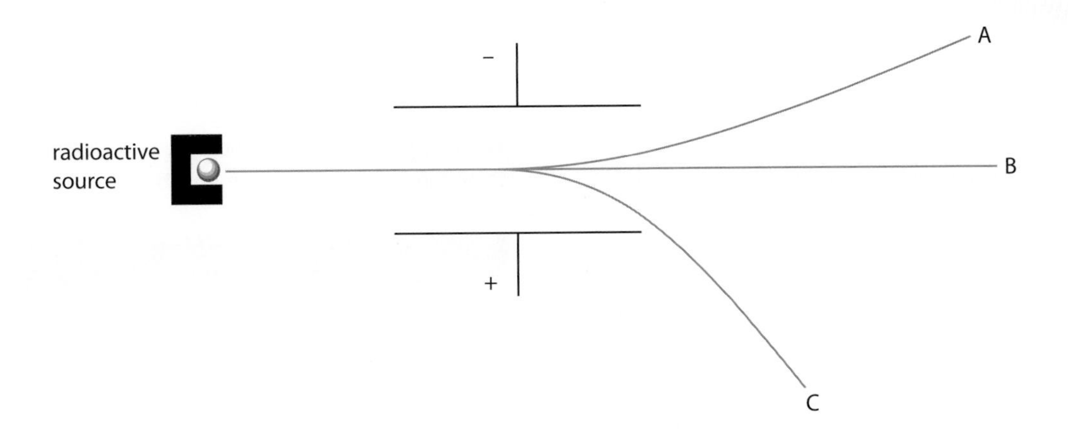

(a) Identify the types of radiation A, B and C. *(2 marks)*

(b) Using the properties of the three types of radiation, explain the different paths. *(6 marks)*

7 **(a)** In nuclear reactors, the nuclei of uranium-235 atoms are bombarded with neutrons.

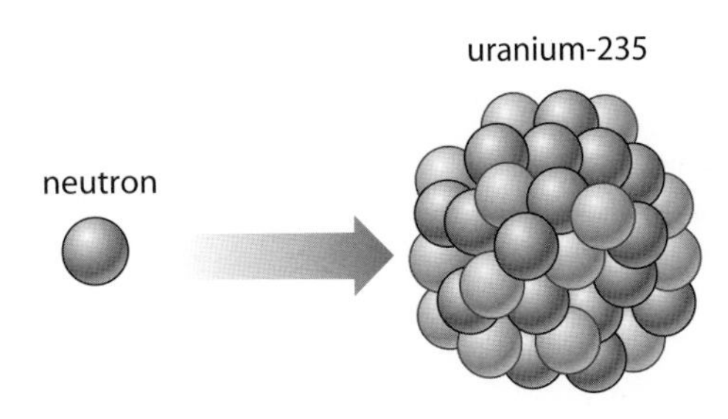

(i) What is the name of the nuclear process that leads to the relaese of large amounts of energy when neutrons bombard uranium nuclei?

fission fusion emission radioactivity *(1 mark)*

(ii) Describe what happens when the neutron hits the nucleus and how the process is used in the generation of electricity.

In this question you will be assessed on using good English, organising information clearly and using specialist terms where appropriate. *(6 marks)*

(b) The nuclear process that releases energy in stars is different from the process in part (a). Describe the process that occurs in stars. *(3 marks)*

Glossary

A

acceleration – The rate of change of velocity over time; measured in m/s^2.

accurate – An accurate measurement is close to the true value.

acidic – Having the properties of an acid.

action force – The force exerted by one object on another, which is matched by a reaction force that the second object exerts on the first object.

activation energy – The minimum amount of energy particles must have to react.

adult stem cell – A stem cell extracted from differentiated tissue, which can normally differentiate only into a limited range of specialised cells.

aerobic respiration – Respiration that requires the presence of oxygen to release energy from glucose, producing carbon dioxide and water.

alkali – A base that dissolves in water to form hydroxide ions and give a solution of pH greater than 7.

alkaline – Having the properties of an alkali.

allele – One form of a gene. Different alleles of the same gene produce slightly different characteristics, such as different eye colours.

allergic reaction – A reaction of the body to a compound to which the immune system has been sensitised.

alloy – A mixture of metals blended to have specific properties.

alternating current (a.c.) – A current which changes direction. Current supplied to UK homes is 50 Hz a.c.

alveolus (plural alveoli) – A tiny air sac in the lungs. The alveoli have a large total surface area over which oxygen diffuses into the blood and carbon dioxide diffuses out.

amino acid – A simple compound which, when combined with other amino acids in chains, makes proteins. There are 20 types of amino acids commonly found in living cells.

amylase – A type of digestive enzyme that catalyses the breakdown of starch into glucose.

anaerobic respiration – Respiration which doesn't need oxygen – the release of energy from glucose without oxygen. In muscle cells this produces lactic acid as a waste product. The process releases less energy than aerobic respiration.

anomalous – An anomalous result is a result that is too far away from the rest of the results to be reliable.

asexual reproduction – Reproduction from one cell of a parent organism by simple cell division, without fertilisation of gametes, that is, without fusion of male and female sex cells.

atom – A particle with no overall electric charge, made up of a combination of protons, neutrons and electrons; an atom is the smallest particle of an element that retains the properties of that element.

atomic number – The number of protons in the nucleus of atoms of a particular element.

attract – Exert a force on another object towards the attractor.

autotrophic – Able to produce its own food from simple starting compounds. Plants are autotrophic, because they can photosynthesise.

B

background radiation – The constant, low-level radiation around us at all times, from radioactive elements in rock, cosmic radiation from space, and human sources such as nuclear weapons tests.

base – A compound, usually a metal oxide or metal hydroxide, that reacts with an acid to neutralise it.

baseline – A baseline measurement is one taken before an experiment starts to establish the usual value.

battery – A number of electrical cells in series.

bauxite – An ore containing a high concentration of aluminium oxide.

Big Bang theory – The theory that the entire Universe originated in an explosion from a single point about 14 billion years ago.

bile – A mixture of chemicals produced in the pancreas that emulsifies fats, and so helps digest them.

binding site – The site in an enzyme to which the substrate binds.

black dwarf – An object expected to be formed when a white dwarf star cools to a cinder after billions of years.

black hole – A relatively small ball of very dense matter. Its gravity is so strong that even light cannot escape from it.

bone marrow – Fatty tissue inside long bones that contains adult stem cells capable of producing all types of blood cell.

braking distance – The distance taken to stop from the moment braking begins.

breathing rate – The number of breaths taken in a measured time, for example per minute.

burette – A long, thin, calibrated glass tube with a tap at the base, used to measure volumes of liquid accurately, for example in a titration.

by-product – A substance produced in a chemical process that is not the main product being made.

C

calibration – Setting the points on a measuring scale correctly by comparison with fixed points, so that accurate readings can be taken.

carbohydrase – An enzyme that catalyses the breakdown of starch to sugars.

carrier – A person carrying one allele for a recessive disorder, so they do not have the disorder themselves but could pass it on to their children.

cast – A fossil formed when the space left by the decay of a dead organism fills with minerals that harden into the shape of the organism.

catalyst – A chemical compound that speeds up a reaction but is not itself used up.

categoric variable – A variable that can only take particular values, for example days of the week or different types of food.

cathode ray oscilloscope (CRO) – An electronic device used to show how voltage changes with time.

cell – (biology) The basic unit of all known organisms. Most cells have a nucleus, a cell membrane and other organelles with their own functions.

cell – (electrical) A device for producing electricity, usually from a chemical reaction.

cell membrane – A thin outer layer of a cell that controls what goes into and comes out of a cell.

cell sap – A liquid that contains dissolved sugars and salts and is found in plant cell vacuoles.

cell wall – The outermost layer of a plant cell. It provides shape and support for the cell.

cellulose – A carbohydrate that forms tiny fibres from which plants and algae build cell walls.

chain reaction – In physics, a nuclear reaction in which products from the first reaction go on to cause the same reaction in other nuclei, producing more products. If unchecked, a chain reaction grows exponentially, resulting in a nuclear explosion.

chlorophyll – A green pigment found in the chloroplasts of plant cells. It captures light energy for photosynthesis.

chloroplast – An organelle in plant cells that contains chlorophyll for photosynthesis.

chromosome – An immensely long molecule of DNA containing many regions called genes, each of which carries the genetic information that influences a characteristic of the organism. Chromosomes are found in the nuclei of cells.

circuit – A series of wires and electrical components connected together so that electricity can flow through them.

clone – An individual that is genetically identical to the parent because it is produced by mitosis.

closed system – A sealed system that matter cannot enter or leave. This means that total momentum does not change.

collision theory – In chemistry, a theory that relates the rate of a chemical reaction to the number of collisions of sufficient energy between the reacting particles.

competition – The struggle between individual organisms for a share of a limited resource, such as water or food; for example, competition between two predators hunting the same prey.

component – (In electrical circuits) A bulb, resistor, diode or other electrical device that functions only when connected as part of an electrical circuit.

compound – A substance that contains two or more elements chemically joined together to form a new substance with new properties.

concentrated – Containing a large amount of solute.

concentration – A measure of how much solute is dissolved in a solvent such as water.

concentration gradient – A difference in the concentration of a substance in two different areas, for example two different solutions of the same solute on opposite sides of a partially permeable membrane.

conductivity – The ease with which a material allows electricity to flow through it, or transmits temperature changes.

continuous variable – A variable that can take any value, such as the time taken for a chemical reaction to happen.

control experiment – An experiment in which the variable that is to be investigated in the study is not changed; the control experiment picks up any changes that are not due to changes in the variable.

control measure – An action that reduces a hazard to an acceptable level of risk.

control rod – A rod of a material that absorbs neutrons. It is used in a nuclear reactor to control the chain reaction.

control variable – A variable that must be kept constant throughout an experiment to ensure that it does not affect the dependent variable.

coulomb – The unit of electrical charge. 1 C is just over 6 000 000 000 000 000 000 electrons, or 6×10^{18} electrons.

covalent bond – A pair of electrons shared between two atoms, forming a bond that holds the atoms together within a molecule.

cover – *See* ground cover.

cross-link – A strong covalent bond between polymer chains.

cryolite – A mineral used in aluminium smelting to reduce the melting point of the ore and thus the heating costs of the process.

cystic fibrosis – An inherited disorder caused by a recessive allele that results in production of thick, sticky mucus, affecting the lungs and other parts of the body.

cytoplasm – The substance outside the nuclei in cells, in which much of the cell metabolism (the chemical reactions within cells) takes place.

D

decay – The fission of an unstable atomic nucleus to release radiation and form a different nucleus.

deceleration – Negative acceleration, that is, the rate of change of velocity over time; measured in m/s^2, when the object is slowing down.

decommission – Take out of use.

delocalised – Delocalised electrons are electrons that have dissociated from their individual atoms.

denature – To alter the shape of an enzyme (a protein molecule), usually by heating it, in such a way that it no longer performs its function.

dependent variable – A dependent variable is the variable that is measured for each change in the independent variable in an experiment.

deuterium – An isotope of hydrogen with one neutron in the nucleus.

differentiated – (Of a cell) Specialised to carry out a specific job. Muscle cells, for example, are specialised to be able to contract.

diffusion – Movement of a substance by random motion from a region of high concentration to a region of lower concentration.

digest – Break down into smaller particles, as when food is digested in the gut.

dilute – Containing only a small amount of solute dissolved in a solvent such as water.

direct current (d.c.) – Current flowing in one direction continuously.

directly proportional – A directly proportional relationship between two variables is a simple mathematical relationship: if one variable is doubled, for example, the other is doubled too.

discharge – In electrolysis, to gain or lose electrons at an electrode.

DNA (deoxyribonucleic acid) – The genetic material found in the nucleus of living cells. Chromosomes are made up of DNA, and a gene is a section of a chromosome.

DNA profiling (DNA fingerprinting) – A process that produces an image of variable sections of DNA, used to identify individuals. Previously called DNA fingerprinting.

dominant – In genetics, an allele is dominant if it produces its form of the characteristic in the organism even if only one chromosome of the pair carries that allele (that is, whether the organism is homozygous or heterozygous).

double helix – The shape of the long DNA molecule, with two strands twisting about one another.

double-insulated – An electrical device designed to need no earth connection, with a plastic casing rather than a metal one, is double-insulated.

E

earth – An earth wire is a low-resistance path for electric current to flow through for safety if there is a fault in an appliance.

efficiency – The proportion of the energy supplied to a device that is transformed usefully rather than wasted.

effluent – A waste liquid that flows out of a reactor.

elastic – Able to be deformed when a force is applied but to regain its original shape when released.

electric current – A flow of charge.

electrode – A conductor that acts as a terminal through which electric current passes into a liquid, a solution, a gas or an electric circuit. For current to flow there must be a positive and a negative electrode.

electrode reaction – Reaction occuring at the surface of an electrode during electrolysis.

electrolysis – Decomposition of a molten or dissolved ionic compound by an electric current passed through it.

electrolyte – The molten or dissolved ionic substance that is decomposed during electrolysis.

electron – A tiny particle with a single negative charge that occupies an energy level around an atom's nucleus. Electrons are responsible for the chemical bonds between atoms, and move through metals when a current flows.

electron microscope – A microscope that uses electrons rather than light to create images. It has a much higher magnification than a light microscope.

electron shell – A position that electrons can occupy around an atom, also known as an energy level.

electroplating – Applying a thin layer of metal to an object by electrolysis.

element – A substance containing only one type of atom.

embryo – A new individual in its first stage of development, when a fertilised egg cell starts to divide.

embryo screening – Testing an embryo to see if it carries certain alleles, such as those for a genetic disorder.

embryonic stem cell – A stem cell extracted from an embryo, which can differentiate into all the kinds of specialised cell found in that organism.

emit – Give out, for example when light is emitted by a light bulb.

empirical formula – The simplest whole number ratio of atoms (or ions) of each element in a substance.

endothermic – Cooling the surroundings: refers to chemical reactions in which the products have more stored chemical energy than the reactants.

end-point – The point in a titration at which the two reactants have reacted completely but none of either is left over. For acid/alkali titrations, the end-point is often marked by a colour change in an indicator.

energy level – A position that electrons can occupy around an atom, also known as an electron shell.

energy store – A deposit of starch that a plant can convert back into glucose when it needs more energy from respiration.

energy transfer – Energy is transferred from one store to another; for example energy stored chemically in a handwarmer can be transferred into energy stored in the atmosphere as heat.

enzyme – A protein molecule that acts as biological catalyst to speed up the rate of a reaction taking place within or outside a cell.

enzyme technology – The use of enzymes as industrial catalysts.

epithelial cell – A cell that is part of the outer layer (the epithelium) of a structure or an organ.

ethical question – A question concerning what is morally right or wrong.

evolutionary tree – A diagram that shows how a group of organisms evolved from earlier organisms.

exothermic – Warming the surroundings; refers to chemical reactions in which the products have less stored chemical energy than the reactants.

extension – The extra length of a spring to which a force is applied.

extinct – (A species) having no individuals still living.

extracellular – Outside the cell.

F

family tree – Diagram that shows the members of a family linked by relationship, which can be used to show inheritance of a genetic disorder.

fatigue – Tire and lose level of response, as in muscles during extended vigorous activity.

fertilisation – The fusion of a male gamete with a female gamete to produce a cell with two sets of chromosomes (a zygote).

fertiliser – A compound added to soil to replace the minerals used up by plants.

filament bulb – An incandescent lightbulb containing a tungsten filament that glows whenit is heated by an electric current.

fission – A reaction in which a large, unstable atomic nucleus splits into smaller nuclei, releasing energy.

fixed resistor – A component in an electric circuit that offers a specific amount of resistance to the flow of current.

force – A force causes a body to change its velocity or shape.

forensic science – Scientific methods used to help identify what happened in a crime, including using DNA profiling to identify people involved.

fossil – The remains of an organism that lived in the past found preserved in rock, or evidence of organisms having been there (such as footprints).

frequency – The number of times per second that something happens, for example the number of complete waves or complete swings of a pendulum per second; measured in hertz (Hz).

fuel cell – A device in which a fuel such as hydrogen is oxidised continuously and directly, rather than by burning, to generate electricity.

fullerene – A form of the element carbon; a molecule made up of at least 60 carbon atoms linked together in rings to form a hollow sphere or tube.

fungicide – A chemical compound used to kill or reduce fungal growth.

fuse – A small safety device containing a length of wire that is designed to melt if the current in a circuit gets too high.

fusion – The combination of atomic nuclei to form a larger nucleus and release energy.

G

gall bladder – An organ that stores bile and releases it into the small intestine.

gamete – A specialised sex cell, such as sperm or eggs, involved in sexual reproduction in plants and animals.

gas chromatography – A method that separates chemicals in a very small sample. It can be used to separate fragments of DNA to make a DNA profile.

gel electrophoresis – A procedure that separates chemicals in a liquid mixture. It was used to make DNA profiles, but has been largely replaced by gas chromatography, which can analyse smaller samples.

gene – Small section of DNA in a chromosome. Each gene contains the code for a particular inherited characteristic, that is, to make a particular protein.

genetic – Relating to genes and DNA (the hereditary material).

genetic diagram – A diagram that shows the inheritance of alleles or characteristics from parents by offspring. Punnett squares are a type of genetic diagram.

giant covalent structure – A structure containing billions of atoms in a network linked together by covalent bonds; also called a macromolecular structure.

glucose – A simple sugar (carbohydrate) produced in plants by photosynthesis and from starch by digestion, which is broken down in respiration to release energy.

glycogen – A form of carbohydrate made from glucose in animals. It is stored in muscle and liver cells, then broken down when glucose levels in the blood are low, for example during vigorous exercise.

gradient – The slope of a graph; it shows the relationship between the two variables. For example, the gradient of a distance–time graph shows the way distance changes over time: the steeper it is, the greater the speed.

greenhouse – A glass building, sometimes heated, that is used for growong plants outside their normal growing season.

ground cover – The area of ground covered by a particular plant species.

group 0 – The vertical column in the periodic table containing the noble (inert) gases.

group 1 – The vertical column in the periodic table containing the alkali metals.

group 7 – The vertical column in the periodic table containing the halogens.

H

Haber process – An industrial process named after the German chemist Fritz Haber, used to make ammonia from nitrogen.

half equation – A balanced equation, including electrons, that represents the reaction at one electrode during electrolysis.

half-life – The average time it takes for something to decline by half, for example for the count rate from a sample of radioactive material to fall to half its initial level.

halogen – An element in group 7, such as fluorine, chlorine or iodine. They react with metals to form ionic compounds containing halide ions.

hazard – Something that can go wrong with an experiment and cause injury to people or objects.

heart rate – The number of heartbeats in a measured time, usually one minute.

heavy water – Water containing deuterium in place of one of the hydrogen atoms.

hydrogen ion – A hydrogen atom without its electron; a proton. Acids release hydrogen ions when they dissolve in water.

hydroxide ion – An negatively charged ion made up of a hydrogen atom and an oxygen atom, OH^-. Alkalis release hydroxide ions when they dissolve in water.

hypothesis – An idea that is suggested to explain a set of observations. It is used to make predictions that can be tested scientifically.

I

independent variable – A variable that is changed or selected by the investigator.

indicator – In chemistry, a substance that changes colour depending on its pH.

infrared radiation (IR) – Electromagnetic radiation that we can feel as heat. IR has a longer wavelength than visible light, but a shorter wavelength than microwaves.

insoluble – Unable to dissolve in a particular solvent, usually water.

instrumental technique – An automated method of performing a chemical analysis.

insulating – Acting as a barrier to the transfer of energy by heating, or to the conduction of electricity.

intermolecular force – A weak force between simple molecules.

interval – The gap between planned measurements in an experiment.

ion – An electrically charged particle, containing different numbers of protons and electrons. An ion is an atom or molecule that has either lost (positively charged) or gained (negatively charged) one or more electrons.

ionic bond – A chemical bond in which oppositely charged ions are held together by mutual attraction.

ionic compound – An ionically bonded compound.

ionic equation – An equation that shows what happens to ions in a reaction.

isotope – Two atoms of the same element with different numbers of neutrons in the nucleus are isotopes of the element; for example, ^{35}Cl and ^{37}Cl; both have 17 protons, but one has 18 neutrons and one has 20 neutrons.

IVF (*in vitro* fertilisation) – A procedure carried out in the laboratory involving mixing eggs and sperm removed from the reproductive organs to encourage fertilisation.

K

kinetic energy – The energy an object has because it is moving.

L

lactic acid – A breakdown product of anaerobic respiration in muscle cells.

large hadron collider (LHC) – The world's largest particle accelerator. In the LHC subatomic particles are smashed together in high-energy collisions to give physicists information about the structure of matter.

lattice – A regular, continuous structure of atoms or ions, for example in a crystal.

law of conservation of momentum – If two objects collide in a closed system, the total momentum is the same before and after the collision.

LDR – A light-dependent resistor.

light intensity – The amount of light energy falling on a measured surface area, for example one square metre.

limiting factor – An environmental variable, such as light intensity, that limits the rate of a process, such as a chemical reaction.

lipase – An enzyme that catalyses the digestion of lipids (fats or oils) into fatty acids and glycerol.

lipid – A fat or oil.

live – A live wire is the wire that carries the oscillating voltage of an a.c. supply.

liver – An organ that produces bile and many other compounds necessary for life.

M

macromolecular – A macromolecular substance has a giant covalent structure.

magnetic resonance imaging – *See* MRI scanner.

main-sequence star – A star in the main part of its life cycle, when it is using hydrogen fuel.

mass extinction – An event in which a large number of species become extinct during the same period.

mass number – The number of protons and neutrons in the nucleus of an atom.

mass spectroscopy – An analytical technique that involves breaking molecules into charged fragments and measuring their mass/charge ratios. Also known as mass spectrometry.

mean – The arithmetical average of a set of data.

median – The middle value of a set of values arranged in number order. The median of the values 3, 4, 4, 6, 7, 7, 9 is 6.

meiosis – A type of cell division to produce gametes. Two divisions of the original cell produce four cells with half the normal number of chromosomes.

metallic structure – A giant structure of close-packed, positively charged metal ions surrounded by delocalised electrons.

microbe – A microorganism.

microorganism – An organism that is too small to be seen without a microscope.

mitochondrion (plural mitochondria) – An organelle (structure) within a cell in which aerobic respiration takes place.

mitosis – A form of cell division that produces two cells genetically identical to the parent cell.

mode – The value that occurs most often among a set of data.

moderator – In a nuclear reactor, a subtance that slows down fast neutrons. Without the moderator, the fission chain reaction woud break down.

mole – The mass of a mole of particles equals the relative formula mass (M_r) in grams.

molecular formula – The chemical formula showing the different elements and the number of atoms of each element in a molecule, for example CH_4 (methane).

molecular ion – The ion formed by the otherwise unfragmented molecule in mass spectroscopy.

molecule – A particle made of two or more atoms joined through covalent bonds.

molten – Melted and in the liquid state.

momentum – The mass of an object multiplied by its velocity.

mould – The shape formed by a dead organism pressing into soft sediment.

MRI scanner – An imaging technique, especially useful for visualising the soft tissues of the body, that works by measuring the effects of a very strong magnetic field upon the body tissues.

mummy – A type of fossil in which soft tissue is well preserved.

N

nanometre – 1 nm = 10^{-9} m, one millionth of a millimetre.

nanoparticle – A particle between 1 and 100 nm in size.

nanoscience – The study of nanoparticles.

nanotube – A tiny tubular structure formed by a giant lattice of linked carbon rings.

natural selection – A natural process whereby the organisms with genetic charateristics best suited to their environment survive to reproduce and pass on their genes to the next generation.

negative linear – A straight-line plot in which one variable decreases in proportion to the increase in the other variable is a negative linear plot.

net movement – The overall movement of particles from an area of higher concentration to one of lower concentration.

neutral – A neutral wire is held at or near earth potential while the voltage in the live wire cycles between poitive and negative in an a.c. supply.

neutralisation reaction – A chemical reaction between an acid and alkali (or base) that produces water and a neutral salt.

neutralise – Make a solution neither acid nor alkaline. In a neutralisation rection an acid and a base combine to make a salt.

neutron – A subatomic particle found in the nucleus that has the same mass as a proton, but no overall charge.

neutron star – The final stage of the life of a star up to about three solar masses in size; it is about the size of the Earth, but extremely dense.

nitinol – A shape-memory alloy made of of nickel and titanium.

nitrate – A compound containing ions with the formula NO_3^-.

nuclear fission – A reaction in which a large, unstable atomic nucleus splits into smaller nuclei, releasing energy.

nuclear fuel – An element that contains a high proportion of unstable nuclei that split easily so it can be used to fuel nuclear reactors.

nuclear fusion – The combination of atomic nuclei to form a larger nucleus and release energy.

nucleus – (biology) The large, membrane-bound organelle inside a cell that contains genetic material.

nucleus – (chemistry and physics) The central part of an atom, containing most of the mass. It is made up of protons and neutrons.

O

observation – A measurement or note made during an experiment.

ore – Rock containing a high concentration of a particular metal or metal compound.

organ – A body structure that has a specific function and is made up of several different types of tissue.

organelle – A structure found within a cell that has a particular function, such as the nucleus or a chloroplast.

origin – The point at which the axes of a graph intersect, (0, 0). (Take care not to call the intersection the origin of the axes do not both start at 0.)

ovary – Female reproductive organ in humans and other animals. Ovaries produce eggs.

oxidation – A type of chemical reaction. When a compound is oxidised it gains oxygen, loses hydrogen or loses electrons.

oxygen debt – The extra oxygen that the body needs after vigorous exercise.

P

palisade mesophyll cell – A leaf cell packed with chloroplasts. Most photosynthesis takes place in palisade mesophyll cells.

pancreas – An organ that produces digestive enzymes and controls blood sugar concentration by producing insulin.

paper chromatography – An analytical technique that separates compounds by their relative speeds in a solvent as it spreads through paper.

parallel circuit – A parallel circuit contains at least one place where the current splits into two or more separate currents.

partially permeable membrane – A thin membrane containing tiny pores that allow the smallest particles to pass through, but not others.

penetration – The distance that radiation can travel through matter before it is stopped.

percentage yield – % yield = (mass of product obtained)/(maximum theoretical mass of product) × 100.

period – The time taken for one complete oscillation, for example the swing of a pendulum or the passage of a wave.

pH scale – A measure of the acidity or alkalinity of a solution. pH 1 is strongly acidic, pH 7 is neutral and pH 14 is strongly alkaline.

photosynthesis – The process by which green plant cells produce sugars and oxygen out of carbon dioxide from the air, water from the soil and energy from sunlight.

physiotherapy – Treatment for a problem or disorder that involves physical activity, such as firm patting on the back for cystic fibrosis suffers to loosen sticky mucus.

pigment – A solid, coloured substance.

pipette – A glass tube for precisely measuring and dispensing volumes of liquid. Many pipettes are made to measure and dispense a single volume, but some are graduated and can dispense a variety of volumes.

plankton – Microscopic single-celled organisms found in lakes, rivers and the sea.

polydactyly – An inherited condition that causes a person to have more fingers, thumbs or toes than usual, in some cases caused by a dominant allele.

polymer – A long-chain molecule made by joining many short molecules (monomers) together.

polypeptide – A long-chain molecule made by joining many amino acids together. Proteins are long polypeptides.

polytunnel – A long arched structure of poly(ethene) sheet used as a greenhouse in agriculture.

positive linear – A straight-line plot in which one variable increases in proportion to the other variable is a positive linear plot.

potential difference (pd) – The difference in the energy carried by electrons before and after they have flowed through an electrical component. Measured in volts (V).

potential energy – The energy stored in a mass because of its position in a gravitational field.

power – The amount of energy transferred per second; measured in watts (W).

precipitate – An insoluble solid formed by a chemical reaction, such as the reaction between two soluble salts.

precipitation reaction – A reaction in which an insoluble solid is formed.

precision – The closer experimental measurements are to each other and the mean of the results, the greater the precision of the results. Precision is not the same as accuracy.

predation – The killing and eating of one kind of organism by another kind of organism, usually in terms of animal predators and their prey.

prediction – A statement of what is expected to happen in the future under specified conditions. A prediction on the basis of a hypothesis provides a way to test the hypothesis.

preliminary work – Work carried out before the main series of tests in an experiment, in order to ensure that the measurements to be made cover a useful range at useful intervals, and the design of the experiment is valid.

pressure – The force exerted divided by the area on which it is exerted. Usually measured in pascals (Pa).

product – A compound formed during a chemical reaction.

proportional – Two variables are proportionally related if changing one by a particular factor changes the other by the same factor (direct proportionality) or by the reciprocal of the factor (inverse proportionality). For example, doubling the force on a spring doubles its extension, so the extension is directly proportional to the force.

protease – A digestive enzyme that catalyses the breakdown proteins into amino acids.

protein – A large molecule made of amino acids linked together; proteins play important roles in all living things and are an important part of the human diet.

proton – A subatomic particle found within the atomic nucleus, with a single positive charge and a relative mass of 1.

protostar – A spinning disc of hot gas in the process of condensing into a star.

Punnett square – A table used to predict the probability that a characteristic will be inherited by the offspring of two organisms with known combinations of alleles. One form of genetic diagram.

Q

quadrat – A frame used for sampling the distribution of species in an area.

R

radioactive – Having some atoms with unstable nuclei, which may spontaneously break down and emit radiation.

radioactive dating – Measuring the ratios of radioactive elements in rock to work out when the rock formed.

radioactivity – A property of elements whose atoms have unstable nuclei, which may spontaneously break down and emit radiation.

random error (uncertainty) – An unpredictable variation around the true value, causing each reading to be slightly different.

random sampling – Sampling by a scheme generated randomly rather than according to a pattern.

range – The spread between maximum and minimum values in a set of experimental results.

reactant – A compound that takes part in a chemical reaction.

reaction – A chemical process in which compounds react together to yield different compounds as products.

reaction force – The force exerted by one object on another in response to an action force that the second object exerts on the first object.

reactivity series – A list of elements in order of their reactivity

recessive – In genetics, an allele is recessive if it produces its form of the characteristic in the organism only if both chromosomes of the pair carry that allele (that is, the organism is homozygous).

red giant – A giant star late in the stellar lifecycle, which has expanded and cooled so that it has a red appearance.

reduction – A type of chemical reaction. When a compound is reduced it loses oxygen, gains hydrogen or gains electrons.

relative atomic mass (A_r) – The average mass of the atoms in an element (their individual mass numbers will differ because of the existence of isotopes).

relative formula mass (M_r) – The sum of the relative atomic masses of all the atoms in the formula.

reliable – A measurement is reliable if you get almost the same value every time it is repeated.

repeatable – Results are repeatable if on repeating the investigation you get the same or similar results.

repel – Exert a force on another object away from the repeller.

reproducible – Results are reproducible if, when you change the method or use different equipment, or if someone else does the investigation, the results are still similar.

resistance – The degree to which a component resists a current flowing through it; the higher the resistance, the greater the potential difference necessary to make a given current flow. The unit of resistance is the ohm (Ω).

resolution – The smallest change in value that an instrument can detect.

respiration – The breakdown of glucose in cells to release energy, carbon dioxide and water.

resultant force – The overall force resulting from a combination of separate forces.

retention time – The time taken for a substance to reach the detector at the end of a gas chromatography column.

reversible – A reaction is reversible if both the forward and reverse reactiosn can take place; for example, if you cool brown NO_2 gas, it reacts with itself to give colourless N_2O_4 gas, but if you heat N_2O_4 it decomposes to give NO_2.

ribosome – Tiny organelles inside a cell where protein synthesis occurs.

risk – The chance of something (often a hazard) happening.

root hair cell – A cell in the root of a plant with long narrow projections that extend into the soil to absorb water and dissolved mineral ions.

root tip squash – The preparation of a root tip by gently squashing it, so that the individual cells can be seen clearly under a microscope.

S

salivary gland – One of several glands in the mouth that produce the digestive enzymes amylases.

sampling – Examining a small portion of an area and using the results to estimate the value for the total area.

series circuit – An electrical circuit in which the current can follow only one route, through one component after another in sequence; there are no branches.

sex chromosome – A chromosome that determines the sex of the individual: human females have two X chromosomes, human males have one X and one Y chromosome.

shape-memory alloy – An alloy that reverts to its original shape when it is heated after being deformed.

small intestine – A section of the digestive tract that produces protease, amylase and lipase enzymes.

soluble – Able to dissolve in a particular solvent, usually water.

solute – A substance that is dissolved in a liquid to make a solution.

specialised cell – A cell that has special features that are required for its function. For example retina cells are specialised to sense light. Also called differentiated cells.

speciation – The evolution of new species from one original species.

staple – A staple food is one that forms the main part of the diet. Staple foods are high in carbohydrate, for example wheat, potatoes and maize.

starch – The form of carbohydrate stored by most plants. Starch is made from glucose molecules joined together.

state symbol – Symbols used in balanced equations to show the physical state of each reactant and product: (s) solid, (l) liquid, (g) gas or (aq) aqueous solution.

static electricity – A charge on an object caused by the addition or removal of electrons.

stem cell – A cell that, unlike most body cells, can divide and differentiate into other cell types.

stomach – An organ in the digestive tract that produces protease enzymes and acid.

stopping distance – Thinking distance + braking distance.

strong acid – An acid that splits completely into ions in solution, producing a high concentration of hydrogen ions, $H^+(aq)$.

strong alkali – An alkali that splits completely into ions in solution and produces a high concentration of hydroxide ions, $OH^-(aq)$.

substrate – The molecule upon which an enzyme acts.

succession – The gradual, predictable process by which the species in a habitat change over time until a stable, 'climax' ecosystem is produced.

superconductor – A metal or other material that has no electrical resistance at very low temperatures and can carry huge currents.

supercooled liquid – A liquid cooled below its freezing point without solidifying.

supergiant – An enormous star up to 70 times more massive than the Sun that has expanded near the end of its life and become red as it cools.

supernova – An explosion near the end of the life of a star of more than about 1.5 solar masses.

symptom – A sign of a disease or disorder noticed by the patient, such as high temperature and runny nose.

systematic error (uncertainty) – An error that produces a spread of readings around a value other than the true value. Systematic errors can be caused by the measuring instrument, the recording method or something in the environment.

systematic sampling – Sampling according to a pattern rather than randomly.

T

terminal velocity – The velocity at which the air resistance on a falling object exactly balances the force of gravity so that the resultant force is zero and no further acceleration occurs.

testes – The male reproductive organs in humans and other animals. The testes produce sperm.

theory – When data from testing predictions support a hypothesis, it becomes a theory that is accepted by most (but not necessarily all) scientists.

theory of evolution – The theory that explains how all living things today have been produced by the accumulation of inherited changes in the characteristics of populations through successive generations.

thermistor – A temperature-dependent resistor.

thermoelectric generator – A device that generates electricity from the heat produced by the decay of radioactive elements.

thermosetting – Unable to soften and melt on heating. Applies to polymers with strong covalent bonds (cross-links) between the polymer chains. These polymers cannot be recycled by melting and remoulding.

thermosoftening – Able to soften and melt on heating. Applies to polymers with only weak forces between the polymer chains. These polymers can be recycled by melting and remoulding.

thinking distance – The time it takes for a driver to recognise a hazard and apply the brake.

tissue – A mass of similar cells, for example muscle tissue.

titration – A method used to find the volumes of solutions that react completely with one another with no excess left over. From the results, concentrations can be calculated.

titre – The volume of solution needed to react exactly with another solution in a titration.

trace fossil – A fossil that provides evidence of the presence of organisms, such as footprints, burrows or root shapes, rather than a fossil of the organism itself.

tracer – A substance that is injected into a body or structure and detected from outside using a scanner. Tracers can be used in humans for medical diagnosis, or in engineering to detect faults.

transect – A line across an area along which the species are sampled in a field study.

transfer – Energy is transferred from one store to another; for example energy stored chemically in a handwarmer can be transferred into energy stored in the atmosphere as heat.

tritium – An isotope of hydrogen with two neutrons in the nucleus.

tungsten – A heavy metal used to make the filaments in incandescent bulbs.

turbulence – Rough flow.

U

uncertainty – The range of possible values of a measurement around the true value.

unicellular organism – An organism that has only one cell, such as a bacterium.

unstable nucleus – A nucleus that may split into a smaller nucleus and release nuclear radiation.

V

vacuole – A sac filled with cell sap found in most plant cells.

validity – An experiment is valid if it has genuinely investigated the hypothesis and the prediction it was intended to investigate.

variable resistor – An electrical component with a resistance that can be altered.

variegated – Of more than one colour; a variegated leaf has green areas containing chlorophyll and yellow/white areas with no chlorophyll.

W

weak acid – An acid that does not completely split into its component ions in solution, and produces a low concentration of hydrogen ions, $H^+(aq)$.

weak alkali – An alkali that does not completely split into its component ions in solution, and produces a low concentration of hydroxide ions $OH^-(aq)$.

white dwarf – A small, dim star that forms from stars of about one solar mass or less toward the end of the star's life.

work – Work is a measure of energy transferred.
$W = F \times d$ (units Nm or J).

X

xylem – A plant tissue made up of long hollow tubes that act as vessels for the transport of water.

Y

yield – The amount of a crop that a plant produces.

Z

zero error (uncertainty) – A systematic error that arises when a measuring instrument does not read exactly zero when there is no input.

zygote – The cell formed when two gametes fuse to create a new individual in sexual reproduction.

Index

A

B

C

D

E

F

G

H

I

K

L

M

N

O

P

Q

R

S

T

U

V

W

X

Y

Z